*Applied Probability and Statistics (Continued)*

BROWNLEE · Statistical Theory and Metho
Engineering, *Second Edition*
BUSH and MOSTELLER · Stochastic Models for Learning
CHAKRAVARTI, LAHA and ROY · Handbook of Methods of Applied Statistics, Vol. I
CHAKRAVARTI, LAHA and ROY · Handbook of Methods of Applied Statistics, Vol. II
CHERNOFF and MOSES · Elementary Decision Theory
CHEW · Experimental Designs in Industry
CHIANG · Introduction to Stochastic Processes in Biostatistics
CLELLAND, deCANI, BROWN, BURSK, and MURRAY · Basic Statistics with Business Applications
COCHRAN · Sampling Techniques, *Second Edition*
COCHRAN and COX · Experimental Designs, *Second Edition*
COX · Planning of Experiments
COX and MILLER · The Theory of Stochastic Processes
DEMING · Sample Design in Business Research
DODGE and ROMIG · Sampling Inspection Tables, *Second Edition*
DRAPER and SMITH · Applied Regression Analysis
GOLDBERGER · Econometric Theory
GUTTMAN and WILKS · Introductory Engineering Statistics
HALD · Statistical Tables and Formulas
HALD · Statistical Theory with Engineering Applications
HANSEN, HURWITZ, and MADOW · Sample Survey Methods and Theory, Volume I
HOEL · Elementary Statistics, *Second Edition*
JOHNSON and LEONE · Statistics and Experimental Design: In Engineering and the Physical Sciences, Volumes I and II
KEMPTHORNE · An Introduction to Genetic Statistics
MEYER · Symposium on Monte Carlo Methods
PRABHU · Queues and Inventories: A Study of Their Basic Stochastic Processes
SARHAN and GREENBERG · Contributions to Order Statistics
SEAL · Stochastic Theory of a Risk Business
TIPPETT · Technological Applications of Statistics
WILLIAMS · Regression Analysis
WOLD and JURÉEN · Demand Analysis
YOUDEN · Statistical Methods for Chemists

*Tracts on Probability and Statistics*

BILLINGSLEY · Ergodic Theory and Information
BILLINGSLEY · Convergence of Probability Measures
CRAMÉR and LEADBETTER · Stationary and Related Stochastic Processes
RIORDAN · Combinatorial Identities
TAKÁCS · Combinatorial Methods in the Theory of Stochastic Processes

*Introduction to Probability Theory and Statistical Inference*

# INTRODUCTION TO PROBABILITY THEORY AND STATISTICAL INFERENCE

*HAROLD J. LARSON*
*Naval Postgraduate School*
*Monterey, California*

*Library of Congress Catalog Card Number: 68-56163*
*SBN 471 51780 1*
*Printed in the United States of America*

2 3 4 5 6 7 8 9

*For Marie*

# Preface

I have written this book to give the sophomore-junior student a more rigorous (but not more difficult) introduction to probability theory and statistical inference than is commonly available from other texts. Since the calculus frequently is being covered in the freshman year, the development here is based upon a prerequisite of a one-year course in calculus. I have not slanted the coverage and examples toward any particular subject area because the domain of statistical methods is unlimited. I believe that both physical science and social science students, as well as students in many engineering disciplines, will profit from this type of coverage.

The first five chapters give an introduction to set theory and probability theory, both discrete and continuous, and to the concept of random variables, with specialized discussion of many standard types of distributions. The modern approach of carefully distinguishing between random variables and their observed values has been used throughout the text. In particular, this notation, perhaps uniquely, has been continued in the treatment of statistical inference in the final five chapters. This approach makes it much easier for the student to grasp the distinctions between the parameter being estimated, the estimate computed from a particular sample, and the estimator defined on the space of all possible outcomes.

The first five chapters can be used in a four-hour one-quarter or three-hour one-semester introduction to probability theory and random variables. The last five chapters treat random sampling and classical statistical inference, especially point and interval estimation, tests of hypotheses, Bayesian approaches, and an introduction to least squares; these chapters can be covered in an additional quarter or semester.

I feel that no student should be expected to grasp the ideas of probability theory or statistics without attempting a large number of problems. For this reason the reader will find 198 worked examples and 432 problems scattered throughout the text; answers for all problems are provided at the back of the book. Also located at the back of the book are five appendices. The first of these deals with the summation and product operations, topics frequently bothersome to students at this level. The second appendix derives the binomial theorem and the values of some special sums connected with it. The third and fourth appendices briefly discuss infinite sums, geometric progressions, improper integrals, and the gamma and beta functions. The fifth appendix gives tabulations of the distribution functions for the binomial, Poisson, standard normal, $\chi^2$ and $t$ random variables. I would like to thank D. Van Nostrand Co., Inc., the editor of *Biometika*, and Oliver & Boyd, Ltd., for permission to reproduce parts of previously published tables.

None of the methods and techniques discussed in this book is original with the author; I would like to acknowledge personally all those who did originally devise them but that would prove an overwhelming task.

My family has proved most helpful and patient during the writing of this book; I very gratefully acknowledge their assistance. I would also like to thank Lt. Col. Richard W. Diller for preparing answers to the exercises in the first half of the book and Mrs. Sylvia Vaden for an excellent job of typing most of the manuscript.

HAROLD J. LARSON

November 1968

# Contents

## 4 Some Standard Distributions



## 5 Jointly Distributed Random Variables



## 6 Sampling and Statistics



## 7 Statistical Inference: Estimation



## 8 Statistical Inference: Tests of Hypotheses

## 9 Bayesian Methods



## 10 Least Squares and Regression Theory



## Appendices

# 1

# Set Theory

In the study of probability theory and statistics an exact medium of communication is extremely important; if the meaning of the question that is asked is confused by semantics, the solution is all the more difficult, if not impossible, to find. The usual exact language employed to state and solve probability problems clearly is that of set theory. The amount of set theory that is required for relative ease and comfort in probability manipulations is easily acquired. We shall take a brief look at some of the simpler definitions, operations, and concepts of set theory, not because these ideas are necessarily a part of probability theory but rather because the time needed to master them is more than compensated for by later simplifications in the study of probability.

## 1.1. Set Notation, Equality, and Subsets

A set is a collection of objects. The objects themselves can be anything from numbers to battleships. An object that belongs to a particular set is called an element of that set. We shall commonly use capital letters from the beginning of the alphabet to denote sets ($A$, $B$, $C$, etc.) and lower case letters from the end of the alphabet to denote elements of sets ($x$, $y$, $z$, etc.).

To specify that certain objects belong to a given set, we shall use braces $\{\ \}$ (commonly called set builders) and either the roster (complete listing of all elements) or the rule method. For example, if we want to write that the set $A$ consists of the letters $a$, $b$, $c$ and that the set $B$ consists of the first 10 integers, we may write

$$A = \{a, b, c\} \text{ (roster method of specification)}$$

$$B = \{x: x = 1, 2, 3, \ldots, 10\} \text{ (rule method of specification).}$$

The above two sets can easily be read as "$A$ is the set of elements $a$, $b$, $c$" and "$B$ is the set of elements $x$ such that $x = 1$ or $x = 2$ or $x = 3$ and so on up to or $x = 10$." We shall use the symbol $\in$ as shorthand for "belongs to" and thus can write for the two sets defined above that $a \in A$, $7 \in B$. Just as a line drawn through an equals sign is taken as negation of the equality, we shall use $\notin$ to mean "does not belong to"; thus $a \notin B$, $9 \notin A$, $f \notin A$, $102 \notin B$, etc., where $A$ and $B$ are the sets defined above.

DEFINITION 1.1.1. Two sets $A$ and $B$ are *equal* if and only if every element that belongs to $A$ also belongs to $B$ and every element that belongs to $B$ also belongs to $A$.

Then two sets are equal only if both contain exactly the same elements. Notice in particular that the order of listing of elements of a set is of no importance and that the sets

$$A = \{1, 2, a, 3\}, \qquad B = \{a, 1, 2, 3\}$$

are equal. Also, the number of times that an element is listed in the roster specifying the set is of no concern; the two sets

$$C = \{1, 2, 3\}, \qquad D = \{1, 2, 2, 1, 3, 1\}$$

are equal. We shall have no use for the redundancy exhibited in the roster of $D$ and thus shall generally assume that if the roster of a particular set contains $n$ elements, the elements are distinguishable and that the same element does not occur more than once.

DEFINITION 1.1.2. $A$ is a *subset* of $B$ (written $A \subset B$) if and only if every element that belongs to $A$ also belongs to $B$.

In a sense, $A$ is a subset of $B$ if $A$ is contained in $B$. For example, $\{a\}$ is a subset of $\{a, z\}$ and both of these are subsets of $\{a, b, z\}$. The set of all people residing in California is a subset of the set of all people residing in the United States; the set of all pine trees is a subset of the set of all trees.

A word regarding the difference between something belonging to a set $A$ and being a subset of $A$ may be in order. For example, define

$$A = \{1, 2, 3\}, \qquad B = \{1, 3\}, \qquad C = \{1\}.$$

Then it is correct to say $B \subset A$ and $C \subset A$ since every element that belongs to $B$ also belongs to $A$, as does every element that belongs to $C$. However, it is not correct to say that $B \in A$ or that $C \in A$ since $B$ and $C$ are not specified in the roster of elements that belong to $A$. Similarly, it is correct to say that $1 \in A$ and $1 \in C$, not $1 \subset A$ and $1 \subset C$ since 1 is not a set. Thus we can say that the set of all married people is a subset of the set of all people, but we do not say that the set of all married people belongs to the set of all people. A particular married person belongs to the set of all married people (as well as to the set of all people), but this person is not a subset of either (since a person is not a set).

An alternative definition of set equality can be given by saying that $A = B$ if and only if $A \subset B$ and $B \subset A$. This, in fact, provides a very useful way of demonstrating that two sets are equal, as we shall see. Most simple set equalities can be rigorously and easily proved in this way.

From the definitions, it is easily seen that every set is equal to itself and that every set is a subset of itself. That is, we can say

$$A = A, \qquad \text{for all } A,$$
$$A \subset A, \qquad \text{for all } A.$$

In textbooks on algebra, these two statements are summarized by saying that both the equality and the subset relations are reflexive.

**EXERCISE 1.1.**

**1.** Show that the subset relationship is transitive; that is, $A \subset B$ and $B \subset C$ imply $A \subset C$.

**2.** Show that the subset relationship is not symmetric; that is, $A \subset B$ and $B \subset A$ are logically different statements and one is not implied by the other. Give an example illustrating this nonsymmetry.

**3.** Give an example showing that if $E \subset F$ and $D \subset F$, we do not necessarily have $D \subset E$ or $E \subset D$.

**4.** Show that set equality is transitive. (Transitivity is defined in problem 1 above.)

**5.** If $A = \{x: x = 1, 0, 1\}$, $B = \{-1, 0, 1\}$, $C = \{0, 1, -1\}$, $D = \{-1, 1, -1\}$ mark each of the following either true or false.

(a) $A = B$

(b) $A = C$

(c) $B = C$

(d) $A \subset B$

(e) $A \subset C$

(f) $B \subset C$

(g) $C \subset A$

(h) $C \subset B$

(i) $0 \in B$

(j) $0 \subset C$

(k) $A = D$

(l) $D \subset A$

(m) $B \subset D$

(n) $D \in C$.

**6.** Define $E = \{x: 0 \leq x \leq 1\}$, $F = \{y: 0 \leq y \leq 1\}$, $G = \{x: 0 < x < 1\}$, $H = \{1\}$, Mark each of the following either true or false.

(a) $E = F$

(b) $F = G$

(c) $F \subset G$

(d) $G \subset F$

(c) $H \subset G$

(f) $H \subset E$

(g) $H \in F$.

**7.** Is the set of all students a subset of the set of all people? Is it a subset of the set of all people under 36 years of age? Under 50 years of age?

### 1.2. Union and Intersection

We shall now study some of the algebra of sets. Just as the algebra of real numbers is concerned with operations performed with numbers and their consequences, the algebra of sets is concerned with operations performed with sets and their consequences. The first set operation we define is called the union of two sets.

DEFINITION 1.2.1. The *union* of $A$ and $B$ (written $A \cup B$) is the set which consists of all the elements that belong to $A$ or to $B$ or to both; i.e., $A \cup B = \{x: x \in A \text{ or } x \in B\}$.

For example, if $A = \{1, 2\}$, $B = \{1, 3\}$, $C = \{0\}$, then $A \cup B = \{1, 2, 3\}$, $A \cup C = \{0, 1, 2\}$, and $B \cup C = \{1, 3, 0\}$. Notice from the definition of the union that $A \cup B$ and $B \cup A$ are identical sets (the operation is commutative), since the collection of objects that belong to $A$ or to $B$ is the same; it does not matter whether we first list those that belong to $A$ or those that belong to $B$. We also note that the operation of forming unions is associative; i.e.,

$$(A \cup B) \cup C = A \cup (B \cup C).$$

For this reason it is not necessary to place parentheses about pairs of sets when we take the union of several sets; it doesn't matter which union is formed first.

We notice a number of interesting relationships linking unions of sets and the concept of subsets. For example, for any $A$ and $B$ we would know that $A \subset A \cup B$ and $B \subset A \cup B$ (verification of these facts is asked for in problem 1 below). Not quite so obviously, $A \cup B = A$ if $B \subset A$, since the set which consists of all those elements that belong to $A$ or to $B$ or to both would merely consist of those elements that belong to $A$ originally if every element belonging to $B$ also belongs to $A$. Thus the union of the set of people living in California with the set of people living in the United States is the set of people living in the United States; the union of the set of all cows with the set of all animals is the set of all animals.

A second set operation which we shall find useful is called the intersection of two sets.

DEFINITION 1.2.2. The *intersection* of $A$ and $B$ (written $A \cap B$) is that set which consists of all elements that belong both to $A$ and to $B$; i.e., $A \cap B = \{x: x \in A \text{ and } x \in B\}$.

For example, if $A = \{0, 1\}$, $B = \{1, 3\}$, $C = \{0, 1, 3\}$, then $A \cap B = \{1\}$, $A \cap C = \{0, 1\}$, $B \cap C = \{1, 3\}$. If $A$ is the set of married people and $B$ is the set of people living in California, then $A \cap B$ is the set of married people living in California.

It is easy to see that the intersection operation is also commutative and associative; that is,

$$A \cap B = B \cap A \qquad \text{and} \qquad A \cap (B \cap C) = (A \cap B) \cap C.$$

We note as well that

$$A \cap B \subset B, \qquad A \cap B \subset A,$$

no matter what the sets $A$ and $B$ may be. Similarly, if $B \subset A$, then

$$A \cap B = B.$$

Thus the intersection of the set of all buildings in California with the set of all buildings in the United States is the set of all buildings in California.

A handy way of picturing sets, relationships between sets, and operations with sets is to employ a Venn diagram. Various geometric shapes can be drawn on a plane and, if we assume that the points interior to and on the boundary of each figure constitute a set, many of the concepts already discussed can be visually displayed. In Figure 1.1 are given sets $A$, $B$, $C$, $D$, and $A \cup B$ and $C \cap D$.

There are two distributive laws linking the operations of unions and

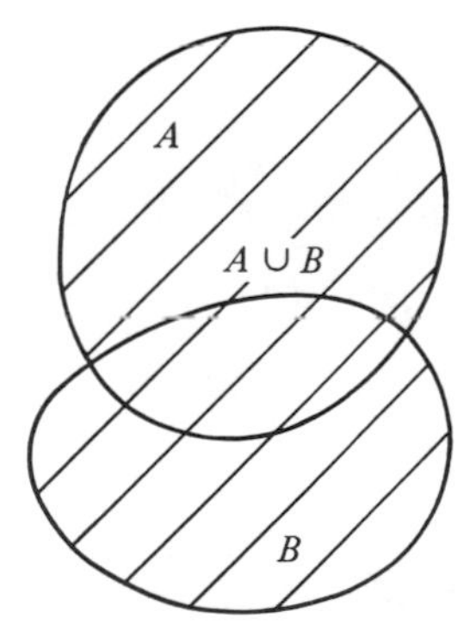


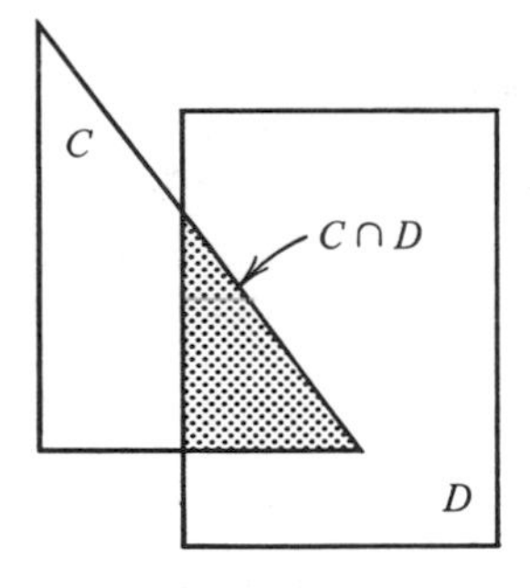


**Figure 1.1.**

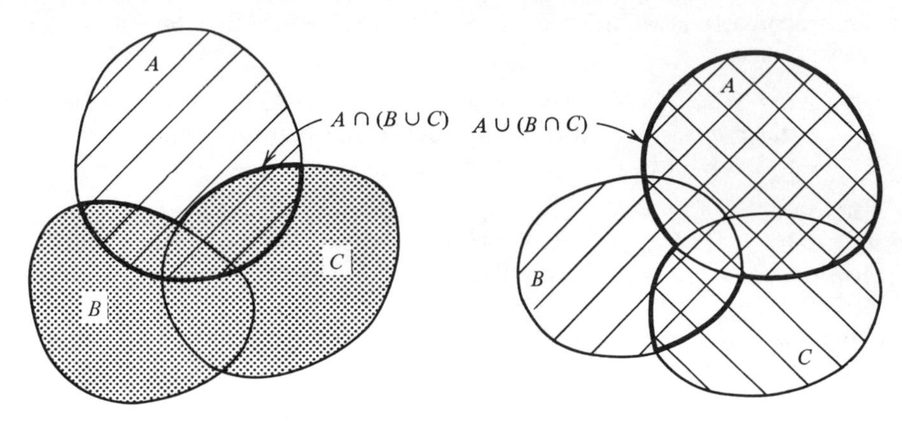


Figure 1.2.

intersections. These are:

$$A \cap (B \cup C) = (A \cap B) \cup (A \cap C) \tag{1.1}$$

$$A \cup (B \cap C) = (A \cup B) \cap (A \cup C). \tag{1.2}$$

Figure 1.2 gives a Venn diagram which the reader can use to verify that both of these statements are true.

Note that there exists a rather natural analog between unions of sets and addition of numbers since, in some senses, the set $A \cup B$ is the result of "adding" the set $B$ to the set $A$. If we also pretend that there is an analog between set intersection and multiplication of numbers, then distributive law 1.1 above is analogous to $a(b + c) = ab + ac$, which is true for all real numbers $a$, $b$, and $c$. However, distributive law 1.2 would be analogous to $a + b \cdot c = (a + b) \cdot (a + c)$, which is not true for all $a$, $b$, and $c$.

**EXERCISE 1.2.**

**1.** Show that $A \cap B \subset A \subset A \cup B$ and that $A \cap B \subset B \subset A \cup B$.

**2.** If $A = \{1, 0\}$, $B = \{x: 0 < x < 1\}$, $C = \{\frac{1}{2}\}$, compute $A \cup B$, $A \cup C$, $B \cup C$, $A \cap B$, $A \cap C$, $B \cap C$.

**3.** Show that $B \cup B = B \cap B = B$ for all $B$.

**4.** Prove distributive law number 1.2.

**5.** What is the implication of the equation $E \cap F = F$?

**6.** What is the implication of the equation $E \cup F = E$?

**7.** Define:

$$A = \{x: x = 1, 2, 3, \ldots, 10\}$$
$$B = \{x: 1 \leq x \leq 10\}$$
$$C = \{x: x = 0, 1, 2, 3, 4, 5, 6\}$$
$$D = \{0, 10, 20, 30\}$$

and compute

(a) $A \cup B$
(b) $A \cap B$
(c) $A \cup C$
(d) $A \cap C$
(e) $A \cup D$
(f) $A \cap D$
(g) $B \cup C$
(h) $B \cap C$
(i) $B \cup D$
(j) $B \cap D$
(k) $C \cup D$
(l) $C \cap D$
(m) $A \cup B \cup C$
(n) $A \cap (B \cup C)$
(o) $A \cup (B \cap C)$
(p) $A \cap B \cap C$
(q) $C \cup (A \cap D)$
(r) $(A \cup B) \cap (C \cup B)$.

### 1.3. Universal Set, Complement, and Cartesian Product

Generally, all of the sets entering into particular discussions will have certain things in common. For example, a group of botanists might discuss sets of trees, in which event all of the sets entering into their discussion would have trees as elements. A group of mathematicians might discuss sets of real numbers, in which event all of the sets entering into their discussion would have real numbers as elements. The universal set for a particular discussion is that set which has as elements all elements of every set entering into the discussion. Thus the universal set for the botanists' discussion could be the set of all trees; the universal set for the mathematicians' discussion might be the set of all real numbers. Note that different discussions may have different universal sets.

If we have in mind a particular universal set, then definition of a set $A$ automatically also specifies a second set $\bar{A}$, called the complement of $A$, which is defined as follows.

Definition 1.3.1. The *complement* of $A$ (with respect to a given universal set) is the set of all elements not belonging to $A$; i.e., $\bar{A} = \{x: x \notin A\}$.

If, for example, our universal set is

$$U = \{x: 1 \leq x \leq 10\}$$

and we define

$$A = \{x: 1 \leq x \leq 2\}$$
$$B = \{1, 10\}$$

then

$$\bar{A} = \{x: 2 < x \leq 10\}$$
$$\bar{B} = \{x: 1 < x < 10\}.$$

Or, if our universal set is the set of all people living in the United States and we define $A$ as the set of people living in California and $B$ as the set of people living in San Francisco, then $\bar{A}$ is the set of people living in the United States outside of California and $\bar{B}$ is the set of people living in the United States outside of San Francisco.

Figure 1.3 gives a Venn diagram illustrating a universal set $U$, a set $A$, a set $B \subset A$, $\bar{A}$, and $\bar{B}$. Note that $A \cup \bar{A} = U$ and, as is always the case, that if $B \subset A$, then $\bar{A} \subset \bar{B}$.

The null (or empty) set is frequently of use and is defined as follows:

DEFINITION 1.3.2. The *null set* $\varnothing$, is the set with no elements; i.e., $\varnothing = \{\ \}$.

$\varnothing$ occurs as the intersection of two sets that have no elements in common. For example, if $A$ is the set of all frogs and $B$ is the set of all people, then $A \cap B = \varnothing$. More generally, from the definition of $\bar{A}$, $A \cap \bar{A} = \varnothing$ for any $A$.

The empty set plays a role analogous to the real number 0 in some (not all) manipulations. If $a$ is any real number, we know that

$$a + 0 = a, \qquad a \cdot 0 = 0.$$

If $A$ is any set, we can easily see that

$$A \cup \varnothing = A, \qquad A \cap \varnothing = \varnothing.$$

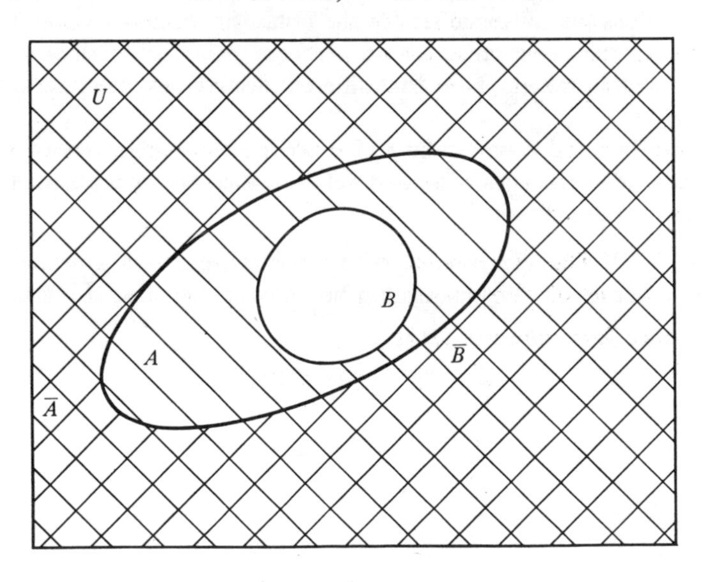


**Figure 1.3.**

Note, however, that we can have $A \cap B = \varnothing$ where neither $A$ nor $B$ is empty, whereas $a \cdot b = 0$ implies that $a$ or $b$ (or both) is 0. From our definition of subsets, we must admit that $\varnothing \subset A$ for all $A$, since every element belonging to $\varnothing$ (none) also belongs to $A$. Alternatively, every $x$ belonging to $\varnothing$ also belongs to $A$ is a true statement, vacuously, since there is no $x$ belonging to $\varnothing$ to make the statement false. Unfortunately, $\varnothing$ cannot be pictured on a Venn diagram like other sets, because if we draw any shape at all we don't have a picture of $\varnothing$ since our shape is not empty.

As was discussed earlier, a set of elements is unchanged if we merely rearrange the ordering of its elements. To have at our disposal collections of $n$ elements where order is of importance, we define an $n$-tuple as follows.

DEFINITION 1.3.3. An $n$-tuple is an ordered array of $n$ elements written $(x_1, x_2, \ldots, x_n)$.

For example, (1, 2), (2, 1), (0, 100), $(a, b)$ are all 2-tuples; $(a, b, c)$, $(b, a, c)$, (1, 1, 1), (2, 1, 2) are all 3-tuples; etc. Two $n$-tuples are different, even if they contain the same elements, if they are written in a different order and, of course, an $n$-tuple and an $(n + 1)$-tuple cannot be equal since they have different numbers of elements. Thus

$$(1, 2) \neq (2, 1)$$

$$(1, 3, 1) \neq (1, 1, 3)$$

$$(1, 1) \neq (1, 1, 1).$$

We shall have use for sets of $n$-tuples as we proceed. For example, if we want to discuss the set of married couples living in Texas, then we want each couple to be an element of the set; for clarity we list the husband's name first when specifying the couple. Each element of the set then is a couple or a 2-tuple. Note that a married individual living in Texas does not belong to the set; he is a component of one of the 2-tuples that belongs to the set. As a second example, the set of points lying in the first quadrant of the usual Cartesian plane is a set of 2-tuples. Each point is represented by a 2-tuple $(x, y)$ where $x$ is the horizontal coordinate and $y$ the vertical coordinate.

One other set operation, the Cartesian product of two sets will be of interest to us. It is defined as follows.

DEFINITION 1.3.4. The *Cartesian product* $A \times B$ of $A$ and $B$ is the set of all possible 2-tuples $(x_1, x_2)$ where $x_1 \in A$, $x_2 \in B$. That is, $A \times B = \{(x_1, x_2): x_1 \in A, x_2 \in B\}$.

Let us consider a number of examples of Cartesian products. Suppose that $A = \{0, 1, 2\}$, $B = \{3, 5\}$, $C = \{0\}$. Then

$$
\begin{aligned}
A \times B &= \{(0, 3), (0, 5), (1, 3), (1, 5), (2, 3), (2, 5)\} \\
A \times C &= \{(0, 0), (1, 0), (2, 0)\} \\
B \times C &= \{(3, 0), (5, 0)\} \\
B \times A &= \{(3, 0), (3, 1), (3, 2), (5, 0), (5, 1), (5, 2)\} \\
C \times A &= \{(0, 0), (0, 1), (0, 2)\} \\
C \times B &= \{(0, 3), (0, 5)\} \\
A \times A &= \{(0, 0), (0, 1), (0, 2), (1, 0), (1, 1), (1, 2), (2, 0), (2, 1), (2, 2)\} \\
B \times B &= \{(3, 3), (3, 5), (5, 3), (5, 5)\} \\
C \times C &= \{(0, 0)\}.
\end{aligned}
$$

If $D$ is the set of all positive real numbers, then $D \times D$ is the set of points in the first quadrant of the plane.

Notice that $A \times B$ and $B \times A$ are not equal in general. The operation is not commutative. Problem 9 below asks you to specify when these two would be equal. We can form the Cartesian product of any number of sets; $A \times B \times C$ is a set of 3-tuples, $A_1 \times A_2 \times \cdots \times A_n$ is a set of $n$-tuples. Taking $A$, $B$, and $C$ as defined above,

$$
\begin{aligned}
A \times B \times C &= \{(0, 3, 0), (0, 5, 0), (1, 3, 0), (1, 5, 0), (2, 3, 0), (2, 5, 0)\} \\
A \times C \times B &= \{(0, 0, 3), (0, 0, 5), (1, 0, 3), (1, 0, 5), (2, 0, 3), (2, 0, 5)\}.
\end{aligned}
$$

Strictly speaking, we should denote the first element listed in $A \times B \times C$ above as $((0, 3), 0)$ if we are thinking of $(A \times B) \times C$, or as $(0, (3, 0))$ if we are thinking of $A \times (B \times C)$. We have no reason to distinguish between $((0, 3), 0)$, $(0, (3, 0))$, and $(0, 3, 0)$, so we simply define $A \times B \times C$, as noted, to be the collection of 3-tuples which results when the innermost pair of parentheses is deleted. Thus Cartesian products are associative.

## EXERCISE 1.3.

**1.** Name two distinct universal sets for a discussion involving sets whose elements are all people residing in a particular town.

**2.** Given the universal set

$$U = \{1, 2, 3, \ldots, n\}$$

and the sets

$$A = \{1, 2, 3\}, \qquad B = \{2, 3, n\},$$

compute $\bar{A}$, $\bar{B}$, and $\bar{A} \cup \bar{B}$. Compare this latter set with $\overline{A \cap B}$.

**3.** Can we define a set $A$ such that $A = \bar{A}$? Such that $A \subset \bar{A}$?

**4.** Show that $\bar{\bar{A}} = A$.

**5.** Show that $\overline{A \cup B} = \bar{A} \cap \bar{B}$. (This is one of De Morgan's laws.)

**6.** Show that $\overline{A \cap B} = \bar{A} \cup \bar{B}$. (This is also one of De Morgan's laws.)
**Hint:** In problem 5, replace $A$ and $B$ by their complements and then take the complement of both sides.

**7.** Draw Venn diagrams illustrating the sets $A \cap \bar{B}$, $\bar{A} \cap \bar{B}$, $A \cup B \cup C$, $A \cap B \cap C$, $(A \cup B) \cap C$, $(A \cap C) \cup (B \cap C)$.

**8.** Define $A = \{1, 2\}$, $B = \{2, 1\}$, $C = \{10, 12\}$, and form the Cartesian products $A \times B$, $A \times C$, $B \times C$, $B \times A$, $C \times A$, $C \times B$, $A \times A$, $B \times B$, $C \times C$, $A \times B \times C$, $C \times B \times A$, $C \times A \times B$.

**9.** What are the conditions under which $A \times B = B \times A$?

**10.** If $A$ is the set of married men living in Texas and $B$ is the set of married women living in Texas, is $A \times B$ the set of married couples living in Texas?

**11.** If $A$ is the set of people with United States citizenship and $B$ is the set of people with Canadian citizenship, does $A \cap B = \varnothing$?

### 1.4. Element Functions and Set Functions

When discussing variables whose values are determined by an experiment or some chance mechanism, we shall have use for the definition of a function of the elements of a set.

DEFINITION 1.4.1. A (real-valued) *element function* $f$ defined on a set $S$ is a rule which associates a (real) number with every element of the set. The number associated with a particular element is called the value of the function for that element and is denoted by $f(\omega)$ for $\omega \in S$.

Many different functions could be defined for the elements of the same set (that is, many different rules could be used). Note that the same number could be associated with more than one element.

For example, consider the set of people residing in a certain town. There are many different rules that we could use to associate a real number with each element of the set, such as the age of the individual, the weight of the individual, the height of the individual, the street number of the individual's residence, the distance of the individual's residence from the city hall, etc. Each of these rules—age, height, weight, etc.—is called an element function defined on the set.

As a second example, suppose that we are given the set of 2-tuples

$$S = \{(x_1, x_2): x_1 = 1, 2, 3;\ \ x_2 = 1, 2, 3\}.$$

The rule $f[(x_1, x_2)] = x_1 + x_2$ for $(x_1, x_2) \in S$ associates the sum of the two numbers making up the 2-tuple with each element of $S$. Thus,

$$f[(1, 1)] = 1 + 1 = 2, \qquad f[(2, 3)] = 2 + 3 = 5,$$

etc. A second-element function which we might define for the same set $S$ is

$$g[(x_1, x_2)] = \frac{x_1}{x_2} \qquad \text{for} \qquad (x_1, x_2) \in S.$$

Then

$$g[(1, 1)] = \tfrac{1}{1} = 1, \qquad g[(2, 2)] = \tfrac{2}{2} = 1,$$
$$g[(1, 3)] = \tfrac{1}{3}, \qquad g[(3, 1)] = \tfrac{3}{1} = 3,$$

and so on.

Two sets which are of interest for any function are its domain of definition and its range. These are defined as follows.

DEFINITION 1.4.2. The *domain of definition* for an element function defined on a set $S$ is simply the set $S$ itself; the *range* of an element function defined on a set $S$ is the collection of real numbers it associates with the elements of $S$ (the range is the set of values of the function).

Thus, for the first example discussed above, the domain of definition of each of the functions given is the set of people residing in the town; the range of the function which associates ages with people is the set of ages of the residents of the town. The range of the weight function is the set of weights of the residents; the range of the height function is the set of heights of residents; the range of the street number function is the set of street numbers in the town; the range of the distance function is the set of distances. In the second example, the domain of definition for both $f$ and $g$ is the set $S$; the range of $f$ is the set $\{2, 3, 4, 5, 6\}$ and the range of $g$ is the set $\{\frac{1}{3}, \frac{1}{2}, \frac{2}{3}, 1, \frac{3}{2}, 2, 3\}$.

Let us now briefly discuss sets, each of whose elements is a set. Such a set is called a class.

DEFINITION 1.4.3. A *class* is a set, each of whose elements is a set.

Generally we shall use script letters to denote classes. Thus,

$$\mathcal{H} = \{\{1\}, \{2\}\}$$
$$\mathcal{K} = \{\varnothing, \{1\}, \{2\}, \{1, 2\}\}$$
$$\mathcal{A} = \{\{a\}, \{b\}, \{a, b\}\}$$

are all classes. The set $\mathcal{B}$ which has as elements all the subsets of $S = \{1, 2, 3\}$ is a class.

DEFINITION 1.4.4. A class $\mathcal{F}$ is *closed* with respect to unions and intersections if and only if $A \cup B \in \mathcal{F}$ and $A \cap B \in \mathcal{F}$ for all $A$ and $B$ belonging to $\mathcal{F}$.

The class $\mathcal{H}$ defined above is not closed since $\{1\} \cup \{2\} = \{1, 2\} \notin \mathcal{H}$. The class $\mathcal{K}$ defined above is closed since the union or the intersection of any two of its elements again belongs to $\mathcal{K}$. The class $\mathcal{A}$ is closed with respect to unions since the union of any two of its elements again belongs to $\mathcal{A}$, but it is not closed with respect to intersections since $\{a\} \cap \{b\} = \varnothing$ does not belong to $\mathcal{A}$. Thus we do not say that $\mathcal{A}$ is closed. The class $\mathcal{B}$ is closed since the union of any two subsets of $S$ is again a subset of $S$, as is the intersection of any two subsets of $S$.

Given a class $\mathcal{F}$ that is closed with respect to unions and intersections, a set function on $\mathcal{F}$ is defined as follows.

DEFINITION 1.4.5. A rule $f$ which associates a real number (denoted by $f(A)$) with each $A \in \mathcal{F}$ is called a (real-valued) *set function* defined on $\mathcal{F}$.

Consider the class $\mathcal{F}$ of subsets of $S = \{0, 1, 2\}$. Then

$$\mathcal{F} = \{\varnothing, \{0\}, \{1\}, \{0, 1\}, \{2\}, \{0, 2\}, \{1, 2\}, \{0, 1, 2\}\}.$$

As in the case of element functions, there are many different set functions that could be defined on $\mathcal{F}$. For example,

$$\begin{aligned} f(A) &= 0, \qquad \text{if } 1 \in A \\ &= 1, \qquad \text{if } 1 \notin A \end{aligned}$$

is a rule which can be used to associate a number with every set $A \in \mathcal{F}$. Thus $f$ is called a set function defined on $\mathcal{F}$. Similarly,

$$\begin{aligned} n(A) &= \text{number of elements in } A,\ A \in \mathcal{F} \\ s(A) &= \text{square of the number of elements belonging to } A,\ A \in \mathcal{F} \end{aligned}$$

are each set functions defined on $\mathcal{F}$. The domain of definition of each of them then is $\mathcal{F}$. The range of $f$ is the set $\{0, 1\}$, the range of $n$ is the set $\{0, 1, 2, 3\}$, and the range of $s$ is the set $\{0, 1, 4, 9\}$.

**EXERCISE 1.4.**

**1.** Let $A$ be the set of individuals in your family and define the element function $f(\omega) = \text{age of } \omega$ for $\omega \in A$. Specify the range of $f$.

**2.** Let

$$\begin{aligned} B &= \{1, 2, 3, 4, 5, 6\} \\ C &= B \times B \end{aligned}$$

and define the element function

$$g[(x_1, x_2)] = x_1 + x_2 \qquad \text{for} \qquad (x_1, x_2) \in C.$$

What is the range of $g$?

**3.** Let $D = \{(x_1, x_2, x_3): x_1 = 0, 1;\ x_2 = 0, 1;\ x_3 = 0, 1\}$ and define

$$h[(x_1, x_2, x_3)] = x_1 + x_2 - x_3.$$

What is the range of $h$?

**4.** Define $C = \{(x_1, x_2): x_1 = 1, 2, 3, 4, 5, 6;\ x_2 = 1, 2, 3, 4, 5, 6\}$ and let

$$g[(x_1, x_2)] = x_1 + x_2, \qquad \text{for } (x_1, x_2) \in C.$$

Specify

$$A_7 = \{(x_1, x_2): g[(x_1, x_2)] = 7\}$$
$$A_3 = \{(x_1, x_2): g[(x_1, x_2)] = 3\}$$
$$A_{10} = \{(x_1, x_2): g[x_1, x_2)] = 10\}.$$

**5.** Let $\mathcal{F}$ be the class of all subsets of $S = \{1, 2, 3, \ldots, k\}$. Define $n(A)$ = number of elements in $A$ for $A \in \mathcal{F}$. What is the range of $n$?

**6.** Let $\mathcal{H}$ be the class of all subsets of $S = \{1, 2, 3, \ldots, k\}$. Define $p(A) = 1/k$ times the number of elements in $A$ for $A \in \mathcal{H}$. What is the range of $p$?

**7.** Let $S$ and $n$ be as defined in problem 5. Show that

$$n(A) \leq n(B), \qquad \text{if } A \subset B,$$
$$n(A \cup B) = n(A) + n(B), \qquad \text{if } A \cap B = \varnothing.$$

# 2

# Probability

We are now ready to begin our study of probability theory itself. The earliest known applications of probability theory occurred in the seventeenth century. A French nobleman of that time was interested in several games then played at Monte Carlo; he tried unsuccessfully to describe mathematically the relative proportion of the time that certain bets would be won. He was acquainted with two of the best mathematicians of the day, Pascal and Fermat, and mentioned his difficulties to them. This began a famous exchange of letters between the two mathematicians concerning the correct application of mathematics to the measurement of relative frequencies of occurrences in simple gambling games. Historians generally agree that this exchange of letters was the beginning of probability theory as we now know it.

For many years a simple relative frequency definition of probability was all that was known and was all that many felt was necessary. This definition proceeds roughly as follows. Suppose that a chance experiment is to be performed (some operation whose outcome cannot be predicted in advance); thus there are several possible outcomes which can occur when the experiment is performed. If an event $A$ occurs with $m$ of these outcomes, then the probability of $A$ occurring is the ratio $m/n$ where $n$ is the total number of outcomes possible. Thus, if the experiment consists of one roll of a fair die and $A$ is the occurrence of an even number, the probability of $A$ is $\frac{3}{6}$.

There are many problems for which this definition is appropriate, but such a heuristic approach is not conducive to a mathematical treatment of the theory of probability. The mathematical advances in probability theory were relatively limited and difficult to establish on a firm basis until the Russian mathematician A. N. Kolmogorov gave a simple set of three axioms or rules which probabilities are assumed to obey. Since the establishment of this firm axiomatic basis, there have been great strides made in the theory of probability and in the number of practical problems to which it is applied.

In this chapter we shall see what these three axioms are and why we might reasonably adopt these rules for probabilities to follow. The axioms do not give any unique value which the probability of an event must equal; rather they express internal rules which ensure consistency in our arbitrary assignment of probabilities. The relative frequency definition of probability which was already mentioned is only one arbitrary way in which probabilities can be computed; it is discussed more fully in Section 2.3 and, as is shown there, does satisfy the axioms. This definition has built within it certain assumptions about the outcomes of the experiment which are not always appropriate. Thus, this arbitrary way of assigning probabilities is not always applicable.

It might aid the reader to discuss briefly at this time the notion of a probability model. An experiment is a physical operation which, in the real world, can result in one of many possible outcomes. For example, if we roll a pair of dice one time the two numbers we might observe can range anywhere from a pair of ones to a pair of sixes. Or, if a particular individual is going to run one hundred yards as fast as he can, the elapsed time from when he starts until when he finishes might be anywhere from nine seconds to thirty seconds. Or, if a particular person lives in a given way in terms of rest and habits of various kinds, his total life span might lie anywhere from zero years to one hundred years. In each of these cases the particular outcome we might observe cannot be predicted in advance, but the total collection of outcomes can be assumed to be known (one of which will be observed when the experiment is completed). When building a probability model for an experiment, we are concerned with specifying: (1) what the total collection of outcomes could be and (2) the relative frequency of occurrence of these outcomes, based on an analysis of the experiment. We are in many senses idealizing the physical situation by restricting the set of possible outcomes; but, to the extent that this idealization does not affect the relative frequency of events of interest, we can profitably use the results of our computations as descriptions of the actual physical experiment. The probability model then consists of the assumed collection of possible outcomes and the assigned relative frequencies or probabilities of these outcomes. The axioms are used to assure consistency in this assignment of probabilities.

It is easy here to give also an indication of the subject matter of statistics that we shall study in the latter portions of this book. Probability theory is concerned with the consistency of assumed probabilities of various outcomes of an experiment and its implications. One problem of statistics, on the other hand, is concerned with whether or not the actual observed outcomes of the physical experiment (in a series of experiments) are consistent with an assumed probability model; another problem of statistics might be that of using a set of actual experimental outcomes to construct a probability model of the physical experiment. Thus in statistics we shall generally be concerned with problems of inference from a set of observed sample outcomes; the theory of probability will prove very useful in measuring the correctness of these inferences.

It is hoped that these brief statements will help orient the reader for our studies of probability and of statistics.

## 2.1. Sample Space; Events

As mentioned in Chapter 1, set theory provides a language ideally suited for the efficient study of probability theory. After the following two basic definitions, we shall see in what manner set theory can be utilized.

DEFINITION 2.1.1. An *experiment* is any operation whose outcome cannot be predicted with certainty.

DEFINITION 2.1.2. The *sample space* $S$ of an experiment is the set of all possible outcomes for the experiment.

Let us examine some particular examples which utilize these two definitions.

*Example 2.1.1.* We roll a single die one time (dice is the plural of die). Then the experiment is the roll of the die. A sample space for this experiment could be

$$S = \{1, 2, 3, 4, 5, 6\}$$

where each of the integers 1 through 6 is meant to represent the face having that many spots being uppermost when the die stops rolling.

*Example 2.1.2.* We select 1 card at random from a standard deck of 52. The experiment is the selection of a card. We might assume that we have numbered the cards (in some specified order) from 1 to 52; then a sample space for this experiment would be

$$S = \{1, 2, 3, \ldots, 52\}$$

since the particular card that we select in performing the experiment must correspond to exactly one of these integers.

*Example 2.1.3.* We select a person at random from the student body of UCLA. The experiment is the selection of a student from the total student body. A sample space for the experiment would be the set of all students that could be selected; i.e., $S$ would simply be the roster of students at UCLA at the time the selection is made.

Frequently the performance of an experiment naturally gives rise to more than one piece of information which we may want to record. If we observe two pieces of information every time the experiment is performed, we would reasonably want a sample space that is a collection of 2-tuples, the two positions corresponding to the two pieces of information. Or, if we observe three pieces of information, we would want a sample space of 3-tuples. Or, more generally, if we observe $r$ pieces of information, we would want a sample space of $r$-tuples. In each of the three examples given above, a single piece of information was generated when performing the experiment; thus each of these sample spaces had 1-tuples as elements. The next three examples discuss experiments in which more than one piece of information is derived.

*Example 2.1.4.* Suppose our experiment consists of one roll of two dice, one red and the other green. A reasonable sample space for the experiment would be the collection of all possible 2-tuples $(x_1, x_2)$ that could occur where the number in the first position of any 2-tuple corresponds to the number on the red die and the number in the second position corresponds to the number on the green die. Thus, we might use as our sample space

$$S = \{(x_1, x_2): x_1 = 1, 2, 3, \ldots, 6;\ x_2 = 1, 2, 3, \ldots, 6\}.$$

*Example 2.1.5.* Doug, Joe, and Hugh match coins. The experiment they perform is one flip of three coins. A reasonable sample space for this experiment is the set of 3-tuples, each of which has $H$ (head) or $T$ (tail) in every position. The first position in the 3-tuple corresponds to the face on Doug's coin, the second position to the face on Joe's coin, and the third position to the face on Hugh's coin. This sample space $S$ can be written

$$S = \{(x_1, x_2, x_3): x_1 = H \text{ or } T, x_2 = H \text{ or } T, x_3 = H \text{ or } T\}.$$

*Example 2.1.6.* We select 10 students, at random, from the student body of the University of Chicago. We would normally restrict the sampling method so that the same student does not occur more than once in the sample (this is called sampling without replacement; see Chapter 6 for a more complete discussion of sampling and its relation to probability). The experiment then is the selection of 10 students. A sample space for the experiment would be the total collection of 10-tuples that could occur, the first element in any 10-tuple corresponding to the first student selected, the second element corresponding to the second student selected, the third element to the third student, etc. For convenience, let us assume that the student body at the University of Chicago consists of 30,000 students and that they have been numbered from 1 to 30,000. Then this sample space may be written

$$S = \{(x_1, x_2, \ldots, x_{10}): x_i = 1, 2, 3, \ldots, 30{,}000;\quad i = 1, 2, \ldots, 10;\ x_i \neq x_j \text{ for all } i \neq j\}.$$

One thing to notice about the sample space of an experiment is that it is not unique; that is, there generally is more than one reasonable way of specifying all possible outcomes of the experiment. For example, if we roll a pair of dice we could use the set of 2-tuples mentioned in Example 2.1.4 as the sample space, or we could reason that every time we roll two dice the sum of the two numbers that occur could be any of the integers 2 through 12 and adopt this set as the sample space. In Example 2.1.5, Doug, Joe, and Hugh will observe 0 or 1 or 2 or 3 heads when they flip their coins; therefore we could adopt the set having these integers as elements as our sample space rather than the set of 3-tuples mentioned above. It would be equally easy to mention an alternative sample space for Example 2.1.6. As we shall see in the sequel, one sample space is generally easier to use than another for a given experiment. Of course, we shall try to use the one that is easiest for our purposes. Which sample space is used has no effect on the values of probabilities of interest but does affect the ease of computation of such probabilities.

The following two definitions are very basic to the sequel. Misunderstanding of the definition of an event and of when an event has or has not occurred can lead to a great deal of difficulty in many problems.

DEFINITION 2.1.3. An *event* is a subset of the sample space. Every subset is an event.

DEFINITION 2.1.4. An event *occurs* if any one of its elements is the outcome of the experiment.

The sample space used for Example 2.1.1 was

$$S = \{1, 2, 3, 4, 5, 6\}.$$

Then each of the sets

$$A = \{1\}, \quad B = \{1, 3, 5\}, \quad C = \{2, 4, 6\},$$
$$D = \{4, 5, 6\}, \quad E = \{1, 3, 4, 6\}, \quad F = \{2\},$$

is an event (these are not the only events since they are not the only subsets of $S$). These are all distinct (different) events because no two of these subsets are equal. If we actually were to perform the experiment (roll the die) and we got a 1, then events $A$, $B$, and $E$ are said to have occurred since each of these has 1 as an element. Events $C$ and $D$ did not occur since $1 \notin C$ and $1 \notin D$. If we got a 4 when the die was rolled, then we would say that events $C$, $D$, and $E$ occurred since 4 is an element of each of them. Notice that no matter which outcome we observe when the experiment is performed, many different events have each occurred (as we shall see, exactly half the possible events occur for any particular outcome).

In Example 2.1.3, which consists of selecting one student at random from the student body of UCLA, the sample space consists of the roster of students at the school. Assuming there are thirty-five thousand students at UCLA, we can represent the sample space by

$$S = \{x\colon x = 1, 2, 3, \ldots, 35{,}000\}.$$

The sets

$$A_1 = \{1, 50\}, \qquad B = \{2, 76, 140, 64\}, \qquad C = \{10\}, \qquad D = \{x\colon x = 101, 102, \ldots, 960\},$$

etc., are all events since each is a subset of $S$. If, when we performed the experiment, we happened to select the student numbered 176, then $D$ occurred since $176 \in D$, but none of the other three events listed occurred. If we happened to select the student numbered 9999, then none of the four events listed occurred since this number doesn't belong to any of them.

We shall frequently want to take a word description of an event and translate it into a subset of the sample space. For example, suppose that in Example 2.1.2 we had numbered the 52 cards in the following way: A, 2, 3, . . . , K of hearts are numbered 1 through 13, respectively; A, 2, 3, . . . , K of diamonds are numbered 14 through 26, respectively; A, 2, 3, . . . , K of spades are numbered 27 through 39, respectively; A, 2, 3, . . . , K of clubs are numbered 40 through 52, respectively. Then we might define the following events in words:

$A$: a red card is drawn

$B$: a spade is drawn

$C$: an ace is drawn

$D$: a face card is drawn (aces are not counted as face cards).

Then, since an event occurs if any of its elements occurs, we have

$$A = \{1, 2, \ldots, 26\}$$
$$B = \{27, 28, \ldots, 39\}$$
$$C = \{1, 14, 27, 40\}$$
$$D = \{11, 12, 13, 24, 25, 26, 37, 38, 39, 50, 51, 52\}.$$

In Example 2.1.4 a pair of dice is rolled one time. The sample space $S$ is the set of 2-tuples

$$S = \{(x_1, x_2)\colon x_1 = 1, 2, \ldots, 6;\ x_2 = 1, 2, \ldots, 6\}.$$

The events:

$A$: the sum of the two dice is 3

$B$: the sum of the two dice is 7

$C$: the two dice show the same number

are as subsets,

$$A = \{(1, 2), (2, 1)\}$$
$$B = \{(1, 6), (2, 5), (3, 4), (4, 3), (5, 2), (6, 1)\}$$
$$C = \{(1, 1), (2, 2), (3, 3), (4, 4), (5, 5), (6, 6)\}.$$

Suppose that in Example 2.1.5, Doug, Joe, and Hugh are playing a game called "odd man loses." That is, if two of the coins' faces match and the third person's does not, the third person loses. The sample space is

$$S = \{(x_1, x_2, x_3): x_1 = H \text{ or } T, x_2 = H \text{ or } T, x_3 = H \text{ or } T\},$$

where the first position corresponds to Doug's coin, the second to Joe's, the third to Hugh's. Define the events:

$A$: Doug loses
$B$: Doug doesn't lose
$C$: Joe loses
$D$: No one loses.

Then, written as subsets, we have

$$A = \{(H, T, T), (T, H, H)\}$$
$$B = \{(H, H, T), (H, T, H), (H, H, H), (T, T, H), (T, H, T), (T, T, T)\}$$
$$C = \{(T, H, T), (H, T, H)\}$$
$$D = \{(H, H, H), (T, T, T)\}.$$

*Example 2.1.7.* Suppose our experiment consists of a hundred-yard dash involving four college-age sprinters. It is clear that there are many facets of the experiment that we might be interested in such as the name of the winner, the winning time, the order in which the four cross the finish line, the time of the second-place man, or the time of the third-place man. Which facets were of interest would determine the sample space to be used. For example, if we were going to refer only to the winning time we could use

$$S_1 = \{t: 0 \le t \le 15\}$$

(measuring time in seconds). Or, if we were interested in the times of both the first- and second-place men, we could use

$$S_2 = \{(t_1, t_2): 0 \le t_1 \le 15, t_1 < t_2 \le 20\}.$$

In this latter case $t_1$ is the time of the winner and $t_2$ is the time of the second-place man; thus the requirement $t_2 > t_1$. For either of these sample spaces we could define the event $A$: winning time is between 9.45 and 9.65 seconds; using $S_1$ we would have

$$A = \{t: 9.45 \le t \le 9.65\}.$$

And using $S_2$ we would have

$$A = \{(t_1, t_2): 9.45 \leq t_1 \leq 9.65, t_1 < t_2 \leq 15\}.$$

Again, then, the sample space is not unique. If $A$ were the only event of interest, we would undoubtedly decide to use $S_1$ rather than $S_2$ since this would require keeping track of only the single time of the first-place finisher.

## EXERCISE 2.1.

**1.** Specify a sample space for the experiment which consists of drawing 1 ball from an urn containing 10 balls of which 4 are white and 6 are red. (Assume that the balls are numbered 1 through 10.)

**2.** Specify a sample space for the experiment which consists of drawing 2 balls with replacement from the urn containing 10 balls (that is, the first ball removed is replaced in the urn before the second is drawn out). Again assume that they are numbered.

**3.** Specify a sample space for the experiment which consists of drawing 2 balls without replacement from the urn containing 10 balls (that is, the first ball removed is not replaced in the urn before the second is drawn out). Assume that they are numbered.

**4.** For the sample space given in problem 1, define the events (as subsets):

$A$: a white ball is drawn

$B$: a red ball is drawn.

**5.** For the sample space given in problem 2, define the events (as subsets):

$C$: the first ball is white

$D$: the second ball is white

$E$: both balls are white.

Does $C \cap D = E$?

**6.** A cigarette company packs 1 of 5 different slips, labelled $a, b, c, d, e$, respectively, with each pack it produces. Suppose that you buy 2 packs of cigarettes of this brand. What is a good sample space for the experiment whose outcome is the pair of slips you receive with the 2 packs?

**7.** Suppose that all of the residents of a particular town are bald or have brown hair or have black hair. Furthermore, each resident has blue eyes or brown eyes. We select one resident at random. Give a sample space $S$ for this experiment and define, as subsets, these events:

$A$: the selected resident is bald

$B$: the selected resident has blue eyes

$C$: the selected resident has brown hair and brown eyes.

**8.** Three girls, Marie, Sandy, and Tina, enter a beauty contest. Prizes are awarded for first and second place. Specify a sample space for the experiment which consists of the choice of the two winners. Define, as subsets, the events:

$A$: Marie wins

$B$: Marie gets second prize

$C$: Tina and Sandy get the prizes.

**9.** Three cards are selected at random without replacement from a deck which contains 3 red, 3 blue, 3 green, and 3 black cards. Give a sample space for this experiment and define the events:

$A$: all the selected cards are red

$B$: 1 card is red, 1 green, and 1 blue

$C$: 3 different colors occur

$D$: all 4 colors occur.

**10.** A small town contains 3 grocery stores (call them 1, 2, 3). Four ladies living in this town each randomly and independently pick a store in which to shop (in this town). Give a sample space for the experiment which consists of the selection of stores by the ladies and define the events:

$A$: all the ladies choose store 1

$B$: half the ladies choose store 1 and half choose store 2

$C$: all the stores are chosen (by at least one lady).

## 2.2. Probability Axioms

As we shall see in this section, the theory of probability is concerned with consistent ways of assigning numbers to events (subsets of the sample space $S$) which are called the probabilities of occurrence of these events. It is because we want to have the ability to compute the probability of occurrence of any subset of the sample space that we permit every subset to be called an event. In almost every problem, then, we shall be aware that there are many events which we could define and whose probabilities we could compute in addition to the particular few events that we shall have interest in. It seems much more satisfying to have the ability to compute probabilities for any conceivable event (most of which are not of interest) than suddenly to come across a problem for which we cannot derive an answer because the quantity of interest is not an event.

Most people with an intuitive feeling for what probability should be give a relative frequency interpretation to numbers called probabilities. For example, most would be quick to agree that the probability of a head occurring if we flip a fair coin is one-half, meaning that if the coin is fair then half the time we should observe a head. There are two immediate consequences to a relative frequency motivation for probability. First, the relative frequency of occurrence of something we are sure will occur should be 1; thus its probability should be 1. For example, if we flip a coin one time the relative frequency of the event of observing a head or a tail should be 1, thus the probability of this event should be 1. Second, a relative frequency can never be negative; thus the probability of any event should be nonnegative. These two rules are the first two axioms of an abstract probability theory in order that assignments of probabilities will have these intuitive properties. We shall insist that the probabilities of occurrence of any events satisfy these two requirements.

There is only one other rule or axiom which we shall insist is always satisfied—the additivity property of probability. If we refer to the experiment which consists of one roll of one die, it is only reasonable to expect that the probability that we observe a 1 or a 2 should be equal to the sum of the probability of observing a 1 plus that of observing a 2 since relative frequencies of occurrence have this property. More generally, we would expect the probability of any event $A$ to be equal to the sum of the probabilities of any two nonoverlapping events which together constitute the event $A$ (their union is $A$). This is the additivity requirement.

Probability can be called a measure applied to the events that can occur when an experiment is performed. To the extent that the assumptions that generate the numerical values of the measure are "correct" for the given problem, the probabilities coincide with the relative frequency notions just discussed. To ensure that this is the case, the probability measures must satisfy the three axioms given below.

Formally, a probability function is a real-valued set function defined on the class of all subsets of the sample space $S$; the value that is associated with a subset $A$ is denoted by $P(A)$. The assignment of probabilities must satisfy the following three rules (in order that the set function may be called a probability function):

1. $P(S) = 1$
2. $P(A) \geq 0 \qquad \text{for all } A \subset S$
3. $P(A_1 \cup A_2 \cup A_3 \cup \cdots) = P(A_1) + P(A_2) + P(A_3) + \cdots$

$$\text{if } A_i \cap A_j = \varnothing \qquad \text{for all } i \neq j.$$

Note that these are the three axioms just discussed. For any experiment, the sample space $S$ plays the role of the universal set; thus any complements which we refer to are taken with respect to $S$.

Many consequences or theorems can be derived for any probability function. Let us take a look at some of these now. It will be recalled that $\varnothing \subset S$. Thus our probability measure must assign some number to this event. The number that must always be assigned is 0, as is proved in Theorem 2.2.1.

***Theorem 2.2.1.*** $P(\varnothing) = 0$ for any $S$.

*Proof:* $S \cup \varnothing = S$ and thus $P(S \cup \varnothing) = P(S) = 1$ by axiom 1. But $S \cap \varnothing = \varnothing$ so that $P(S \cup \varnothing) = P(S) + P(\varnothing) = 1 + P(\varnothing)$ by axiom 3. Thus $1 + P(\varnothing) = 1$; that is, $P(\varnothing) = 0$. ◀

A second consequence of the assumed axioms is given as Theorem 2.2.2. We shall see many instances in which it saves a great deal of effort in computing probabilities.

***Theorem 2.2.2.*** $P(\bar{A}) = 1 - P(A)$, where $\bar{A}$ is the complement of $A$ with respect to $S$.

*Proof:* $A \cup \bar{A} = S$ so $P(A \cup \bar{A}) = P(S) = 1$ by axiom 1. But $A \cap \bar{A} = \varnothing$ and thus $P(A \cup \bar{A}) = P(A) + P(\bar{A})$ by axiom 3. Thus we have established that $P(A) + P(\bar{A}) = 1$, from which the result follows immediately. ◀

Axiom 3 tells us that if two events $A$ and $B$ have no elements in common, then the probability of their union is the sum of their individual probabilities. Theorem 2.2.3 derives a preliminary result which is used to establish Theorem 2.2.4, regarding $P(A \cup B)$ when $A$ and $B$ have elements in common.

***Theorem 2.2.3.*** $P(\bar{A} \cap B) = P(B) - P(A \cap B)$.

*Proof:* By referring to Figure 2.1, it can be seen that

$$B = (\bar{A} \cap B) \cup (A \cap B).$$

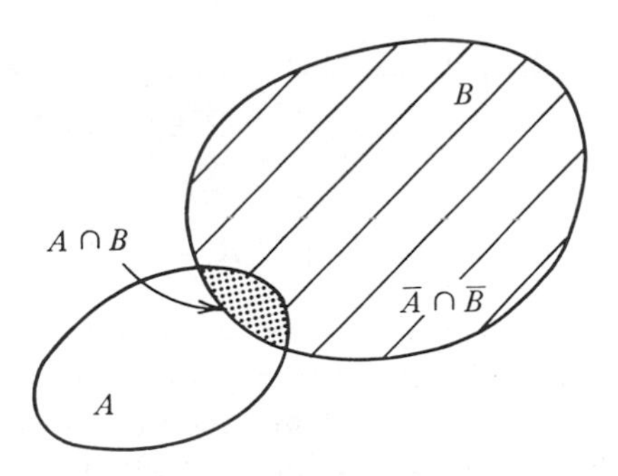


**Figure 2.1.**

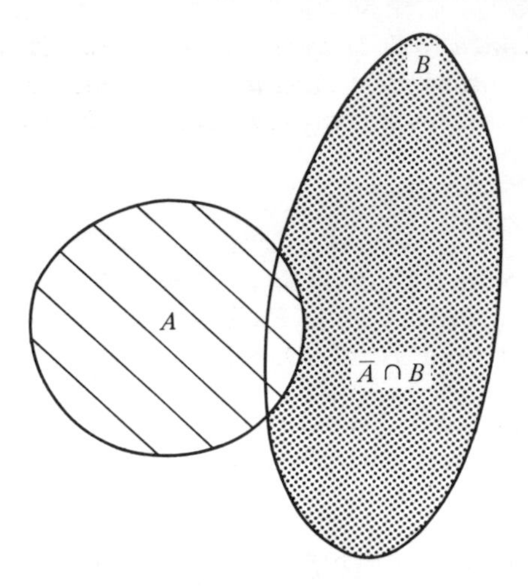


**Figure 2.2.**

Then $P(B) = P((\bar{A} \cap B) \cup (A \cap B))$. Furthermore, $(\bar{A} \cap B) \cap (A \cap B) = \varnothing$; thus, $P((\bar{A} \cap B) \cup (A \cap B)) = P(\bar{A} \cap B) + P(A \cap B)$, and we have established that

$$P(B) = P(\bar{A} \cap B) + P(A \cap B)$$

from which the desired result follows immediately. ◀

***Theorem 2.2.4.*** $P(A \cup B) = P(A) + P(B) - P(A \cap B)$.

*Proof:* By referring to Figure 2.2, it can be seen that we can write

$$A \cup B = A \cup (\bar{A} \cap B),$$

and thus $P(A \cup B) = P(A \cup (\bar{A} \cap B))$. Furthermore, $A \cap (\bar{A} \cap B) = \varnothing$ so that

$$\begin{aligned} P(A \cup (\bar{A} \cap B)) &= P(A) + P(\bar{A} \cap B) \\ &= P(A) + P(B) - P(A \cap B), \end{aligned}$$

by Theorem 2.2.3. Thus

$$P(A \cup B) = P(A) + P(B) - P(A \cap B).$$ ◀

Notice that the probability function we have discussed in this section is just a special sort of (real-valued) set function. For every experiment that can be performed we assume that we can define the sample space $S$ which has as elements all possible outcomes; the class of all subsets of $S$ can then be defined and probabilities are just the values assumed by a set function defined on the class of all subsets. The set function whose values are called probabilities is distinguished by the fact that the three axioms just discussed must be satisfied.

It is important to realize that these axioms will not give a unique assignment of probabilities to events; rather the axioms simply clarify relationships between probabilities that we assign so that we shall be consistent with our intuitive notions of probability. For example, if a rocket has been designed to take a man to the moon, then the experiment which consists of firing the rocket and the man at the moon can be thought of as having two outcomes: success and failure. Success would be the safe arrival of the man on the moon, failure anything else that might occur. Then the axioms do not imply that the probability of the event {success} must be $\frac{1}{2}$ or $\frac{3}{4}$ or .99 or any other particular value. If we denote this probability by $p$, they do imply that $0 \leq p \leq 1$ and that the probability of the event {failure} must be $1 - p$. Beyond this, $p$ is still unspecified.

Actual specification of the value of $p$ must come from analytical considerations of the experiment performed and the mechanism behind it. For the rocket example just mentioned, this would consist of detailed examination of the rocket design and conditions under which it is to be fired, in addition to any prior test firings or performance data available. Generally, considerations of prior data and their implications regarding the value of $p$ fall into the realm of statistics, the topic for the latter half of this volume.

## EXERCISE 2.2.

**1.** Given $S = \{1, 2, 3\}$, $A = \{1\}$, $B = \{3\}$, $C = \{2\}$, $P(A) = \frac{1}{3}$, $P(B) = \frac{1}{3}$, find:

(a) $P(C)$

(b) $P(A \cup B)$

(c) $P(\bar{A})$

(d) $P(\bar{A} \cap \bar{B})$

(e) $P(\bar{A} \cup \bar{B})$

(f) $P(B \cup C)$.

**2.** Let $S$, $A$, $B$, $C$ be defined as in problem 1, but now let $P(A) = \frac{1}{2}$, $P(B) = \frac{1}{5}$. Compute the probabilities asked for in (a) through (f).

**3.** Let $S$, $A$, $B$, $C$ be defined as in problem 1 and let $P(A) = 1$. Compute the probabilities asked for in (a) through (f). Could we let $P(A)$ be 2?

**4.** Define $S = \{a, b\}$, $B = \{b\}$. Give three different assignments of probabilities to the subsets of $S$.

**5.** Prove, from the axioms, that probabilities are monotonic; that is, $P(A) \leq P(B)$ if $A \subset B$.

**6.** Prove, from the axioms, that $P(A) \leq 1$ for all $A$.

**7.** Given an experiment such that $P(A) = \frac{1}{2}$, $P(B) = \frac{1}{2}$, $P(A \cup B) = \frac{2}{3}$, compute:

(a) $P(\bar{A})$
(b) $P(\bar{B})$
(c) $P(A \cap B)$
(d) $P(\bar{A} \cap \bar{B})$
(e) $P(\bar{A} \cup \bar{B})$
(f) $P(A \cap \bar{B})$
(g) $P(\bar{A} \cap B)$
(h) $P(\bar{A} \cup B)$.

**8.** Given an experiment such that $P(A) = \frac{1}{2}$, $P(B) = \frac{1}{3}$, $P(A \cap B) = \frac{1}{4}$, compute:

(a) $P(A \cup B)$
(b) $P(\bar{A} \cup B)$
(c) $P(\bar{A} \cap B)$
(d) $P(A \cap \bar{B})$
(e) $P(\bar{A} \cap \bar{B})$
(f) $P(\bar{A} \cup \bar{B})$.

**9.** Is it possible to have an assignment of probabilities such that $P(A) = \frac{1}{2}$, $P(A \cap B) = \frac{1}{3}$, $P(B) = \frac{1}{4}$?

**10.** If we know that $P(A \cup B) = \frac{2}{3}$ and $P(A \cap B) = \frac{1}{3}$, can we determine $P(A)$ and $P(B)$?

## 2.3. Single-Element Events and the Equally Likely Case

When we are dealing with an experiment that has a finite number of possible outcomes (thus $S$ is a finite set), the concept of a single-element event becomes rather important; as we shall see, the specification of the values of the probability function $P$ for the single-member events then completely specifies the values of $P$ for all events. First, the definition of a single-element event is as follows.

DEFINITION 2.3.1. A *single-element event* $A$ is a subset of $S$ which has only one element of $S$ belonging to it; that is, if there exists only one $x \in S$ such that $x \in A \subset S$, then $A$ is a single-element event.

For the sample space $\{H, T\}$, the single-element events are $\{H\}$, $\{T\}$. For the sample space $\{1, 2, 3, 4, 5, 6\}$, the single-element events are $\{1\}$, $\{2\}$, $\{3\}$, $\{4\}$, $\{5\}$, $\{6\}$. For the sample space $\{(0, 0), (0, 1), (1, 0), (1, 1)\}$, the single-element events are $\{(0, 0)\}$, $\{(0, 1)\}$, $\{(1, 0)\}$, $\{(1, 1)\}$. As can easily be seen, if $S$ has $n$ elements, then there are exactly $n$ distinct single-element events. The following theorem shows how the probabilities of occurrence of the single-element events imply the probabilities of occurrence of any event, no matter how many elements belong to it.

***Theorem 2.3.1.*** Given a sample space $S$ and any event $A \subset S$,

$$P(A) = P(A_1) + P(A_2) + \cdots + P(A_k)$$

where $A_1, A_2, \ldots, A_k$ are distinct single-element events and $A = A_1 \cup A_2 \cup \cdots \cup A_k$.

*Proof:* Suppose that $A_1, A_2, \ldots, A_k$ are distinct single-element events. Then $A_i \cap A_j = \varnothing$ for all $i \neq j$ and, by axiom 3, $P(A_1 \cup A_2 \cup \cdots \cup A_k) = \sum_{i=1}^{k} P(A_i)$. Thus, if $A = A_1 \cup A_2 \cup \cdots \cup A_k$, we have

$$P(A) = \sum_{i=1}^{k} P(A_i).$$

◀

*Example 2.3.1.* Suppose we roll a six-sided die. Then $S = \{1, 2, 3, 4, 5, 6\}$. The single-element events are $\{1\}, \{2\}, \{3\}, \{4\}, \{5\}, \{6\}$. Theorem 2.3.1 tells us that if we know the probability of occurrence for each of these six events, we can use this information to compute the probability of occurrence of any other event of interest. For example,

$$P(\{1, 4, 6\}) = P(\{1\}) + P(\{4\}) + P(\{6\}).$$

In many experiments it is quite reasonable to assume that each single-element event is as likely to occur as any other. For example, if our experiment consists of one flip of a fair coin, it is reasonable to assume that the single-element events of $S = \{H, T\}$ are equally likely to occur. Or, if our experiment consists of one roll of a fair die, it is reasonable to assume that the single-element events of $S = \{1, 2, 3, 4, 5, 6\}$ are equally likely to occur. If we make the assumption that all single-element events of $S$ are equally likely, there is a very simple way in which we can assign probabilities to all the subsets of $S$ and still be consistent with the three axioms.

Assume that we have an experiment with $k$ possible outcomes, and from the description of the experiment we are willing to assume that each of the single-element events is as likely to occur as any other. Then, since the total probability of the whole sample space is to be 1, the common value of the probability for each of the single-element events must be $1/k$. Furthermore, since any event is the union of single-element events (as we just saw in Theorem 2.3.1), the probability of any event $A \subset S$ then is given by the ratio of the number of elements in $A$ (the number of single-element events whose union is $A$) to the number of elements in $S$. That is, we use the rule

$$P(A) = \frac{n(A)}{n(S)} \qquad \text{for } A \subset S$$

where $n(A)$ is the number of elements in $A$ (see the last example in Chapter 1, Section 1.4). That this rule will satisfy the three axioms given in Section 2.2 is proved in the next theorem.

***Theorem 2.3.2.*** If $S$ has $k$ elements the rule

$$P(A) = \frac{n(A)}{n(S)}$$

satisfies the three axioms for a probability function.

*Proof:* If $S$ has $k$ elements, then $n(S) = k$.

$$P(S) = \frac{n(S)}{n(S)} = \frac{k}{k} = 1$$

so axiom 1 is satisfied. If $A$ is any subset of $S$, it contains a nonnegative number of elements; i.e., $n(A) \geq 0$ for all $A \subset S$. Then

$$\frac{n(A)}{k} = \frac{n(A)}{n(S)} = P(A) \geq 0 \qquad \text{for all } A \subset S$$

and axiom 2 is satisfied. If $A \cap B = \varnothing$, then $A$ and $B$ have no elements in common and we would know that $n(A \cup B) = n(A) + n(B)$. Thus

$$\frac{n(A \cup B)}{n(S)} = \frac{n(A)}{n(S)} + \frac{n(B)}{n(S)};$$

that is,

$$P(A \cup B) = P(A) + P(B).$$

Clearly this line of reasoning is valid for the union of any number of non-overlapping events and thus axiom 3 is also satisfied. ◄

Thus, for any problem in which we are justified in assuming equally likely single-element events, we now have a rule which enables us to compute the probability of occurrence of any event.

*Example 2.3.2.* Suppose we roll a fair die one time. What is the probability of getting an even number? What is the probability of getting a number which is greater than 4?

Our sample space is $S = \{1, 2, 3, 4, 5, 6\}$. Since the die is fair, we assume the single-element events to be equally likely to occur; each then has probability $\frac{1}{6}$ of occurrence. Let $A$ be the event that an even number occurs and let $B$ be the event that we get a number greater than 4.

$$A = \{2, 4, 6\}, \qquad B = \{5, 6\},$$

$$n(A) = 3, \qquad n(B) = 2, \qquad n(S) = 6,$$

and we get $P(A) = \frac{3}{6}$, $P(B) = \frac{2}{6}$.

*Example 2.3.3.* We roll a pair of dice one time. What is the probability that the sum of the two numbers is 2? That it is 7? That it is 11?

Our sample space is $S = \{(x_1, x_2)\colon x_1 = 1, 2, \ldots, 6;\ x_2 = 1, 2, \ldots, 6\}$. Since the first die can have a number from 1 through 6 on it and, quite independently, the second die can also have any number 1 through 6 on it, we reason that there are $6 \cdot 6 = 36$ elements belonging to $S$ (or we simply list them all and count them). Thus $n(S) = 36$. Let $A$ be the event that the sum is 2, $B$ the event that the sum is

7, $C$ the event that the sum is 11. Then

$$A = \{(1, 1)\}$$
$$B = \{(1, 6), (2, 5), (3, 4), (4, 3), (5, 2), (6, 1)\}$$
$$C = \{(5, 6), (6, 5)\}$$

and we see that $n(A) = 1$, $n(B) = 6$, $n(C) = 2$. Then, since we are assuming that the dice are fair, $P(A) = \frac{1}{36}$, $P(B) = \frac{6}{36}$, $P(C) = \frac{2}{36}$.

*Example 2.3.4.* A certain class has 20 students. Of the 20, 7 are blue-eyed blond girls, 4 are blue-eyed brunette girls, 5 are blue-eyed blond boys, and the remaining 4 are brown-eyed brunette boys. We select 1 student at random. What is the probability that the selected student is a girl? Has blue eyes? Is a brunette? Is a brown-eyed blond? For ease of notation, we assume that the 20 students are numbered in some specified order. Then our sample space is

$$S = \{1, 2, \ldots, 20\}$$

and the single-element events are equally likely to occur (since the student is selected at random). Define the following events:

$A$: the selected student is a girl

$B$: the selected student has blue eyes

$C$: the selected student is a brunette

$D$: the selected student is a brown-eyed blond.

Then $N(S) = 20$, $N(A) = 11$, $N(B) = 16$, $N(C) = 8$, $N(D) = 0$, and we have

$$P(A) = \tfrac{11}{20}, \qquad P(B) = \tfrac{16}{20}, \qquad P(C) = \tfrac{8}{20}, \qquad P(D) = 0.$$

Many important problems have equally likely single-element events. As we have seen, computations of probabilities in these cases reduce to counting the number of elements in $S$ and the number of elements in the events of interest. Probabilities are then given by the ratio of these quantities. Since counting the number of elements belonging to an event plays a role in many practical problems, the next two sections are devoted to counting techniques.

## EXERCISE 2.3.

**1.** If two fair coins are flipped, what is the probability that the two faces are alike?

**2.** If we draw 1 card at random from a standard deck of 52, what is the probability that it is red? That it is a diamond? That it is an ace? That it is the ace of diamonds?

**3.** Five different colored rubber bowls with the same identical dog food in them are laid out in a row. If a dog chooses a bowl at random from which to eat, what is the probability that he selects the blue one? If a second dog is used, what is the probability that he chooses the blue one? What is the probability that both choose the blue one?

**4.** A pair of fair dice is rolled once. Compute the probability that the sum is equal to each of the integers 2 through 12.

**5.** A one is painted on the head side and a two is painted on the tail side of each of 3 fifty-cent pieces. If the 3 coins are all tossed once (together), compute the probability that the sum of the three numbers occurring is each of the integers 3 through 6.

**6.** Forty people are riding on the same railroad car. Of this number, 5 are Irish ladies with blue coats, 2 are Irish men with green coats, 1 is an Irish man with a black coat, 7 are Norwegian ladies with brown coats, 2 are Norwegian ladies with blue coats, 6 are Norwegian men with black coats, 4 are German men with green coats, 3 are German ladies with black coats, 5 are German ladies with blue coats, and 5 are German men with black coats. If we select one person at random from this car, what is the probability that the selected person is a man? Is wearing a green coat? Is wearing a brown coat? Is a Norwegian? Is a German? Is a German wearing a green coat?

**7.** Suppose that a die has been loaded in such a way that the probability of a particular number occurring is proportional to that number. Compute the probabilities of all the single-element events and use these to compute the probability of occurrence of an even number and of a number greater than 4.

### 2.4. Counting Techniques

As was mentioned in the last section, the solutions of many probability problems depend on being able to count the numbers of elements belonging to particular sets. A number of techniques are invaluable aids in counting the number of elements belonging to certain sets and thus they are of use in solving probability problems in which these sets occur. Unfortunately, there is no general technique that is universally applicable to all counting problems (other than making a complete list and counting the number of items in it); thus it is necessary to attempt to tailor-make counting methods to the particular counting problem in hand, a process that is always frustrating and frequently fruitless. Be that as it may, almost all counting problems seem to yield eventually to a closely reasoned analysis.

The first technique of some generality that we shall discuss is a very simple one frequently referred to as the multiplication principle. This is defined as follows.

DEFINITION 2.4.1. If a first operation can be performed in any of $n_1$ ways and a second operation can then be performed in any of $n_2$ ways, both operations can be performed (the second immediately following the first) in $n_1 \cdot n_2$ ways.

This definition can immediately be extended to the simultaneous performance of any number of operations. For example, if we can travel from town $A$ to town $B$ in 3 ways and from town $B$ to town $C$ in 4 ways, then we can travel from $A$ to $C$ in a total of $3 \cdot 4 = 12$ ways. Or, if the operation of tossing a die gives rise to 1 of 6 possible outcomes and the operation of tossing a second die gives rise to 1 of 6 possible outcomes, then the operation of tossing a pair of dice gives rise to $6 \cdot 6 = 36$ possible outcomes.

*Example 2.4.1.* Suppose that a set $A$ has $n_1$ elements and a second set $B$ has $n_2$ elements. Then the Cartesian product $A \times B$ (see Definition 1.3.4) has $n_1 n_2$ elements. The Cartesian product $A \times A$ has $n_1^2$ elements, $B \times B$ has $n_2^2$ elements.

DEFINITION 2.4.2. An arrangement of $n$ symbols in a definite order is called a *permutation* of the $n$ symbols.

We shall frequently want to know how many different $n$-tuples can be made using $n$ different symbols (that is, how many permutations are possible). The multiplication principle will immediately give us the answer. We can count the number of $n$-tuples by reasoning as follows. In listing all the possible $n$-tuples, we would perform $n$ natural operations. First we must fill the leftmost position of the $n$-tuple. Then we must fill the second leftmost position, the third leftmost position, etc. Since we could put any of the $n$ elements available into the leftmost position, this operation can be performed in $n$ ways. After we fill the leftmost position, we can use any of the remaining $n - 1$ elements to fill the second leftmost position; we can use any of the remaining $n - 2$ elements to fill the third leftmost position; and so on until we finally arrive at the rightmost position and want to count the number of ways it can be filled. We have used $n - 1$ elements at this point to fill the first $n - 1$ positions and thus have left only one element which must be used to fill the final position. Then the total number of ways we can perform all $n$ operations (which is also the total number of $n$-tuples we could make with the given $n$ symbols) is given by the product of the numbers of ways of doing the individual operations. Thus the number of different $n$-tuples is $n(n - 1)(n - 2) \cdots 2 \cdot 1$ which we write $n!$ (read $n$-factorial).

*Example 2.4.2.* Suppose that the same 5 people park their cars on the same side of the street in the same block every night. How many different orderings of the 5 cars parked on the street are possible? The different orderings of the cars parked on the street could be represented by 5-tuples with 5 distinct elements; then if we can count the number of different 5-tuples that are possible, we also know the number of different orderings on the street. The number of 5-tuples possible is of course $5! = 5 \cdot 4 \cdot 3 \cdot 2 \cdot 1 = 120$; thus these 5 people could park their cars on the street in a different order every night for 4 months without repeating an ordering they had already used.

DEFINITION 2.4.3. The number of $r$-tuples we can make ($r \leq n$), using $n$ different symbols (each only once), is called the *number of permutations of n things r at a time* and is denoted by ${}_nP_r$.

How might we compute the value of ${}_nP_r$? Each $r$-tuple has exactly $r$ positions. The leftmost position could be filled by any of the $n$ symbols; the second leftmost position could then be filled by any of the remaining $n - 1$ symbols, etc. By the time we are ready to fill the $r$th position, we have used $(r - 1)$ symbols already and any of the remaining $n - (r - 1)$ symbols could be used in the $r$th position. Thus the total number of $r$-tuples we could construct is $n(n - 1) \cdots (n - r + 1)$ and we have

$$ {}_nP_r = n(n-1)\cdots(n-r+1). $$

If we multiply this number by $(n - r)!/(n - r)!$, we certainly do not change its value; but in so doing we get a form of ${}_nP_r$ that is much easier to remember:

$$ \begin{aligned} {}_nP_r &= n(n-1)(n-2)\cdots(n-r+1)\frac{(n-r)!}{(n-r)!} \\ &= \frac{n(n-1)(n-2)\cdots(n-r+1)(n-r)(n-r-1)\cdots 2\cdot 1}{(n-r)!} \\ &= \frac{n!}{(n-r)!} \end{aligned} $$

*Example 2.4.3.* (a) Fifteen cars enter a race. In how many different ways could trophies for first, second, and third place be awarded? This answer is simply ${}_{15}P_3 = 15!/12! = 2730$ since the question is equivalent to asking how many permutations are there of 15 objects, 3 at a time.

(b) How many of the 3-tuples just counted have car number 15 in the first position? This can be answered in two ways. First we might reason that there are ${}_{14}P_2 = 14!/12! = 182$ ways in which the last two positions could be filled, having already put 15 into the first position of the 3-tuple. Alternatively, there must obviously be equal numbers of 3-tuples (in the totality of all possible) having car 15 in first place as there are having car 14 in first place, 13 in first place, etc. Thus, if we divide the total number of 3-tuples by 15, we should get the number that have car 15 in first place; this gives $2730/15 = 182$—the same answer derived above.

*Example 2.4.4.* (a) How many three-letter "words" can we make using the letters w, i, n, t, e, r (allowing no repetition)? (A "word" is any arrangement of letters, regardless of whether it is in actual fact a word listed in the dictionary for some language.) This is, of course, just ${}_6P_3 = 6!/3! = 120$. The number of four-letter words is ${}_6P_4 = 360$, etc.

(b) Suppose that repetition of a letter is allowed in making three-letter words using w, i, n, t, e, r. How many three-letter words can we make? The answer is $6 \cdot 6 \cdot 6 = 6^3 = 216$, since we now would be able to fill each position with six letters. The number of four-letter words we could make, allowing repetition, then is $6 \cdot 6 \cdot 6 \cdot 6 = 6^4 = 1296$.

(c) How many three-letter words are there with one or more repeated letters? How many four-letter words are there with one or more repeated letters? We know that there are 216 three-letter words if repetitions are allowed and 120 three-letter words if repetition is not allowed. Thus, there are $216 - 120 = 96$ three-letter words with one or more repeated letters. Analogously, there are $1296 - 360 = 936$ four-letter words with one or more repeated letters.

DEFINITION 2.4.4. The number of subsets, each of size $r$, that a set with $n$ elements has is called the number of *combinations* of $n$ things $r$ at a time and is denoted by $\binom{n}{r}$.

Remember that sets are not ordered. If we knew $\binom{n}{r}$ and multiplied by $r!$, we would have to get ${}_nP_r$, the number of permutations of $n$ things $r$ at a time, since each distinct subset of $r$ elements would give rise to $r!$ different $r$-tuples. Thus we get

$$\binom{n}{r} r! = {}_nP_r = \frac{n!}{(n-r)!}$$

which, dividing through by $r!$, gives us

$$\binom{n}{r} = \frac{n!}{r!\,(n-r)!}.$$

The most difficult part of many counting problems is deciding whether ordering should be of importance. If ordering does not matter we want combinations; if ordering is of importance we want permutations.

*Example 2.4.5.* (a) How many distinct 5-card hands can be dealt from a standard 52-card deck? Since the 5-card hand remains unchanged if you received the same 5 cards, but in a different order, the answer is

$$\binom{52}{5} = \frac{52!}{5!\,47!} = 2{,}598{,}960.$$

(b) How many distinct 13-card hands can be dealt from a standard deck?

$$\binom{52}{13} = 6.35 \times 10^{11}.$$

(c) Suppose that 10 boys go out for basketball at a particular school. How many different teams could be fielded from this school?

$$\binom{10}{5} = 252.$$

(d) One of the 10 boys out for basketball is named Joe. How many of the 252 teams include Joe as a member? If we want to count only those teams which include Joe, then we need only count how many ways might we select 4 additional individuals to be on the team. This is

$$\binom{9}{4} = 126.$$

$\binom{n}{k}$ is frequently called a combinatorial coefficient. Tables are available giving the value of $\binom{n}{k}$ for varying values of $n$ and $k$. Let us note a few facts regarding these coefficients. First, $\binom{n}{0} = 1$ for all $n$ as can be seen by simply evaluating the factorials involved. As is also easily seen, $\binom{n}{1} = n$ for all $n$. One further result which may be useful is the identity $\binom{n}{k} = \binom{n}{n-k}$ for any $n$ and $k$. By writing these two coefficients out in factorial form, it is immediately evident that they are always equal. Recalling that $\binom{n}{k}$ is the number of subsets of size $k$ that we can construct for a set having $n$ elements, we realize that every time we write down a subset of $k$ items, we leave behind a subset of size $n - k$. Thus it is not surprising that there are always equal numbers of subsets of these two sizes.

*Example 2.4.6.* (a) How many subsets does a set $S$ with $n$ elements have? (If $\mathcal{F}$ is the class of all subsets of $S$, then how many elements does $\mathcal{F}$ have?) We are able to compute the answer by using the binomial theorem, reviewed in Appendix 2, and the combinatorial coefficients. Clearly, if we add together the number of subsets of $S$ having 0, 1, 2, . . . , $n$ elements, respectively, this sum would be the total number of subsets of $S$. $\binom{n}{k}$ is the number of subsets of $S$, each having $k$ elements, for $k = 0, 1, 2, \ldots, n$. Thus, the total number of subsets is

$$\sum_{k=0}^{n} \binom{n}{k} = \sum_{k=0}^{n} \binom{n}{k} 1^k 1^{n-k}$$
$$= (1 + 1)^n = 2^n.$$

A set with 2 elements has $2^2 = 4$ subsets; a set with 5 elements has $2^5 = 32$ subsets; a set with 100 elements has $2^{100} = 1.27 \times 10^{30}$ subsets.

(b) If $S = \{1, 2, \ldots, n\}$, how many subsets of $S$ have 1 as an element? Clearly, only one single-element event has 1 as an element. The number of subsets of size 2 having 1 as an element is $\binom{n-1}{1}$ since the second element in the subset could be any one of the remaining $n - 1$. The number of subsets of size 3 having 1 as an element is $\binom{n-1}{2}$; the number of subsets of size $r$ having 1 as an element is $\binom{n-1}{r-1}$ where $r = 1, 2, \ldots, n$. Thus, the total number of subsets having 1 as an element is

$$1 + \binom{n-1}{1} + \binom{n-1}{2} + \cdots + \binom{n-1}{n-1}$$

but, by the binomial theorem, this sum equals $2^{n-1}$. Obviously, no matter which single element belonging to $S$ we consider, it belongs to $2^{n-1}$ events. Since there are $2^n$ events in total, then exactly half of all possible events occurs, no matter which element belonging to $S$ is the outcome we observe.

## EXERCISE 2.4.

**1.** How many ways can 3 different books be arranged side by side on a shelf?

**2.** If an item sold by a vending machine costs 40 cents, and the money deposited into the machine must consist of a quarter, a dime, and a nickel, in how many different orders could the money be inserted into the machine?

**3.** Six people are about to enter a cave in single file. In how many ways could they arrange themselves in a row to go through the entrance?

**4.** A bag contains 1 red, 1 black, and 1 green marble. I randomly select 1 of the marbles and record its color. I then replace it in the bag, shake the bag, and randomly select a second marble, again recording its color. The second marble is then replaced and a third marble is randomly selected and its color recorded. How many different samples of 3 colors could occur?

**5.** An ant farm contains both red and black ants. A particular passage in the farm is so narrow that only 1 ant can get through at a time. If 4 ants follow each other through the passage, how many different color patterns (having 4 elements) could be produced (assuming that red ants are indistinguishable from one another, as are black ants)?

**6.** A particular city is going to give 3 awards to outstanding residents. If 4 people are eligible to receive them, in how many different ways could they be distributed among the 4 people (assuming that no person may receive more than 1 award)?

**7.** If a set has 3 elements, how many subsets does it have?

**8.** Could we define a set that has exactly 9 subsets?

**9.** How many selections of 5 dominoes can be made from a regular 28-domino double-6 set?

**10.** In how many ways could 2 teams be chosen from an 8-team league?

**11.** How many committees of 3 people could be chosen from a group of 10?

**12.** How many 5-man squads could be chosen from a company of 20 men?

**13.** Given a set of 15 points in a plane, how many lines would be necessary to connect all possible pairs of points?

**14.** A "complete graph of order 3" is given by connecting 3 points in all possible ways. If 15 points are joined in all possible ways, how many complete graphs of order 3 would be included? Of order $k = 4, 5, 6, \ldots, 15$?

**15.** Given a box with 2 25-watt, 3 40-watt, and 4 100-watt bulbs, in how many ways could 3 bulbs be selected from the box?

**16.** Referring to the light bulbs in problem 15, how many of these selections of three bulbs would include both 25-watt bulbs? How many would include no 25-watt bulbs?

**17.** How many bulb selections defined in problem 15 would include exactly 1 of each of the 3 different wattages?

## 2.5. Some Particular Probability Problems

In this section we shall take up a few problems that should help acquaint the reader with counting techniques and their applications to probability problems.

*Example 2.5.1.* A bag contains 4 red and 2 white marbles. If these are randomly laid out in a row, what is the probability that the 2 end marbles are white? That they are not white? That the 2 white marbles are side by side? For convenience we assume that the white marbles are numbered 1 and 2 and that the red marbles are numbered 3 through 6. Then we might adopt as our sample space $S$ the collection of $6! = 720$ permutations of 6 things; i.e.,

$$S = \{(x_1, x_2, \ldots, x_6): x_i = 1, 2, 3, \ldots, 6, \text{ for all } i \text{ and } x_i \neq x_j \text{ for } i \neq j\}.$$

If the marbles are randomly laid out in a row, then each of these 6-tuples is equally likely to occur and we can use our equally likely formula for computing probabilities. Define $A$ to be the event that the first and last marbles are white (the collection of 6-tuples with marbles 1 and 2 on the ends) and $B$ to be the event that marbles 1 and 2 are side by side. Then the number of elements belonging to $A$ is

$$n(A) = 2 \cdot 4! = 48$$

(the white marbles could be on the ends in 2 ways and, for either of these, the red marbles could be arranged between in 4! ways). We also find that

$$n(B) = 5 \cdot 2 \cdot 4! = 240.$$

(There are 5 side-by-side positions for the white marbles to occupy, namely 12, 23, 34, 45, 56; whichever of these is the one to occur, the white marbles can occupy the selected pair of positions in 2 ways and the red balls can be permuted in the remaining positions in 4! ways.) As we noted above,

$$n(S) = 6! = 720$$

and we have

$$P(A) = \tfrac{48}{720} = \tfrac{1}{15}$$

and

$$P(B) = \tfrac{240}{720} = \tfrac{1}{3}.$$

$\bar{A}$ is the event that the white marbles are not on the 2 end positions; we immediately have

$$P(\bar{A}) = 1 - P(A) = \tfrac{14}{15}.$$

An alternative sample space for this problem can be derived as follows. If we pretend that the marbles are going to be put into numbered spots, then all possible outcomes of the experiment can be recorded by keeping track of just the 2 positions that the white marbles occupy; all remaining positions are, of course, filled with the red marbles. All possible pairs of positions are equally likely to occur if the marbles are put down randomly. Thus,

$$S = \{(x_1, x_2)\colon x_1 = 1, 2, \ldots, 6;\ x_2 = 1, 2, \ldots, 6;\ x_1 < x_2\}.$$

Note that $S$ does list every possible pair of position numbers that we could select for the white marbles and that it lists each one only once. The number of elements belonging to $S$ is equal to the number of subsets of size 2 that a set with 6 elements has; i.e.,

$$n(S) = \binom{6}{2} = 15.$$

If we define $A$ and $B$ as above, exactly 1 of these subsets consists of the largest and the smallest elements of $S$ and exactly 5 of them consist of consecutive pairs. Thus,

$$n(A) = 1, \qquad n(B) = 5$$

and, as above,

$$P(A) = \tfrac{1}{15}, \qquad P(\bar{A}) = \tfrac{14}{15}, \qquad P(B) = \tfrac{1}{3}.$$

In many examples, more than one equally likely sample space is possible; used correctly any one of them will give answers to problems of interest.

*Example 2.5.2.* Suppose that we select a whole number at random between 100 and 999, inclusive. What is the probability that it has at least one 1 in it? What is the probability that it has exactly two 3's in it? For a sample space, we choose

$$S = \{x\colon x = 100, 101, \ldots, 999\}.$$

Then $n(S) = 900$ and, since the number is chosen at random, we assume that all single-element events are equally likely to occur. Define the events:

$A$: the selected number has at least one 1 in it

$B$: the selected number has exactly two 3's in it.

We shall find it easy to compute $n(\bar{A})$, then use this to get $P(\bar{A})$, and finally use Theorem 2.2.2 to compute $P(A) = 1 - P(\bar{A})$. We shall compute $n(B)$ directly. The event $\bar{A}$ would be the collection of 3-digit numbers, each of which contains no

1's. The first position of any such number can be filled in 8 ways (since the first digit can be neither 1 nor 0) and each of the succeeding two positions can be filled in 9 ways (since 1 cannot occur in either). Thus,

$$n(\bar{A}) = 8 \cdot 9 \cdot 9 = 648$$

and

$$P(\bar{A}) = \frac{n(\bar{A})}{n(S)} = .72$$

$$P(A) = 1 - P(\bar{A}) = .28.$$

To compute $n(B)$ we can reason as follows: If the first digit is a 3, then one of the succeeding digits must be a 3 and the other can be any of the remaining 9 digits. These two succeeding digits can occur in either of 2 orders, so there are $9 \cdot 2 = 18$ 3-digit numbers having a 3 in the first position and each containing exactly two 3's. If the first digit is not a 3, then this position can be filled in 8 ways (neither 0 nor 3 can be used). The last two positions must then both be filled with 3's. Thus,

$$n(B) = 18 + 8 = 26$$

and

$$P(B) = \frac{n(B)}{n(S)} \doteq .029.$$

*Example 2.5.3.* Suppose that $n$ people are in a room. If we make a list of their birthdates (month and day of the month), what is the probability that there will be one or more repetitions in the list? (What is the probability that two or more people have the same birthday?) We shall make the assumption that there are only 365 days available for each birthday (ignoring February 29) and that each of these days is equally likely to occur for any individual's birthday. (It can, in fact, be shown that this is the worst possible assumption we might make relative to this probability; that is, if days in March or some other month are more likely to occur as birthdays, then the probability of one or more repeated birthdays is larger than if all days are equally likely.) Our sample space is the collection of all possible $n$-tuples that could occur for the birthdays, numbering the days of the year sequentially from 1 to 365. Thus,

$$S = \{(x_1, x_2, \ldots, x_n) : x_i = 1, 2, \ldots, 365;\ i = 1, 2, \ldots, n\}.$$

The first position in each $n$-tuple gives the first person's birthday; the second position gives the second person's birthday, etc. Assuming that all days of the year are equally likely for each person's birthday implies that each of these $n$-tuples is equally likely to occur. By using the counting techniques presented in Section 2.4, we see that

$$n(S) = 365^n.$$

Define $A$ to be the event that there is one or more repetitions of the same number in the $n$-tuple that occurs. Then $\bar{A}$ is the collection of $n$-tuples which have no repetitions; we can see rather easily that

$$n(\bar{A}) = 365 \cdot 364 \cdot 363 \cdots (365 - n + 1)$$

which gives us

$$P(\bar{A}) = \frac{n(\bar{A})}{n(S)} = \frac{365 \cdot 364 \cdot 363 \cdots (365 - n + 1)}{365^n}.$$

Again from Theorem 2.2.2,

$$P(A) = 1 - P(\bar{A}).$$

The following table gives the values of $P(\bar{A})$ and $P(A)$ for various values of $n$. It is somewhat surprising that the probability of a repetition exceeds $\frac{1}{2}$ for as few as 23 people in the room and that for 60 people it is a virtual certainty.

**Table 2.1**

| $n$ | $P(\bar{A})$ | $P(A)$ |
|---|---|---|
| 10 | .871 | .129 |
| 20 | .589 | .411 |
| 21 | .556 | .444 |
| 22 | .524 | .476 |
| 23 | .493 | .507 |
| 24 | .462 | .538 |
| 25 | .431 | .569 |
| 30 | .294 | .706 |
| 40 | .109 | .891 |
| 50 | .030 | .970 |
| 60 | .006 | .994 |

*Example 2.5.4.* Suppose that Mrs. Riley claims to be a clairvoyant. Specifically she claims that if she is presented with 8 cards, 4 of which are red and 4 of which are black, she will correctly identify the color of at least 6 of the cards without being able to see their colors. If she is guessing and has no special ability, what is the probability that she would correctly identify at least 6 out of the 8 cards? (She will identify 4 of the cards as red and 4 of the cards as black.) We arbitrarily decide to present her with the 4 red cards first, one by one, and then present her with the 4 black cards. The sample space for the experiment is the set of all possible guesses she might give for the colors of the cards; that is,

$$S = \{(x_1, x_2, \ldots, x_8): x_i = R \text{ or } B, \text{ for all } i; \text{ exactly } 4x_i\text{'s are } R\}.$$

If she is guessing, then each of the single-element events is equally likely to occur. Since each 8-tuple belonging to $S$ contains exactly 4 $R$'s and exactly 4 $B$'s, we can compute $n(S)$ by counting how many ways we might select 4 positions from the 8 in which to place the $R$'s; thus,

$$n(S) = \binom{8}{4} = 70.$$

Define

$A$: she identifies at least 6 cards correctly

$B$: she identifies exactly 6 cards correctly

$C$: she identifies all 8 cards correctly.

Since she will call 4 cards red and 4 cards black, it is not possible for her to be correct on exactly 7 cards; thus,

$$A = B \cup C$$

and, since

$$B \cap C = \varnothing,$$
$$P(A) = P(B) + P(C).$$

Clearly, $n(C) = 1$ so $P(C) = \frac{1}{70}$. If $B$ is to occur, she must identify exactly 3 of the 4 red cards correctly and exactly 3 of the 4 black cards correctly. The 1 red card that she is wrong on could be any of the 4 and the 1 black card she is wrong on could be any one of the 4. Thus, the number of 8-tuples having 1 $B$ in the first 4 positions and 1 $R$ in the last 4 positions is

$$n(B) = 4 \cdot 4 = 16$$

and we have

$$P(B) = \tfrac{16}{70}.$$

Thus

$$P(A) = \tfrac{1}{70} + \tfrac{16}{70} = \tfrac{17}{70} \doteq .243;$$

if she only guesses, there is slightly less than 1 chance in 4 of her doing as well as she claims.

## EXERCISE 2.5.

**1.** In how many ways could a dozen oranges be chosen from a table holding 30 oranges? (How many distinct collections of 12 oranges could be made?)

**2.** How many arrangements could be made of 5 red balls and 1 orange ball?

**3.** In how many of the arrangements counted in problem 2 are the red balls all together?

**4.** A certain market uses red boxes and green boxes for displays at Christmas time. In how many ways could the market arrange 20 boxes in a row if 15 of them are red and 5 are green? If there are 10 boxes of each color?

**5.** Ten people in total are nominated for a slate of 3 offices. If every group of 3 people has the same probability of winning, what is the probability that a particular person will be on the winning slate? That a particular pair of people will be on the winning slate?

**6.** Two people are to be selected at random to be set free from a prison with a population of 100. What is the probability that the oldest prisoner is 1 of the 2 selected? That the oldest and youngest are the pair selected?

**7.** In a certain national election year, governors were to be elected in 30 states. Assume that in every state there were only 2 candidates (called the Republican and Democratic candidates, respectively). What is the probability that the Republicans carried all 30 states, assuming that each state was equally likely to elect either party? What is the probability that the same party carried all the states?

**8.** $A$, $B$, and $C$ are going to race. What is the probability that $A$ will finish ahead of $C$, given that all are of equal ability (and no ties can occur)? What is the probability that $A$ will finish ahead of both $B$ and $C$?

**9.** Each of 5 people is asked to distinguish between vanilla ice cream and French vanilla custard (each is given a small sample of both and asked to identify which is ice cream). If all 5 people are guessing, what is the probability that all will correctly identify the ice cream? If all 5 are guessing, what is the probability that at least 4 will identify the ice cream correctly?

**10.** Compute the probability that a group of 5 cards drawn at random from a 52-card deck will contain

(a) exactly 2 pair
(b) a full house (3 of one denomination and 2 of another)
(c) a flush (all 5 from the same suit)
(d) a straight (5 in sequence, beginning with ace or deuce or trey, . . . , or ten).

**11.** $n$ people are in a room. Compute the probability that at least 2 have the same birth month. Evaluate this probability for $n = 3, 4, 5, 6$.

**12.** A student is given a true-false examination with 10 questions. If he gets 8 or more correct, he passes. If he is guessing, what is his probability of passing the examination?

**13.** Eight black and 2 red balls are randomly laid out in a row. What is the probability that the 2 red balls are side by side? That the 2 red balls are occupying the end positions?

**14.** A person is to be presented with 3 red and 3 white cards in a random sequence. He knows that there will be 3 of each color; thus he will identify 3 cards as being of each color. If he is guessing, what is his probability of correctly identifying all 6 cards? Of identifying exactly 5 correctly? Exactly 4?

## 2.6. Conditional Probability

In some applications we shall be given the information that an event $A$ occurred and will be asked the probability that another event $B$ also occurred. For example, we might be given that the card we selected from a regular 52-card bridge deck was red and then might want to know the probability that the card selected was the ace of hearts. Or, when running an opinion poll, we might be given the fact that the person we have selected is a Republican and then might also want to know the probability that he favors our current actions in Viet Nam. Or, when conducting a medical research experiment, we might be given that a randomly selected person has a family history of diabetes and then ask what is the probability that this person also has diabetes. In each of these examples we are given that an event $A$ has occurred and want to know the probability that a second event $B$ also has occurred.

If we are given that $A$ has already occurred, then in effect we now have the event $A$ as our sample space since we know that any element $x \in \bar{A}$ did not occur. Thus it would seem reasonable to measure the probability that $B$ also has occurred by the relative proportion of the time that $A$ and $B$ occur together (relative to the total probability of $A$ occurring). This is in fact the definition of the conditional probability of $B$ occurring, given that $A$ has occurred, as we see in the following definition.

DEFINITION 2.6.1. The *conditional probability* of $B$ occurring, given that $A$ has occurred (written $P(B \mid A)$) is $P(B \mid A) = P(B \cap A)/P(A)$ if $P(A) > 0$. If $P(A) = 0$, we define $P(B \mid A) = 0$.

*Example 2.6.1.* We roll a pair of fair dice 1 time and are given that the 2 numbers that occur are not the same. Compute the probability that the sum is 7 or that the sum is 4 or that the sum is 12. Define the event

$A$: the two numbers that occur are different.

Then we are given that $A$ has occurred. Also define the events

$B$: the sum is 7

$C$: the sum is 4

$D$: the sum is 12.

Then, assuming equally likely single-element events, we find that

$$P(A) = \tfrac{5}{6}, \qquad P(B) = \tfrac{1}{6}, \qquad P(C) = \tfrac{1}{12}, \qquad P(D) = \tfrac{1}{36},$$
$$P(A \cap B) = \tfrac{1}{6}, \qquad P(A \cap C) = \tfrac{1}{18}, \qquad P(A \cap D) = 0$$

and thus we have

$$P(B \mid A) = \frac{\frac{1}{6}}{\frac{5}{6}} = \tfrac{1}{5}$$

$$P(C \mid A) = \frac{\frac{1}{18}}{\frac{5}{6}} = \tfrac{1}{15}$$

$$P(D \mid A) = \frac{0}{\frac{5}{6}} = 0.$$

*Example 2.6.2.* From past experience with the illnesses of his patients, a doctor has gathered the following information: 5% feel that they have cancer and do have cancer, 45% feel that they have cancer and don't have cancer, 10% do not feel that they have cancer and do have it, and finally 40% feel that they do not have cancer and really do not have it. These percentage figures then imply the following probabilities for a randomly selected patient from this doctor's practice. Define the events

$A$: the patient feels he has cancer

$B$: the patient has cancer.

Then

$$P(A \cap B) = .05, \qquad P(A \cap \bar{B}) = .45,$$
$$P(\bar{A} \cap B) = .1, \qquad P(\bar{A} \cap \bar{B}) = .4.$$

Then we have

$$P(A) = P(A \cap B) + P(A \cap \bar{B}) = .5$$
$$P(B) = P(A \cap B) + P(\bar{A} \cap B) = .15.$$

The probability that a patient has cancer then, given that he feels he has it, is

$$P(B \mid A) = \frac{.05}{.5} = .1.$$

The probability he has cancer given that he does not feel that he has it is

$$P(B \mid \bar{A}) = \frac{.1}{.5} = .2.$$

The probability he feels he has cancer given that he does not have it is

$$P(A \mid \bar{B}) = \frac{.45}{.85} = \frac{9}{17}.$$

The probability he feels he has cancer given that he does have it is

$$P(A \mid B) = \frac{.05}{.15} = \frac{1}{3}.$$

One of the major uses of conditional probability is to allow easy computation of the probabilities of intersections of events for certain experiments. The equation

$$P(B \mid A) = \frac{P(B \cap A)}{(PA)}$$

implies that

$$P(B \cap A) = P(A \cap B) = P(A)P(B \mid A). \tag{2.1}$$

The equation

$$P(A \mid B) = \frac{P(A \cap B)}{P(B)}$$

implies that

$$P(A \cap B) = P(B)P(A \mid B). \tag{2.2}$$

One or the other of these formulas for $P(A \cap B)$ is useful in many problems.

*Example 2.6.3.* We select 2 balls at random without replacement from an urn which contains 4 white and 8 black balls. (a) Compute the probability that both balls are white. (b) Compute the probability that the second ball is white.

(a) Define

$A$: the first ball is white

$B$: the second ball is white

$C$: both balls are white.

Then

$$A \cap B = C \text{ and } P(C) = P(A \cap B) = P(A)P(B \mid A)$$
$$= \tfrac{4}{12} \cdot \tfrac{3}{11} = \tfrac{1}{11}.$$

(b) Clearly $B = (A \cap B) \cup (\bar{A} \cap B)$ and

$$(A \cap B) \cap (\bar{A} \cap B) = \varnothing.$$

Then

$$\begin{aligned} P(B) &= P(A \cap B) + P(\bar{A} \cap B) \\ &= P(A)P(B \mid A) + P(\bar{A})P(B \mid \bar{A}) \\ &= \tfrac{4}{12} \cdot \tfrac{3}{11} + \tfrac{8}{12} \cdot \tfrac{4}{11} \\ &= \tfrac{1}{3}. \end{aligned}$$

Notice that the probability of drawing a white ball the second draw (even though the first one is not replaced) is $\frac{1}{3}$, the same as the probability that the first ball drawn is white. This can be shown to be the case generally. $P(B)$ is an unconditional probability and, as indicated by the way it was computed above, it is an average of the two conditional probabilities $P(B \mid A)$ and $P(B \mid \bar{A})$.

*Example 2.6.4.* Box 1 contains 4 defective and 16 nondefective light bulbs. Box 2 contains 1 defective and 1 nondefective light bulb. We roll a fair die 1 time. If we get a 1 or a 2, then we select a bulb at random from box 1. Otherwise we select a bulb from box 2. What is the probability that the selected bulb will be defective? Define

$A$: we select a bulb from box 1

$B$: the selected bulb is defective.

Then $P(A) = \frac{1}{3}$, $P(\bar{A}) = \frac{2}{3}$, $P(B \mid A) = \frac{1}{5}$, and $P(B \mid \bar{A}) = \frac{1}{2}$. Since

$$B = (B \cap A) \cup (B \cap \bar{A})$$

and

$$(B \cap A) \cap (B \cap \bar{A}) = \varnothing,$$

we have

$$\begin{aligned} P(B) &= P(B \cap A) + P(B \cap \bar{A}) \\ &= P(A)P(B \mid A) + P(\bar{A})P(B \mid \bar{A}) \\ &= \tfrac{1}{3} \cdot \tfrac{1}{5} + \tfrac{2}{3} \cdot \tfrac{1}{2} = \tfrac{6}{15} = \tfrac{2}{5}. \end{aligned}$$

The following result, frequently called Bayes' theorem or Bayes' formula, is useful in many applied problems. It is named after the Reverend Thomas

Bayes, one of the early writers on probability theory; it has recently been applied in many different sorts of problems and plays an important role in many branches of applied statistics.

***Theorem 2.6.1.*** (Bayes) Suppose that we are given $k$ events $A_1, A_2, \ldots, A_k$ such that:

1. $A_1 \cup A_2 \cup \cdots \cup A_k = S$
2. $A_i \cap A_j = \varnothing, \qquad \text{for all } i \neq j$

(these events form a *partition* of $S$); then for any event $E \subset S$,

$$P(A_j \mid E) = \frac{P(A_j)P(E \mid A_j)}{\sum_{i=1}^{k} P(A_i)P(E \mid A_i)}, \qquad j = 1, 2, \ldots, k.$$

*Proof:* Because of properties 1 and 2 listed above, we know that for any event $E \subset S$ we can write

$$E = (E \cap A_1) \cup (E \cap A_2) \cup \cdots \cup (E \cap A_k)$$

where

$$(E \cap A_i) \cap (E \cap A_j) = \varnothing, \qquad \text{for all } i \neq j.$$

Thus

$$\begin{aligned} P(E) &= P(E \cap A_1) + P(E \cap A_2) + \cdots + P(E \cap A_k) \\ &= P(A_1)P(E \mid A_1) + P(A_2)P(E \mid A_2) + \cdots + P(A_k)P(E \mid A_k) \end{aligned}$$

by Equation 2.1. By definition

$$P(A_j \mid E) = \frac{P(A_j \cap E)}{P(E)}, \qquad j = 1, 2, \ldots, k.$$

And by using Equation 2.1 and the result above for $P(E)$, we have

$$P(A_j \mid E) = \frac{P(A_j)P(E \mid A_j)}{\sum_{i=1}^{k} P(A_i)P(E \mid A_i)}$$

which is the desired result. ◀

*Example 2.6.5.* Suppose that you are a political prisoner in Russia and are to be exiled in one of two places: Siberia or Mongolia. The probabilities of being sent to these two places are .7 and .3, respectively. It is also known that if you randomly select a resident of Siberia, the probability that he will be wearing a seal-skin coat is .8, whereas this same event has probability .4 in Mongolia. Late one night you are blindfolded and thrown on a truck. Two weeks later (you estimate) the truck stops, you are told you have arrived at your place of exile, and your blindfold is removed. The first person you see is not wearing a seal-skin coat. What is the probability that your place of exile is Siberia? Bayes theorem can be used to answer this question. Define

$A$: you are sent to Siberia.

Then we have

$\bar{A}$: you are sent to Mongolia.

Also define

$B$: randomly selected resident wears seal-skin coat.

The information we have been given then is:

$$P(A) = .7, \qquad P(\bar{A}) = .3, \qquad P(B \mid A) = .8, \qquad P(B \mid \bar{A}) = .4$$

and we are asked to compute $P(A \mid \bar{B})$. Since $A$ and $\bar{A}$ form a partition of the sample space we have, from Bayes theorem,

$$P(A \mid \bar{B}) = \frac{P(A)P(\bar{B} \mid A)}{P(A)P(\bar{B} \mid A) + P(\bar{A})P(\bar{B} \mid \bar{A})}$$

$$= \frac{(.7)(.2)}{(.7)(.2) + (.3)(.6)} = \frac{7}{16}.$$

**EXERCISE 2.6.**

**1.** An urn contains 4 balls numbered 1, 2, 3, 4, respectively. Two balls are drawn without replacement. Let $A$ be the event that the sum is 5 and let $B_i$ be the event that the first ball drawn has an $i$ on it, $i = 1, 2, 3, 4$. Compute $P(A \mid B_i)$, $i = 1, 2, 3, 4$, and $P(B_i \mid A)$, $i = 1, 2, 3, 4$.

**2.** Suppose that the two balls of problem 1 are drawn with replacement. Let $A$ and $B_i$ be defined as above and compute $P(A \mid B_i)$ and $P(B_i \mid A)$, $i = 1, 2, 3, 4$.

**3.** A fair coin is flipped 4 times. What is the probability that the fourth flip is a head, given that each of the first 3 flips resulted in heads?

**4.** A fair coin is flipped 4 times. What is the probability that the fourth flip is a head, given that 3 heads occurred in the 4 flips? Given that 2 heads occurred in the 4 flips?

**5.** Urn 1 contains 2 red and 4 blue balls, urn 2 contains 10 red and 2 blue balls. If an urn is chosen at random and a ball is removed from the chosen urn, what is the probability that the selected ball is blue? That it is red?

**6.** Suppose that in problem 7, instead of selecting an urn at random we roll a die and select the ball from urn 1 if a 1 occurs on the die and otherwise select the ball from urn 2. What is the probability that the selected ball is blue? That it is red?

**7.** Five cards are selected at random without replacement from a 52-card deck. What is the probability that they are all red? That they are all diamonds?

**8.** An urn contains 2 red, 2 white, and 2 blue balls. Two balls are selected at random without replacement from the urn. Compute the probability that the second ball drawn is red.

**9.** An urn contains 2 black and 5 brown balls. A ball is selected at random. If the ball drawn is brown, it is replaced and two additional brown balls are also put into the urn. If the ball drawn is black, it is not replaced in the urn and no additional balls are added. A ball is then drawn from the urn the second time. What is the probability that it is brown?

**10.** The two-stage experiment described in problem 9 was performed and we are given that the ball selected at the second stage was brown. What is the probability that the ball selected at the first stage was also brown?

**11*.** Suppose that medical science has a cancer-diagnostic test that is 95% accurate both on those that do have cancer and on those that do not have cancer. If .005 of the population actually does have cancer, compute the probability that a particular individual has cancer, given that the test says he has cancer.

**12.** In a large midwestern school, 1% of the student body participates in the intercollegiate athletic program; 10% of these people have a grade point of 3 or more (out of 4) whereas 20% of the remainder of the student body have a grade point of 3 or more. What proportion of the total student body have a grade point of 3 or more? Suppose we select 1 student at random from this student body and find that he has a grade point of 3.12. What is the probability that he participates in the intercollegiate athletic program?

### 2.7. Independent Events

It is possible to define and be interested in events $A$ and $B$ such that if we know that $A$ has occurred, then we are also certain that $B$ has occurred (take any example where $A \subset B$). In such a case there is certainly a degree of dependence between $A$ and $B$. As we shall see in this section, it is also possible to have two events $A$ and $B$ such that the knowledge that $A$ has occurred gives no information on whether or not $B$ also has occurred. Two such events will be called independent (also referred to as being statistically independent). The formal definition of independent events is as follows.

DEFINITION 2.7.1. Two events, $A$ and $B$, are *independent* if and only if

$$P(A \cap B) = P(A)P(B).$$

Before looking at some examples of independent events, let us note that if $A$ and $B$ are independent, then

$$P(A \mid B) = \frac{P(A \cap B)}{P(B)} = \frac{P(A)P(B)}{P(B)} = P(A).$$

Thus, the conditional probability of $A$ occurring is the same as the unconditional probability of $A$. The knowledge that $B$ occurred didn't change the

* Adapted from Emanuel Parzen, *Modern Probability Theory and Its Applications*, Wiley, 1960.

probability of $A$'s occurrence. Similarly, it is easily seen that $P(B \mid A) = P(B)$ if $A$ and $B$ are independent.

*Example 2.7.1.* Assume that the numbers given in the cells of Table 2.2 give the probabilities of a randomly selected individual falling into the given cell. That is,

**Table 2.2**

| | Gets Cancer | Does Not Get Cancer |
|---|---|---|
| Smoker | .50 | .20 |
| Nonsmoker | .10 | .20 |

if we let $A$ be the event that the selected individual is a smoker and let $B$ be the event that the selected individual gets cancer, then

$$P(A \cap B) = .5, \qquad P(A \cap \bar{B}) = .2, \qquad P(\bar{A} \cap B) = .1$$

and

$$P(\bar{A} \cap \bar{B}) = .2.$$

Since

$$P(A) = P(A \cap B) + P(A \cap \bar{B}) = .7$$
$$P(B) = P(A \cap B) + P(\bar{A} \cap B) = .6,$$

We see that $P(A \cap B) = 2 \neq (.7)(.6)$, so $A$ and $B$ are not independent.

*Example 2.7.2.* If 2 fair dice are rolled 1 time, show that the 2 events

$A$: the sum of the 2 dice is 7

$B$: the 2 dice have the same number

are not independent. As we have seen before, $P(A) = P(B) = \frac{1}{6}$; $A \cap B = \varnothing$ so $P(A \cap B) = 0$ which is not $\frac{1}{6} \cdot \frac{1}{6}$. Thus, the 2 events are not independent.

Two events which cannot happen simultaneously are said to be mutually exclusive, as is given in the next definition.

DEFINITION 2.7.2. $A$ and $B$ are *mutually exclusive* if and only if $A \cap B = \varnothing$. The definitions of independence and mutually exclusive are frequently confused; this is probably caused by the fact that in common English usage the word independent is frequently used to mean "having nothing to do with." The phrase "having nothing to do with" could be interpreted to mean that two events could not happen together, which we have just defined as being mutually exclusive. The two definitions (2.7.1 and 2.7.2) are certainly not the same in content, as we can see from the following theorem.

***Theorem 2.7.1.*** Assume that $P(A) \neq 0$ and $P(B) \neq 0$. Then $A$ and $B$ independent implies that they are not mutually exclusive and $A$ and $B$ mutually exclusive implies that they are not independent.

*Proof:* Suppose that $A$ and $B$ are independent. Then $P(A \cap B) = P(A)P(B) \neq 0$, since $P(A) \neq 0$ and $P(B) \neq 0$. Thus they are not mutually exclusive. Now suppose $A$ and $B$ are mutually exclusive. Then $A \cap B = \varnothing$ and $P(A \cap B) = 0$. But since $P(A) \neq 0$ and $P(B) \neq 0$, $P(A)P(B) \neq 0$ and they are then not independent. ◀

Independence of three events is defined as follows.

DEFINITION 2.7.3. $A$, $B$, and $C$ are independent if and only if:

1. $P(A \cap B) = P(A)P(B)$
2. $P(A \cap C) = P(A)P(C)$
3. $P(B \cap C) = P(B)P(C)$
4. $P(A \cap B \cap C) = P(A)P(B)P(C)$.

Many examples can be given which show that the first three of these conditions do not imply the fourth and vice versa. The following is an example in which the first three equations hold but the fourth does not.

*Example 2.7.3.* Suppose that we have a bowl with 4 tags in it labelled 000, 110, 101, 011, respectively. We draw 1 tag from the bowl at random and define $A_i$ ($i = 1, 2, 3$) to be the event that a 0 occurs in the $i$th position. Then equations 1, 2, and 3 in Definition 2.7.3 are satisfied by these three events, but equation 4 is not. $A_1$, $A_2$, $A_3$ are called pairwise independent events but not independent events. Problem 6 below shows a case in which equation 4 is satisfied but at least one of 1, 2, and 3 is not.

Many practical experiments consist of independent trials; by independent trials we mean that a particular outcome in one trial has no affect on the outcome observed in another trial. This idea is made more exact in the following definition.

DEFINITION 2.7.4. An experiment consists of $n$ *independent trials* if and only if: (1) $S$ is the Cartesian product of $n$ sets $S_1, S_2, \ldots, S_n$, and (2) the probability of occurrence of a single-element event $A \subset S$ is the product of the probabilities of occurrence of appropriate single-element events $A_i \subset S_i$, $i = 1, 2, \ldots, n$; that is,

$$P(A) = P_1(A_1)P_2(A_2) \cdots P_n(A_n)$$

where $A \subset S$, $A_i \subset S_i$, $i = 1, 2, \ldots, n$ and $A, A_1, \ldots, A_n$ are each single-element events.

Note immediately then that an experiment that consists of $n$ independent trials has $n$-tuples as elements of its sample space. Furthermore, probabilities of single-element events are assigned in a special way; this special way in fact gives an easy method of computing many probabilities.

*Example 2.7.4.* A box contains 15 electron tubes of which 4 are defective. We select 5 tubes at random, with replacement, from the box. We choose as our sample space

$$S = \{(x_1, x_2, \ldots, x_5): x_i = n \text{ or } d, \quad i = 1, 2, \ldots, 5\},$$

where $n$ is meant to stand for nondefective and $d$ for defective. A single-element event then is a 5-tuple; in fact, if we define

$$S_i = \{n, d\}, \qquad i = 1, 2, \ldots, 5,$$

we see that

$$S = S_1 \times S_2 \times S_3 \times S_4 \times S_5.$$

Since the sampling is done with replacement, the composition of the box is the same for each of the draws. Thus, if the first tube drawn is defective, this has no effect on whether the second tube is also defective (or the third, etc.). The draws then are independent and this is an experiment with 5 independent trials. For each $S_i$ we see that

$$P_i(\{n\}) = \tfrac{11}{15}, \qquad P_i(\{d\}) = \tfrac{4}{15}.$$

This then gives us the assignment of probabilities for each single-element event of $S$. For example

$$\begin{aligned}
P(\{(n, n, n, n, n)\}) &= \tfrac{11}{15} \cdot \tfrac{11}{15} \cdot \tfrac{11}{15} \cdot \tfrac{11}{15} \cdot \tfrac{11}{15} = (\tfrac{11}{15})^5 \\
P(\{(d, d, d, d, d)\}) &= \tfrac{4}{15} \cdot \tfrac{4}{15} \cdot \tfrac{4}{15} \cdot \tfrac{4}{15} \cdot \tfrac{4}{15} = (\tfrac{4}{15})^5 \\
P(\{(n, d, n, d, n)\}) &= \tfrac{11}{15} \cdot \tfrac{4}{15} \cdot \tfrac{11}{15} \cdot \tfrac{4}{15} \cdot \tfrac{11}{15} = (\tfrac{4}{15})^2(\tfrac{11}{15})^3 \\
P(\{(d, d, n, n, n)\}) &= \tfrac{4}{15} \cdot \tfrac{4}{15} \cdot \tfrac{11}{15} \cdot \tfrac{11}{15} \cdot \tfrac{11}{15} = (\tfrac{11}{15})^3(\tfrac{4}{15})^2
\end{aligned}$$

and so on. Once we know the probabilities of the single-element events, we can use these to compute the probabilities of any other events of interest. Note that if we sampled tubes from this box without replacement, we would not have independent draws because the number of tubes to be drawn from would not be the same for each draw nor would the relative proportion of defective tubes remain constant.

*Example 2.7.5.* The Iowa State University football team plays 11 different teams on successive Saturdays. If we assume that their performances from one Saturday to the next are independent, then the experiment which consists of their full schedule for a year is made up of 11 independent trials. Define

$$S_i = \{W, L, T\}, \qquad i = 1, 2, \ldots, 11,$$

where $W$ stands for win, $L$ for loss, and $T$ for tie, and the sample space $S$ for the year's games can be written

$$S = S_1 \times S_2 \times \cdots \times S_{11},$$

Then the probability of occurrence of any single-element event belonging to $S$ is defined to be the product of the probabilities of the single-element events for the appropriate trials; that is

$$P(\{(x_1, x_2, \ldots, x_{11})\}) = P_1(\{x_1\})P_2(\{x_2\}) \cdots P_{11}(\{x_{11}\})$$

where

$$(x_1, x_2, \ldots, x_{11}) \in S, \qquad x_1 \in S_1, x_2 \in S_2, \ldots, x_{11} \in S_{11}.$$

The differing subscript oh $P$ for the various trials is meant to indicate that the probability of Iowa State winning (or losing or tying) a game is not necessarily the same for all weeks. The probability that they win all their games is then

$$\prod_{i=1}^{11} P_i(\{W\});$$

the probability that they lose all their games is

$$\prod_{i=1}^{11} P_i(\{L\}).$$

(The product notation is discussed in Appendix 1.)

## EXERCISE 2.7.

**1.** One fair coin is flipped 2 times. Are the 2 events

$A$: a head occurs on the first flip

$B$: a head occurs on the second flip

independent?

**2.** A fair coin is flipped 2 times. Let $A$ be the event that a head occurs on the first flip and let $B$ be the event that the same face does not occur on both flips. Are $A$ and $B$ independent?

**3.** An urn contains 4 balls numbered 1, 2, 3, 4, respectively. Two balls are drawn without replacement. Let $A$ be the event that the first ball drawn has a 1 on it and let $B$ be the event that the second ball has a 1 on it. Are $A$ and $B$ independent?

**4.** If the drawing is done with replacement in problem 3, are $A$ and $B$ independent?

**5.** A pair of dice is rolled 1 time. Let $A$ be the event that the first die has a 1 on it, $B$ the event that the second die has a 6 on it, and $C$ the event that the sum is 7. Are $A$, $B$, and $C$ independent?

**6.** A fair coin is flipped 3 times. Let $A$ be the event that a head occurs on the first flip, let $B$ be the event that at least 2 tails occur, and let $C$ be the event that we get exactly 1 head or that we get tail, head, head in that order. Show that these 3 events satisfy equation 4 of Definition 2.7.3. but not equations 1, 2, or 3.

**7.** Prove that if $A$ and $B$ are independent, so are $\bar{A}$ and $\bar{B}$.

**8.** The probability that a certain basketball player scores on a free throw is .7. If in a game he gets 15 free throws, compute the probability that he makes them all. Compute the probability that he makes 14 of them. What assumptions have you made in deriving your answer?

**9.** Three teams, $A$, $B$, and $C$, enter a round-robin tournament. (Each team plays 2 games, 1 against each of the possible opponents. The winner of the tournament, if there is a winner, is the team winning both its games.) Assume that the game played is one in which a tie is not allowed. We assume the following probabilities:

$$P(A \text{ beats } B) = .7$$
$$P(B \text{ beats } C) = .8$$
$$P(C \text{ beats } A) = .9.$$

Compute the probability that team $A$ wins the tournament; that team $B$ wins the tournament. Compute the probability no one wins the tournament.

## 2.8. Discrete and Continuous Sample Spaces

The reader may have noted that the sample spaces used in all the preceding examples were finite sets. Such sample spaces are special cases of what are called discrete sample spaces. Discrete sample spaces are defined as follows.

DEFINITION 2.8.1. A *discrete* sample space is one which has a finite or a countably infinite number of elements.

Generally, discrete sample spaces are those for which it is meaningful to consider single-element events. That is, knowledge of the probabilities of occurrence of the single-element events is sufficient for computing the probabilities of any event. All of the finite sample spaces we have seen thus far have been of this sort. The following example gives a problem in which the sample space is countably infinite, yet we still are able to compute probabilities from knowledge of the single-element events.

Before looking at this example, let us note what our third axiom is saying when we have an infinite sample space. In an infinite sample space it is possible to have an infinite number of nonoverlapping events. (Consider, for example, the single-element events.) Axiom 3 says then that the probability assigned to the union of an infinite number of nonoverlapping events must be equal to the value to which an infinite series converges (the sum of the individual probabilities assigned to the events). Appendix 3 gives a more complete discussion of orders of infinity.

*Example 2.8.1.* Suppose we flip a fair coin until we get a head. Compute the probability that it takes less than 4 flips to conclude the experiment. Compute the probability that it takes an even number of flips to conclude the experiment. Our sample space is

$$S = \{H, TH, TTH, \ldots\}.$$

$H$ signifies that we get a head on the first flip, $TH$ that we first got a head on the second flip, $TTH$ that we first got a head on the third flip, etc. Clearly we can set up a one-to-one correspondence between the elements of $S$ and the positive integers; thus $S$ has a countably infinite number of elements and is a discrete sample space. Assuming that the coin is fair and that the flips are independent, we can easily assign probabilities to the single-element events as follows.

$$P(\{H\}) = \tfrac{1}{2}, \qquad P(\{TH\}) = \tfrac{1}{2}\cdot\tfrac{1}{2} = \tfrac{1}{4},$$
$$P(\{TTH\}) = \tfrac{1}{2}\cdot\tfrac{1}{2}\cdot\tfrac{1}{2} = \tfrac{1}{8}, \qquad P(\{TTTH\}) = \tfrac{1}{2}\cdot\tfrac{1}{2}\cdot\tfrac{1}{2}\cdot\tfrac{1}{2} = \tfrac{1}{16},$$

etc. First of all we should make sure that this assignment satisfies axiom 1. Clearly

$$\begin{aligned} P(S) &= \tfrac{1}{2} + \tfrac{1}{4} + \tfrac{1}{8} + \tfrac{1}{16} + \cdots \\ &= \tfrac{1}{2}(1 + \tfrac{1}{2} + \tfrac{1}{4} + \tfrac{1}{8} + \cdots) \\ &= \tfrac{1}{2}\cdot\frac{1}{1-\tfrac{1}{2}} = 1, \end{aligned}$$

since the parenthetic expression is just a geometric progression with $r = \tfrac{1}{2}$. Thus we do satisfy axiom 1. Define

$A$: it takes less than 4 flips to conclude the experiment

$B$: it takes an even number of flips to conclude the experiment.

That is,

$$\begin{aligned} A &= \{H, TH, TTH\} \\ B &= \{TH, TTTH, TTTTTH, \ldots\}. \end{aligned}$$

Then

$$\begin{aligned} P(A) &= \tfrac{1}{2} + \tfrac{1}{4} + \tfrac{1}{8} = \tfrac{7}{8} \\ P(B) &= \tfrac{1}{4} + \tfrac{1}{16} + \tfrac{1}{64} + \tfrac{1}{256} + \cdots \\ &= \tfrac{1}{4}(1 + \tfrac{1}{4} + \tfrac{1}{16} + \tfrac{1}{64} + \cdots) \\ &= \tfrac{1}{4}\cdot\frac{1}{1-\tfrac{1}{4}} = \tfrac{1}{3}. \end{aligned}$$

*Example 2.8.2.* Suppose that we select a number at random from the positive integers. What is the probability that it is even? The sample space for this experiment is

$$S = \{1, 2, 3, \ldots\},$$

so $S$ is discrete. However, it is not possible to assign the same probability to each of the single-element events. No matter how small the value of this common probability, the sum of the probabilities of the single-element events is infinite. That is, suppose we say

$$P(\{x\}) = p > 0 \qquad \text{for } x = 1, 2, 3, \ldots$$

Then

$$P(S) = \sum_{x=1}^{\infty} P(\{x\}) = \sum_{x=1}^{\infty} p$$

and this sum diverges, no matter how close to 0 $p$ is. Thus it would appear that there is no way we could satisfactorily describe such an experiment. This conclusion is perfectly correct because, after a little reflection, we would have to conclude that there is absolutely no way in which we could perform the experiment. No matter what physical mechanism we used to "select a number at random from the positive integers," there are too many positive integers to be considered since the integers have no end. The experiment itself makes no sense; this is why we cannot describe it. In spite of the foregoing, it would seem reasonable to give $\frac{1}{2}$ as the probability of selecting an even number since every other integer is even. This conclusion is actually based on reasoning such as the following. If we were to select randomly an integer from the set $\{1, 2, 3, \ldots, M\}$ where $M$ is very large, the probability that the number is even is $\frac{1}{2}$ if $M$ is even and $\frac{1}{2} - \frac{1}{2M}$ if $M$ is odd. In either case, it is about $\frac{1}{2}$ and, as $M$ gets bigger and bigger, it gets closer and closer to $\frac{1}{2}$ (for odd $M$; for even $M$ it is $\frac{1}{2}$).

Many examples give rise to continuous sample spaces. These are defined as follows:

DEFINITION 2.8.2. A *continuous* (one-dimensional) sample space is one which has as elements all of the points in some interval on the real line.

Thus the sets

$$\{x\colon 0 < x < 1\}, \qquad \{x\colon 10 \leq x \leq 20\}, \qquad \{x\colon x > 0\}$$

are all examples of continuous sample spaces. Generally, if the experiment consists of observing something which could lie anywhere along a continuous line, we shall want to use a continuous sample space. Again, subsets of the continuous interval are events; the single-element events would be the sets of single points in the interval. In the case of continuous sample spaces, however, we shall generally have to conclude that the probabilities of occurrence of these individual points must be zero. Otherwise, since there are a noncountable infinity of them in any interval, we would not be able to satisfy our first axiom. Thus, in continuous sample spaces, we cannot compute the probabilities of occurrence of any event from knowledge of the probabilities of the single-element events.

Another difficulty is also encountered. In discrete sample spaces, all subsets are called events and probabilities can consistently be assigned to them. It can be shown that in a continuous sample space $S$, not all subsets are probabilizable; that is, certain examples of subsets can be constructed such that any assignment of probabilities to them is inconsistent with our three axioms. Such subsets are not called events and thus are not assigned probabilities. The reader should rest assured, however, that in any practical problem we shall be able to call any outcomes of interest events. In more advanced courses it is shown that the subsets of $S$ belonging to the class of Borel sets are all probabilizable; furthermore, the class of Borel sets includes all of the continuous-interval subsets of $S$, as well as unions and intersections of such continuous-interval subsets. Without going further into the matter, we merely acknowledge the fact that not all subsets of $S$ are called events if $S$ consists of all the points in some continuous interval; those that are events are actually Borel sets.

Continuous sample spaces, in quite general examples, are most easily studied after we are acquainted with random variables (to be introduced in Chapter 3). Let us content ourselves here with looking at a particular type of continuous sample space and the way in which we assign probabilities to events. The particular type of continuous sample space we shall investigate now is one which has equally likely elements. As we have seen, each of the single-element events (single points) must then have probability 0. Therefore, what we really mean by saying that the elements of $S$ are equally likely to occur is that the probability of the occurring point lying in a continuous subinterval of $S$ is proportional to the length of the subinterval. If, for example, $A$ and $B$ are both continuous subintervals of $S$ and if $A$ is twice as long as $B$, then the probability of $A$ containing the outcome is twice as big as the probability of $B$ containing the outcome.

It can be shown that for continuous sample spaces, we must specify a rule for assigning probabilities to continuous subintervals rather than specifying probabilities for all single-element events. In this section, as has been noted, we shall be concerned only with the particular continuous sample space which has equally likely outcomes. As mentioned above, this implies that the probability of the outcome lying in a continuous subinterval should be proportional to the length of the subinterval. Accordingly, suppose we are given an experiment whose outcome is equally likely to lie anywhere in the real interval from $a$ to $b$ inclusive. Then

$$S = \{x: a \leq x \leq b\}.$$

Suppose $A$ is a continuous subinterval of $S$; i.e.,

$$A = \{x: c \leq x \leq d\},$$

where $c > a$, $d < b$. (See Figure 2.3.) For any continuous subinterval $A$, define the set function

$$L(A) = \text{length of } A.$$

Then, for the particular subset defined above we have

$$L(A) = d - c,$$

and for $S$ we have

$$L(S) = b - a.$$

One rule for assigning probabilities then is

$$P(A) = \frac{L(A)}{L(S)} = \frac{d - c}{b - a}$$

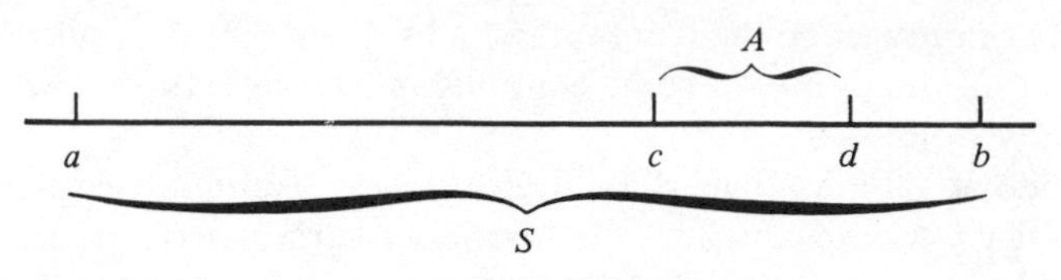


**Figure 2.3.**

if $A$ is any continuous subinterval of $S$. For any subset $B$ of $S$ which is a union of nonoverlapping continuous subintervals of $S$, we define $L(B)$ to be the sum of the lengths of the nonoverlapping subintervals belonging to $B$. That is, if

$$B = C_1 \cup C_2 \cup \cdots, \qquad \text{where } C_i \cap C_j = \varnothing, \qquad \text{for } i \neq j,$$

then

$$L(B) = L(C_1) + L(C_2) + \cdots$$

and we have

$$P(B) = \frac{L(B)}{L(S)} = \frac{L(C_1)}{L(S)} + \frac{L(C_2)}{L(S)} + \cdots = P(C_1) + P(C_2) + \cdots.$$

***Theorem 2.8.1.*** The rule

$$P(A) = \frac{L(A)}{L(S)} \qquad \text{for } A \subset S$$

satisfies the three axioms.

*Proof:* $P(S) = L(S)/L(S) = 1$ so axiom 1 is satisfied. $L(A) \geq 0$ for any $A \subset S$ so $P(A) = L(A)/L(S) \geq 0$ and axiom 2 is satisfied. By the way in which lengths of unions of intervals are defined, we have

$$P(C_1 \cup C_2 \cup \cdots) = \frac{L(C_1 \cup C_2 \cdots)}{L(S)}$$

$$= \frac{L(C_1)}{L(S)} + \frac{L(C_2)}{L(S)} + \cdots = P(C_1) + P(C_2) + \cdots$$

if $C_i \cap C_j = \varnothing$ for all $i \neq j$. ◀

Note that if $A$ is a single-element event then $L(A) = 0$ and we have $P(A) = 0$, as discussed above.

Let us now consider some specific examples.

*Example 2.8.3.* Doug is a 2-year-old boy. From his family history it seems plausible to assume that his adult height is equally likely to lie between 5 feet 9 inches and 6 feet 2 inches. Making this assumption, what is the probability that he will stand at least 6 feet high as an adult? What is the probability that his adult height will lie between 5 feet 10 inches and 5 feet 11 inches?

We use as our sample space

$$S = \{x\colon\ 69 \leq x \leq 74\}$$

where we are recording his achieved adult height in inches. Define

$$A = \{x\colon\ 72 \leq x \leq 74\}$$
$$B = \{x\colon\ 70 \leq x \leq 71\}.$$

Then

$$L(S) = 5, \qquad L(A) = 2, \qquad L(B) = 1$$

and we have

$$P(A) = \tfrac{2}{5}, \qquad P(B) = \tfrac{1}{5}.$$

*Example 2.8.4.* Assume that you daily ride a commuter train from your home in Connecticut into Manhattan. The station you leave from has trains leaving for Manhattan at 7 a.m., 7:13 a.m., 7:20 a.m., 7:25 a.m., 7:32 a.m., 7:45 a.m., and 7:55 a.m. It is your practice to take the first train that leaves after your arrival at the station. Due to the vagaries of your rising time and the traffic you encounter driving to the station, you are equally likely to arrive at the station at any instant between 7:15 a.m. and 7:45 a.m. On a particular day, what is the probability that you have to wait less than 5 minutes at the station? Less than 10 minutes? Suppose the 7:25 a.m. and 7:45 a.m. trains are expresses. What is the probability that you catch an express on a given day?

Let us, for convenience, take as our sample space

$$S = \{x\colon\ 0 \leq x \leq 30\}$$

where the elements of $S$ are actually meant to represent minutes after 7:15 a.m. that might occur until your arrival time. (Refer to Figure 2.4.) Define the events

$A$: you wait less than 5 minutes

$B$: you wait less than 10 minutes

$C$: you catch an express.

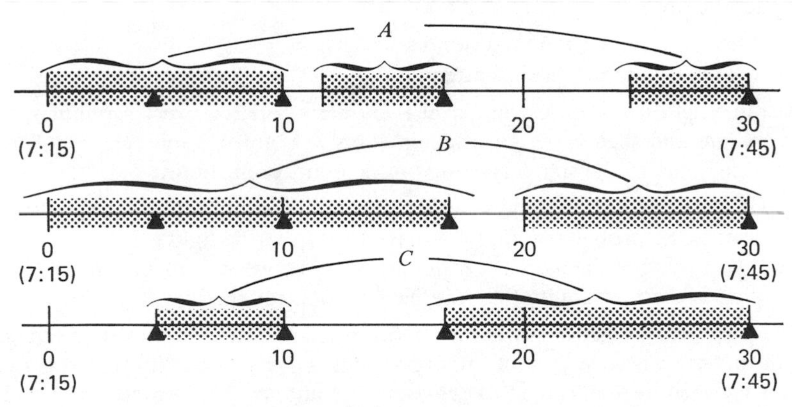


**Figure 2.4.**

Then

$$A = \{x\colon 0 \leq x < 10 \text{ or } 12 \leq x < 17 \text{ or } 25 \leq x < 30\}$$
$$B = \{x\colon 0 \leq x < 17 \text{ or } 20 \leq x < 30\}$$
$$C = \{x\colon 5 \leq x < 10 \text{ or } 17 \leq x < 30\}$$

and

$$L(S) = 30, \qquad L(A) = 20, \qquad L(B) = 27, \qquad L(C) = 18.$$

Thus

$$P(A) = \tfrac{2}{3}, \qquad P(B) = \tfrac{9}{10}, \qquad P(C) = \tfrac{3}{5}.$$

## EXERCISE 2.8.

**1.** A fair die is rolled until a 1 occurs. Compute the probability that:

(a) 10 rolls are needed
(b) less than 4 rolls are needed
(c) an odd number of rolls is needed.

**2.** A fair pair of dice is rolled until a 7 occurs (as the sum of the 2 numbers on the dice). Compute the probability that

(a) 2 rolls are needed
(b) an even number of rolls is needed.

**3.** You fire a rifle at a target until you hit it. Assume the probability that you hit it is .9 for each shot and that the shots are independent. Compute the probability that:

(a) it takes more than 2 shots.
(b) the number of shots required is a multiple of 3.

**4.** Hugh takes a written driver's license test repeatedly until he passes it. Assume the probability that he passes it any given time is .1 and that the tests are independent. Compute the probability that:

(a) it takes him more than 4 attempts
(b) it takes him more than 10 attempts.

**5.** A traffic light on a route you travel every day turns red every 4 minutes, stays red 1 minute and then turns green again (thus it is green 3 minutes, red 1, etc.), with the red part of the signal starting on the hour, every hour.

(a) If you arrive at the light at a random instant between 7:55 a.m. and 8:05 a.m., what is the probability that you have to stop at the light?
(b) If you arrive at the light at a random instant between 7:54 a.m. and 8:04 a.m. what is the probability that you have to stop for the light?

**6.** The plug on an electric clock with a sweep second hand is pulled at a random instant of time within a certain minute. What is the probability that the second hand is between the 4 and the 5? Between the 1 and the 2? Between the 1 and the 6?

**7.** A point is chosen at random between 0 and 1 on the $x$-axis in the $(x, y)$ plane. A circle centered at the origin is then drawn in the plane, with radius determined by the chosen point. Compute the probability that the area of the circle is less than $\pi/2$.

**8.** A 12-inch ruler is broken into 2 pieces at a random point along its length. What is the probability that the longer piece is at least twice the length of the shorter piece?

# 3

# Random Variables and Distribution Functions

Probability theory itself is an interesting mathematical discipline. It can be used in many different contexts; one of its most important applications is in the description of random variables.

Much headway was made in science by assuming an essentially deterministic world—one in which it was assumed that the identical operation or experiment, performed repeatedly under identical conditions, would always give rise to the same result. However, as time passed and technology created more and more precise methods and equipment for observation, it became apparent that the same experiment, performed under identical conditions, does not give exactly the same result each time. It is this variability of observed results that is ideally described with probability theory and random variables. Since the outcomes or results of experiments or observations are usually summarized by numbers, this variability of the results of experiments is represented by a variability of the numbers observed. As we shall see, we can think of random variables as results of an experiment that is affected by some chance mechanism; thus the description of random variables gives us a method of describing the variability of outcomes for the experiment.

### 3.1. Random Variables

Let us first mention some specific examples of random variables. A man who bets 1 dollar in a gambling game is interested in the net change in his capital as a result of this bet. The man firing a rifle at a target is interested in the distance between the point at which his bullet strikes the target and the bull's-eye of the target. A jockey is interested in the total elapsed time from when his horse leaves the starting gates until it crosses the finish line. A variable whose value is determined by the outcome of an experiment is called a random variable. Formally, the definition of a random variable is as follows.

DEFINITION 3.1.1. A *random variable* $X$ is a real-valued function of the elements of a sample space $S$.

We see then that a random variable is an element function whose domain is the sample space for an experiment (see Section 1.4).

Especially now, when we are first studying random variables, we shall use a functional notation to stress the fact that a random variable is a special sort of mathematical function. The Greek letter $\omega$ will be used to represent a generic element of the sample space (regardless of whether the elements are 1-tuples or $n$-tuples) and $X(\omega)$ will be the functional representation of the random variable (the rule) $X$. We shall reserve capital letters from the end of the alphabet ($X$, $Y$, $Z$, $U$, $V$, $W$, etc.) to represent random variables and lower case letters ($x$, $y$, $z$, $u$, $v$, $w$, etc.) to stand for particular values in the range of the random variable.

We shall concern ourselves with two different sorts of random variables, called discrete and continuous random variables (actually, a mixture of these two types could also be defined but it occurs so rarely in practice that we shall not do so). The difference between these two is essentially the same as the difference between discrete and continuous sample spaces, as discussed in Section 2.8. Since every random variable $X(\omega)$ is a real-valued function of the elements of $S$, the range of $X(\omega)$ is always a set of real numbers. It is a basic difference between the ranges of discrete and continuous random variables that forces us to differentiate between them now. The formal definition follows.

DEFINITION 3.1.2. A random variable $X$ is *discrete* if its range forms a discrete (countable) set of real numbers. A random variable $X$ is *continuous* if its range forms a continuous (uncountable) set of real numbers and the probability of $X$ equalling any single value in its range is 0.

Let us consider some specific examples.

*Example 3.1.1.* We roll a pair of fair dice 1 time. Let $X$ be the sum of the 2 numbers that occur. Then the sample space is

$$S = \{(x_1, x_2)\colon x_1 = 1, 2, \ldots, 6; x_2 = 1, 2, \ldots, 6\}$$

and we have defined

$$X(\omega) = x_1 + x_2 \qquad \text{for } \omega = (x_1, x_2) \in S.$$

The range of $X$ is the set $\{2, 3, \ldots, 12\}$ so $X$ is a discrete random variable.

*Example 3.1.2.* A random sample of 3 people is selected from the list of registered voters in the county of Monterey, California. Let $Y$ be the number of Republicans that occur in the sample. For convenience, we use as our sample space

$$S = \{(x_1, x_2, x_3): x_1 = 0, 1; x_2 = 0, 1; x_3 = 0, 1\}$$

where the 3 elements in the 3-tuples represent the people selected in the sample as follows: $x_1$ is 1 if the first person is a Republican and 0 otherwise; $x_2$ is 1 if the second person selected is a Republican and 0 otherwise; $x_3$ is defined similarly for the third person. With our 3-tuples defined in this way, we can express $Y$ as

$$Y(\omega) = x_1 + x_2 + x_3, \qquad \text{for } \omega = (x_1, x_2, x_3) \in S.$$

The range of $Y$ is the set $\{0, 1, 2, 3\}$, so $Y$ is a discrete random variable.

*Example 3.1.3.* Suppose the instant at which you leave your residence for your day's activities is equally likely to fall anywhere in the interval from 7:30 a.m. to 8:00 a.m. inclusive. Let $Z$ be the number of minutes (and fractions thereof) before 8:00 a.m. at which you leave your residence. A reasonable sample space can be defined by calling 7:30 a.m. zero on our time scale and using a minute for our unit. Thus we have

$$S = \{x: 0 \leq x \leq 30\}.$$

Using this sample space the random variable $Z$ can be written in functional notation as

$$Z(\omega) = 30 - \omega, \qquad \text{for } \omega \in S.$$

Then the range of $Z$ is the continuous set $\{x: 0 \leq x \leq 30\}$ so $Z$ is a continuous random variable.

*Example 3.1.4.* We select 1 student at random from those registered at the University of Washington. Let $U$ be the weight of the selected student. We assume that there are 25,000 students registered and that they are numbered from 1 through 25,000. A reasonable sample space then would be

$$S = \{x: x = 1, 2, 3, \ldots, 25{,}000\}.$$

The random variable $U$ then is defined to be

$$U(\omega) = \text{weight of student } \omega, \qquad \text{for } \omega \in S.$$

In actual fact, no more than 25,000 distinct values of $U$ could occur. We might, however, find it convenient to think of $U$ as being a continuous random variable and to use continuous methods in describing its behavior. If we thought of using such a continuous approach, we might assume that no student at Washington weighs less than 50 pounds or more than 500 pounds, in which case the range of $U$

would be the set $\{x: 50 \leq x \leq 500\}$ and $U$ then would be a continuous random variable.

If $X$ is a discrete random variable, we can use the probability measure defined on the subsets of $S$ to define the probability function for $X$. This can be used to evaluate probability statements about $X$ and is defined as follows

DEFINITION 3.1.3. The *probability function* for $X$ is a function (denoted by $p_X(x)$) of a real variable $x$ and is defined to be

$$p_X(x) = P(X(\omega) = x), \qquad \text{for all real } x.$$

If we have a given experiment and a sample space $S$ for that experiment and have defined the discrete random variable $X$ on the elements of $S$, then we can find $p_X(x)$ as follows. Let

$$A(x) = \{\omega: X(\omega) = x\}.$$

Then $p_X(x) = P(A(x))$ for all real $x$. Note immediately that $A(x)$ may equal $\varnothing$ for many $x$ and thus $p_X(x)$ is 0 for any such $x$.

*Example 3.1.5.* Consider the discrete random variable $X$ defined in Example 3.1.1. ($X$ is the sum of the 2 numbers that occur when a pair of fair dice is rolled.) The range of $x$ is the set $\{2, 3, \ldots, 12\}$. Then

$$\begin{aligned}
A(2) &= \{(1, 1)\} \\
A(3) &= \{(1, 2), (2, 1)\} \\
A(4) &= \{(1, 3), (2, 2), (3, 1)\} \\
&\vdots \\
A(11) &= \{(5, 6), (6, 5)\} \\
A(12) &= \{(6, 6)\} \\
A(x) &= \varnothing, \qquad \text{for all other } x.
\end{aligned}$$

Then we have

$$\begin{aligned}
p_X(x) &= \tfrac{1}{36}, & x &= 2 \\
&= \tfrac{2}{36}, & x &= 3 \\
&= \tfrac{3}{36}, & x &= 4 \\
&\vdots & &\vdots \\
&= \tfrac{2}{36}, & x &= 11 \\
&= \tfrac{1}{36}, & x &= 12 \\
&= 0, & &\text{otherwise.}
\end{aligned}$$

We can, of course, use $p_X(x)$ to compute any probabilities of interest regarding $X$

when $X$ is a discrete random variable. For example, the probability that $X$ is an odd number is

$$p_X(3) + p_X(5) + p_X(7) + p_X(9) + p_X(11) = \tfrac{1}{2}.$$

We compute the probability that $X$ lies in any interval by summing the probability function over the interval (strictly speaking, we are summing only over those points in the interval that have positive probability). Thus,

$$P(4 \le X(\omega) \le 8) = \sum_{x=4}^{8} p_X(x) = \tfrac{23}{36}$$

$$P(0 \le X(\omega) \le 3) = \sum_{x=2}^{3} p_X(x) = \tfrac{1}{12}.$$

*Example 3.1.6.* In Example 3.1.2. we selected 3 people at random from the list of registered voters in Monterey county. We assume that the proportion of registered voters who are Republicans is .4 and that the number of registered voters is large enough to consider the selection of the 3 as being independent trials with probability .4 of selecting a Republican. That is, the probabilities of the single-element events can be computed as though we have independent trials. Thus,

$$P(\{(0, 0, 0)\}) = (.6)^3 = .216$$
$$P(\{(1, 0, 0)\}) = (.4)(.6)^2 = .144$$
$$P(\{(0, 1, 0)\}) = (.6)(.4)(.6) = .144,$$

etc. The range of $Y$, the number of Republicans who appear in the sample, is $\{0, 1, 2, 3\}$. We define

$$A(0) = \{(0, 0, 0)\}$$
$$A(1) = \{(1, 0, 0), (0, 1, 0), (0, 0, 1)\}$$
$$A(2) = \{(1, 1, 0), (1, 0, 1), (0, 1, 1)\}$$
$$A(3) = \{(1, 1, 1)\}.$$

The probability function for $Y$ then is

$$\begin{aligned} p_Y(y) &= .216, && \text{for } y = 0 \\ &= .432, && \text{for } y = 1 \\ &= .288, && \text{for } y = 2 \\ &= .064, && \text{for } y = 3. \end{aligned}$$

*Example 3.1.7.* The game of Chuck-a-Luck is played as follows. Three fair dice are rolled. You as the bettor are allowed to bet 1 dollar (or some other amount) on the occurrence of one of the integers 1, 2, 3, 4, 5, 6. Suppose you bet on the occurrence of a 5. Then if one 5 occurs (on the 3 dice) you win 1 dollar, if two 5's occur you win 2 dollars, and if three 5's occur you win 3 dollars. If no 5's occur you lose your 1 dollar. Let $V$ be the net amount you win in one play of this game. The range of $V$ is the set $\{-1, 1, 2, 3\}$ so $V$ is a discrete random variable. We choose as our sample space

$$S = \{(x_1, x_2, x_3): x_i = 1, 2, \ldots, 6;\ i = 1, 2, 3\}.$$

Since we are assuming the dice to be fair, the 3 dice constitute 3 independent trials, each with probability $\frac{1}{6}$ of giving a 5. We immediately then are able to assign probabilities to all of the single-element events (see Section 2.7) and can use these to find the probability function for $V$:

$$\begin{aligned} p_V(v) &= \tfrac{125}{216}, &\quad \text{for } v &= -1 \\ &= \tfrac{75}{216}, &\quad \text{for } v &= 1 \\ &= \tfrac{15}{216}, &\quad \text{for } v &= 2 \\ &= \tfrac{1}{216}, &\quad \text{for } v &= 3. \end{aligned}$$

As you might gather, it is quite significant that $p_V(-1) > \frac{1}{2}$. We shall soon see exactly why this should play a role in Las Vegas casinos' offers to play this game.

Since the probability function for a discrete random variable gives the probabilities of the random variable being equal to various values in its range (and since a random variable assigns a real number to every $\omega \in S$), we know that

$$\sum_{\substack{\text{range} \\ \text{of } X}} p_X(x) = 1, \qquad \text{for any random variable } X. \tag{3.1}$$

Furthermore, since probabilities cannot be negative,

$$p_X(x) \geq 0, \qquad \text{for all } x. \tag{3.2}$$

Any function of $x$ that satisfies Equations 3.1 and 3.2 could in fact be looked at as being the probability function for some random variable $X$. We have purposely not mentioned a probability function for a continuous random variable, because it would be zero for all $x$. We shall have an alternative function to evaluate probabilities about continuous random variables.

A word about future notation. From now on we shall write $P(X = x)$ for $P(X(\omega) = x)$. That is, the functional dependence of $X$ upon the elements $\omega \in S$ will be suppressed. Unless necessary, we shall not even need to refer explicitly to the sample space $S$. We can always in theory make such a reference, but as we gain skill and practice in deriving probability functions it represents a superfluous step which will be deleted.

## EXERCISE 3.1.

**1.** An urn contains 4 balls numbered 1, 2, 3, 4, respectively. Let $Y$ be the number that occurs if 1 ball is drawn at random from the urn. What is the probability function for $Y$?

**2.** Consider the urn defined in problem 1. Two balls are drawn from the urn without replacement. Let $Z$ be the sum of the 2 numbers that occur. Derive the probability function for $Z$.

**3.** Assume that the sampling in problem 2 is done with replacement and define $Z$ in the same way. Derive the new probability function for $Z$.

**4.** Two balls are drawn with replacement from the urn mentioned in problem 1. Let $X$ be the sum of the squares of the 2 numbers drawn and derive the probability function for $X$.

**5.** A class in statistics contains 10 students, 3 of whom are 19, 4 are 20, 1 is 21, 1 is 24, and 1 is 26. Two students are selected at random without replacement from this class. Let $X$ be the average age of the 2 selected students and derive the probability function for $X$.

## 3.2. Distribution Functions and Density Functions

We saw several particular examples in the last section of discrete random variables and the probability function of discrete random variables. The probability function can be derived very naturally from the probability measure of the subsets of the sample space and can be used to evaluate probability statements about the random variable. The distribution function for the random variable $X$ can also be used to evaluate probability statements about $X$; the advantage of the distribution function is that it is appropriate for either discrete or continuous random variables. The distribution function is defined as follows.

DEFINITION 3.2.1. The *distribution function* for a random variable $X$ (denoted by $F_X(t)$) is a function of a real variable $t$ such that: (1) the domain of definition of $F_X$ is the whole real line, and (2) for any real $t$

$$F_X(t) = P(X \leq t).$$

Notice that $F_X(t)$ gives the total accumulation of probability for $X$ equalling any number less than or equal to $t$. The distribution function is a powerful theoretical tool in many applications, as we shall see.

We can immediately derive the distribution function of a discrete random variable $X$ if we know the probability function of $X$. Clearly,

$$F_X(t) = \sum_{x \leq t} p_X(x), \tag{3.3}$$

where the summation is over all $x$ from the range of $X$ that satisfy $x \leq t$.

*Example 3.2.1.* In Example 3.1.6 we found that

$$\begin{aligned} p_Y(y) &= .216, && \text{for } y = 0 \\ &= .432, && \text{for } y = 1 \\ &= .288, && \text{for } y = 2 \\ &= .064, && \text{for } y = 3 \\ &= 0, && \text{otherwise.} \end{aligned}$$

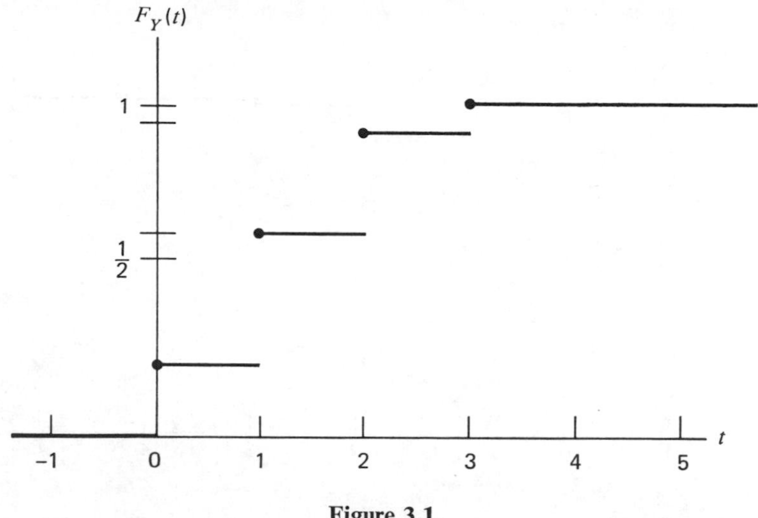


**Figure 3.1.**

The distribution function for $Y$ then is

$$\begin{aligned} F_Y(t) &= 0, & t &< 0 \\ &= .216, & 0 &\le t < 1 \\ &= .648, & 1 &\le t < 2 \\ &= .936, & 2 &\le t < 3 \\ &= 1, & t &\ge 3. \end{aligned}$$

(See Figure 3.1.)

*Example 3.2.2.* In Example 3.1.7 we found

$$\begin{aligned} p_V(v) &= \tfrac{125}{216}, & &\text{for } v = -1 \\ &= \tfrac{75}{216}, & &\text{for } v = 1 \\ &= \tfrac{15}{216}, & &\text{for } v = 2 \\ &= \tfrac{1}{216}, & &\text{for } v = 3. \end{aligned}$$

Then the distribution function for $V$ is

$$\begin{aligned} F_V(t) &= 0, & t &< -1 \\ &= \tfrac{125}{216}, & -1 &\le t < 1 \\ &= \tfrac{200}{216}, & 1 &\le t < 2 \\ &= \tfrac{215}{216}, & 2 &\le t < 3 \\ &= 1, & t &\ge 3. \end{aligned}$$

(See Figure 3.2.)

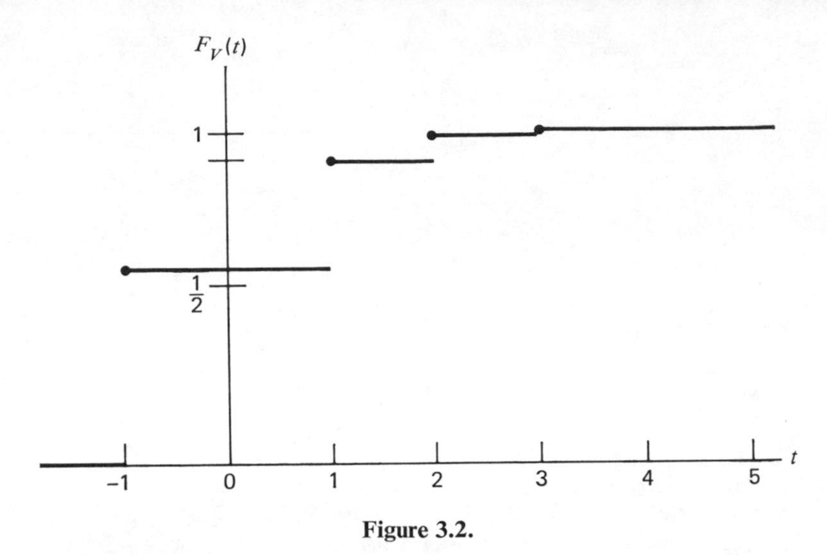


**Figure 3.2.**

The distribution function $F_X$ can be used to evaluate probability statements of any kind regarding $X$; in particular, if we define the two events

$$A: \ X \leq a$$

$$B: \ a < X \leq b$$

where $a$ and $b$ are any two real numbers such that $a < b$, we obviously have

$$A \cap B = \varnothing$$

and thus

$$P(A \cup B) = P(A) + P(B).$$

That is,

$$P(X \leq b) = P(X \leq a) + P(a < X \leq b),$$

or

$$F_X(b) = F_X(a) + P(a < X \leq b).$$

Thus

$$P(a < X \leq b) = F_X(b) - F_X(a), \tag{3.4}$$

which shows us how to evaluate the probability that $X$ lies in a particular interval, given $F_X$.

*Example 3.2.3.* Given the random variable $V$ discussed in Example 3.2.2 (refer to Figure 3.2), use $F_V$ to compute

$$P(0 < V \leq 3), \qquad P(V \leq 0), \qquad P(-1 < V \leq 0).$$

Using Equation 3.4 (and Definition 3.2.1),

$$\begin{aligned}
P(0 < V \leq 3) &= F_V(3) - F_V(0) \\
&= 1 - \tfrac{125}{216} = \tfrac{91}{216} \\
P(V \leq 0) &= F_V(0) = \tfrac{125}{216}, \\
P(-1 < V \leq 0) &= F_V(0) - F_V(1) \\
&= \tfrac{125}{216} - \tfrac{125}{216} = 0.
\end{aligned}$$

This last probability is zero since the distribution function $F_V$ is constant or level on the interval $-1 < t \leq 0$. This means that there are no values of $t$ in the interval $-1 < t \leq 0$ which $V$ will equal with positive probability. This is the case in general; if the distribution function of a random variable $X$ is constant over any interval, then the probability of $X$ taking on values in that interval is zero.

We have seen how we might derive the distribution function for a discrete random variable $X$ from the probability function for $X$. We might logically then ask how we could get the probability function for $X$ from the distribution function for $X$. To answer this, consider the real numbers $t$ in the interval $b - h < t \leq b$ where $h$ is some positive number. As $h$ tends to zero, this interval clearly tends to the single point $t = b$; that is, the limit of any such interval as $h \to 0$ is the single point $t = b$ since $b - h \to b$. Then, we might expect, as is indeed the case, that

$$\lim_{h \to 0} P(b - h < X \leq b) = P(X = b).$$

But

$$\begin{aligned}
\lim_{h \to 0} P(b - h < X \leq b) &= \lim_{h \to 0} [F_X(b) - F_X(b - h)] \\
&= F_X(b) - \lim_{h \to 0} F_X(b - h) \\
&= F_X(b) - F_X(b -),
\end{aligned}$$

where we have written $F_X(b-)$ for $\lim_{h \to 0} F_X(b - h)$. Thus, if $b$ is a point of discontinuity of $F_X$, then $b$ is a value that $X$ takes on with positive probability; the probability that $X = b$ is the size of the jump at $F_X(b)$. (Refer to Figure 3.3.)

*Example 3.2.4.* Suppose that we are given a random variable $U$ with distribution function

$$\begin{aligned}
F_U(t) &= 0, & t &< 0 \\
&= \tfrac{1}{2}, & 0 &\leq t < 2 \\
&= \tfrac{5}{6}, & 2 &\leq t < 3 \\
&= 1, & t &\geq 3.
\end{aligned}$$

Then we see that there are three points at which $F_U$ is discontinuous, namely

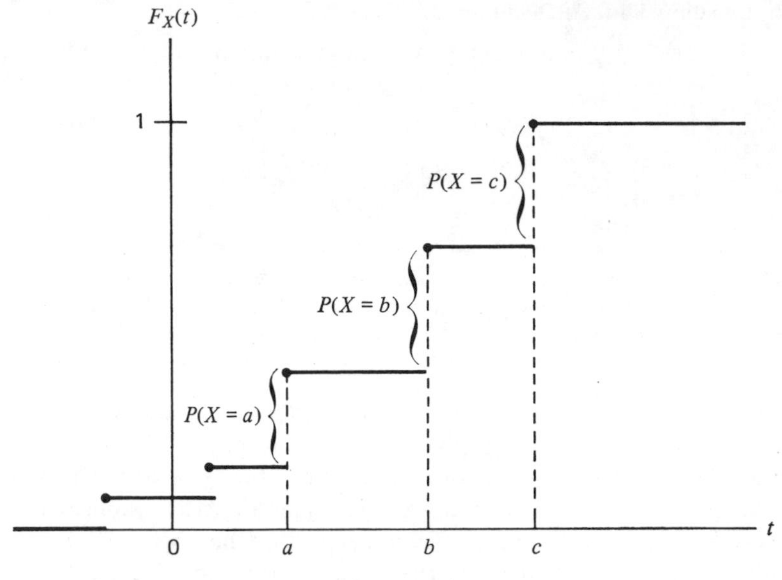


**Figure 3.3.**

$t = 0, 2, 3$. These then are the values which $U$ will equal with positive probability, and we have

$$\begin{aligned} p_U(u) &= F_U(0) - F_U(0-) = \tfrac{1}{2}, &\quad \text{at } u = 0 \\ &= F_U(2) - F_U(2-) = \tfrac{1}{3}, &\quad \text{at } u = 2 \\ &= F_U(3) - F_U(3-) = \tfrac{1}{6}, &\quad \text{at } u = 3. \end{aligned}$$

Since there are no other points of discontinuity of $U$, $p_U(u)$ is 0, otherwise.

The distribution function for any random variable $X$ is always a function of a real variable. We might inquire whether there are definite rules which a function must satisfy in order that it could be the distribution function of some random variable $X$. The answer is yes, of course. A function $H(t)$ of a real variable must satisfy the following four requirements in order to be a distribution function. First,

$$0 \leq H(t) \leq 1, \qquad \text{for all } t. \tag{3.5}$$

This is necessary since $H(t)$ must be a probability for any real $t$ and probabilities are bounded by 0 and 1, as we know. Second,

$$\lim_{t \to -\infty} H(t) = 0, \qquad \lim_{t \to \infty} H(t) = 1. \tag{3.6}$$

This requirement is analogous to axiom 1; it is equivalent to saying that the total probability of value 1 for the sample space is accounted for if we look at all real $t$. The third requirement is

$$H(a) \leq H(b), \qquad \text{for all } a \leq b, \tag{3.7}$$

which is necessary because

$$P(a < X \leq b) = H(b) - H(a).$$

If we did not insist that Equation 3.7 be satisfied, then we would be allowing a negative probability of the event

$$a < X \leq b.$$

Finally, we require that

$$\lim_{h \to 0} H(b + h) = H(b), \qquad \text{for all } b; \tag{3.8}$$

that is, that $H$ is continuous from the right at all points. If this were not required and we had, for example,

$$\lim_{h \to 0} H(b + h) = H(b) + \tfrac{1}{4},$$

then we would not be able to attribute the probability of $\frac{1}{4}$ that has been added to $H(b)$ to either specific values that $X$ might take on or to specific intervals it might lie in. To summarize, any function $H(t)$ which satisfies Equations 3.5 through 3.8 can be thought of as the distribution function of some random variable; conversely, it is easily shown that any distribution function $F_X$ will satisfy all four statements.

We noted in Definition 3.1.2 that the probability was zero for a continuous random variable $X$ being equal to any value in its range. Since,

$$P(X = b) = F_X(b) - F_X(b-),$$

we see that, if $X$ is continuous, then

$$F_X(b) - F_X(b-) = 0, \qquad \text{for all } b.$$

That is,

$$F_X(b) = F_X(b-)$$

and we see that the distribution function of $X$ has no jumps in it. If $X$ is a continuous random variable, then $F_X(t)$ must be a continuous function of $t$ for all $t$.

*Example 3.2.5.* The function

$$\begin{aligned} H(t) &= 0, && t < 0 \\ &= t, && 0 \leq t \leq 1 \\ &= 1, && t > 1 \end{aligned}$$

rather obviously satisfies Equations 3.5 through 3.8. Thus it may be thought of as the distribution function for some random variable $X$. Since $H(t)$ is a continuous function of $t$, the corresponding random variable $X$ is continuous.

*Example 3.2.6.* The function

$$\begin{aligned} H(t) &= 0, & t < 0 \\ &= 1 - e^{-t}, & t \geq 0 \end{aligned}$$

satisfies Equations 3.5 through 3.8 and thus is the distribution function for some random variable $Y$. Since $H(t)$ is a continuous function of $t$, the corresponding random variable $Y$ is also continuous.

It has been pointed out that the probability function, $P(Y = b)$, must be zero for all $b$ if $Y$ is a continuous random variable. Thus, the probability function cannot be used (as an alternative to the distribution function) to compute the probability that $a \leq Y \leq b$ if $Y$ is continuous. We might then inquire whether we might not be able to find some function, similar to the probability function for discrete random variables, which could be used to evaluate the probability that $a \leq Y \leq b$ when $Y$ is continuous. Specifically, if $X$ is discrete, then from Equation 3.3

$$F_X(t) = \sum_{x \leq t} p_X(x).$$

Since integration is the continuous analog to summation, for a continuous random variable $Y$ we might try to find a function $f_Y(y)$ such that

$$F_Y(t) = \int_{-\infty}^{t} f_Y(y)\, dy.$$

For all continuous random variables with which we shall be concerned, such a function $f_Y$ does always exist and is called the probability density function for the continuous random variable $Y$. Note that

$$\frac{d}{dt} F_Y(t) = f_Y(t);$$

that is, we can derive the density function by taking the derivative of $F_Y(t)$. (This derivative may not exist at a finite number of points; since the probability is zero that $Y$ would equal these points, we shall encounter no difficulties if we simply define $f_Y$ to be zero at such points.) We formally define the probability density function for a continuous random variable $Y$ as follows.

DEFINITION 3.2.2. The *probability density function* for a continuous random variable $Y$ (denoted by $f_Y(y)$) is a function of a real variable $y$ such that: (1) the domain of $f_Y$ is the whole real line and (2) for any real number $t$

$$F_Y(t) = \int_{-\infty}^{t} f_Y(y)\, dy.$$

Let us make some comments regarding density functions in general before looking at some specific examples. First, since

$$f_Y(t) = \frac{d}{dt} F_Y(t),$$

we see that the density function at any point $t$ gives the slope of the distribution function $F_Y(t)$; since $F_Y(t)$ is the total accumulation of probability up to and including $t$, we may think of $f_Y(t)$ as measuring the relative rate at which the probability is accumulating at any point $t$. Second, since

$$P(a < Y \leq b) = F_Y(b) - F_Y(a),$$

we see that

$$\begin{aligned} P(a < Y \leq b) &= \int_{-\infty}^{b} f_Y(y)\,dy - \int_{-\infty}^{a} f_Y(y)\,dy \\ &= \int_{a}^{b} f_Y(y)\,dy. \end{aligned}$$

Thus we can compute the probability that $Y$ lies in the interval from $a$ to $b$ by integrating $f_Y(y)$ between these limits. Since the probability is zero that $Y$ would equal any particular number, it doesn't matter whether the end points of the interval are included or not. Thus

$$P(a < Y < b) = P(a \leq Y \leq b) = \int_{a}^{b} f_Y(y)\,dy.$$

Finally, since definite integrals can be thought of as giving the area between the function being integrated and the horizontal axis, between the definite limits, we can graphically think of the probability that $a < Y < b$ as being the area under $f_Y$ between $a$ and $b$.

*Example 3.2.7.* Consider the distribution function given in Example 3.2.5:

$$\begin{aligned} F_X(t) &= 0, \qquad t < 0 \\ &= t, \qquad 0 \leq t \leq 1 \\ &= 1, \qquad t > 1. \end{aligned}$$

The derivative of $F_X$ is

$$\begin{aligned} \frac{d}{dt} F_X(t) &= 0, \qquad \text{for } t < 0 \text{ and } t > 1 \\ &= 1, \qquad \text{for } 0 < t < 1. \end{aligned}$$

The derivative doesn't exist at $t = 0$ and $t = 1$, but we can conveniently define

$$\begin{aligned} f_X(x) &= 1, \qquad 0 < x < 1 \\ &= 0, \qquad \text{otherwise,} \end{aligned}$$

and use $f_X$ to compute

$$P(\tfrac{1}{4} < X < \tfrac{3}{4}) = \int_{1/4}^{3/4} 1\, dx = \tfrac{3}{4} - \tfrac{1}{4} = \tfrac{1}{2}$$

$$\begin{aligned} P(-1 < X < \tfrac{1}{2}) &= \int_{-1}^{1/2} f_X(x)\, dx \\ &= \int_{-1}^{0} f_X(x)\, dx + \int_{0}^{1/2} f_X(x)\, dx \\ &= \int_{-1}^{0} 0\, dx + \int_{0}^{1/2} 1\, dx = \tfrac{1}{2}. \end{aligned}$$

(Refer to Figure 3.4.)

*Example 3.2.8.* The continuous distribution function discussed in Example 3.2.6 was

$$\begin{aligned} F_Y(t) &= 0, & t < 0 \\ &= 1 - e^{-t}, & t \geq 0. \end{aligned}$$

The derivative of $F_Y$ is

$$\begin{aligned} \frac{d}{dt} F_Y(t) &= 0, & t < 0 \\ &= e^{-t}, & t > 0. \end{aligned}$$

We then can define

$$\begin{aligned} f_Y(y) &= e^{-y}, & y > 0 \\ &= 0, & \text{otherwise}, \end{aligned}$$

and use $f_Y$ to compute

$$P(1 < Y < 3) = \int_{1}^{3} e^{-y}\, dy = e^{-1} - e^{-3}$$

$$P(Y > 4) = \int_{4}^{\infty} e^{-y}\, dy = e^{-4}$$

$$P(Y < 2) = \int_{-\infty}^{2} f_Y(y)\, dy = \int_{0}^{2} e^{-y}\, dy = 1 - e^{-2}.$$

Let us conclude this section by commenting that any function $h(t)$ can be considered the probability density function of some continuous random variable if

$$h(t) \geq 0, \qquad \text{for all } t, \tag{3.9}$$

and

$$\int_{-\infty}^{\infty} h(t)\, dt = 1. \tag{3.10}$$

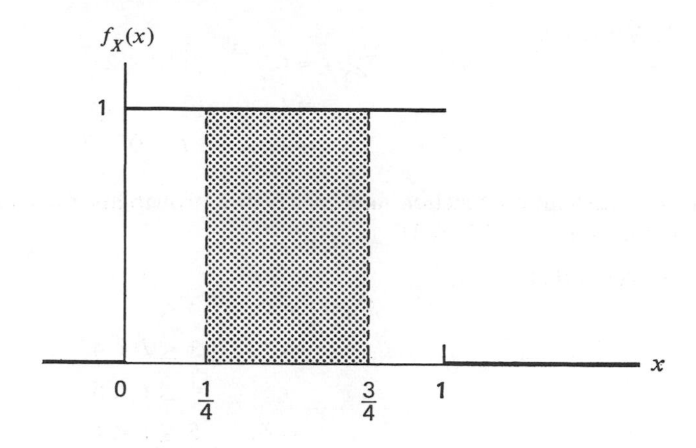


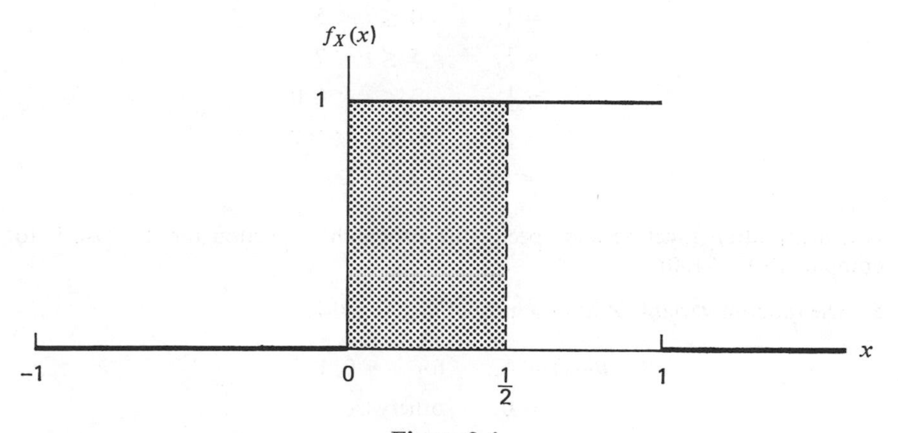


**Figure 3.4.**

**EXERCISE 3.2.**

**1.** Verify that

$$\begin{aligned} F_X(t) &= 0, & t &< -3 \\ &= \tfrac{1}{3}, & -3 &\le t < -1 \\ &= \tfrac{2}{3}, & -1 &\le t < 0 \\ &= 1, & t &\ge 0 \end{aligned}$$

is a distribution function and derive the probability function for $X$.

**2.** Verify that

$$\begin{aligned} F_Z(t) &= 0, & t &< -2 \\ &= \tfrac{1}{2}, & -2 \le t &< 0 \\ &= 1, & t &\ge 0 \end{aligned}$$

is a distribution function and specify the probability function for $Z$. Use it to compute $P(-1 \le Z \le 1)$.

**3.** Verify that

$$\begin{aligned} F_W(t) &= 0, & t &< 3 \\ &= \tfrac{1}{3}, & 3 \le t &< 4 \\ &= \tfrac{1}{2}, & 4 \le t &< 5 \\ &= \tfrac{2}{3}, & 5 \le t &< 6 \\ &= 1, & t &\ge 6 \end{aligned}$$

is a distribution function and specify the probability function for $W$. Use it to compute $P(3 < W \le 5)$.

**4.** Verify that

$$\begin{aligned} F_Y(t) &= 0, & t &< 0 \\ &= \tfrac{1}{4}, & 0 \le t &< 5 \\ &= \tfrac{1}{3}, & 5 \le t &< 7 \\ &= \tfrac{1}{2}, & 7 \le t &< 100 \\ &= \tfrac{5}{6}, & 100 \le t &< 102 \\ &= 1, & t &\le 102 \end{aligned}$$

is a distribution function and specify the probability function for $Y$. Use it to compute $P(Y \le 100)$.

**5.** The random variable $Z$ has the probability function

$$\begin{aligned} p_Z(x) &= \tfrac{1}{3}, & &\text{for } x = 0, 1, 2 \\ &= 0, & &\text{otherwise} \end{aligned}$$

What is the distribution function for $Z$?

**6.** The random variable $U$ has the probability function

$$\begin{aligned} p_U(-3) &= \tfrac{1}{2} \\ p_U(0) &= \tfrac{1}{6} \\ p_U(4) &= \tfrac{1}{3}. \end{aligned}$$

What is the distribution function for $U$?

**7.** Verify that

$$\begin{aligned} F_X(t) &= 0, & t &< -1 \\ &= \frac{t+1}{2}, & -1 &\le t \le 1 \\ &= 1, & t &> 1 \end{aligned}$$

is a distribution function and specify the probability density function for $X$. Use it to compute $P(-\frac{1}{2} \le X \le \frac{1}{2})$.

**8.** Verify that

$$\begin{aligned} F_Y(t) &= 0, & t &< 0 \\ &= \sqrt{t}, & 0 &\le t \le 1 \\ &= 1, & t &> 1 \end{aligned}$$

is a distribution function and specify the probability density function for $Y$. Use it to compute $P(\frac{1}{4} < Y < \frac{3}{4})$.

**9.** Verify that

$$\begin{aligned} F_Z(t) &= 0, & t &< 0 \\ &= t^2, & 0 &\le t < \tfrac{1}{2} \\ &= 1 - 3(1-t)^2, & \tfrac{1}{2} &\le t < 1 \\ &= 1, & t &\ge 1 \end{aligned}$$

is a distribution function and derive the density function for $Z$.

**10.** Is

$$\begin{aligned} H(t) &= 0, & t &< -1 \\ &= 1 - t^2, & -1 &\le t < \tfrac{1}{2} \\ &= \tfrac{1}{2} + t^2, & \tfrac{1}{2} &\le t < 1 \\ &= 1, & t &\ge 1 \end{aligned}$$

a distribution function?

**11.** Verify that

$$\begin{aligned} H(t) &= 0, & t &< 0 \\ &= \tfrac{1}{2}, & 0 &\le t < \tfrac{1}{2} \\ &= t, & \tfrac{1}{2} &\le t \le 1 \\ &= 1, & t &> 1 \end{aligned}$$

is a distribution function. This is the distribution function of what is called a mixed random variable. It has one point of discontinuity (at $t = 0$) and thus there is a positive probability ($\frac{1}{2}$) of the random variable equalling zero. Note that the distribution function is continuous and increasing for $\frac{1}{2} \le t \le 1$. Thus the random variable takes on particular values in this interval with probability zero; we could define a pseudo density function for the random variable lying in subintervals of this interval.

**12.** Given $X$ has probability density function

$$f_X(x) = 1, \qquad 99 < x < 100$$
$$= 0, \qquad \text{otherwise,}$$

derive $F_X(t)$.

**13.** $Y$ is a continuous random variable with

$$f_Y(y) = 2(1 - y), \qquad 0 < y < 1$$
$$= 0, \qquad \text{otherwise.}$$

Derive $F_Y(t)$.

**14.** $Z$ is a continuous random variable with probability density function

$$f_Z(z) = 10e^{-10z}, \qquad z > 0$$
$$= 0, \qquad \text{otherwise.}$$

Derive $F_Z(t)$.

### 3.3. Expected Values and Summary Measures

We have seen that if $X$ is a discrete random variable, we can use either the distribution function $F_X$ or the probability function $p_X$ to evaluate probability statements about $X$. If $X$ is a continuous random variable, we can use the distribution function $F_X$ or the density function $f_X$ to evaluate probability statements about $X$. Frequently, problems are phrased that require the notion of the average or expected value of $X$, not merely a statement of the probability that $X$ will lie in a certain interval. For example, if we bet 1 dollar on the occurrence of a certain outcome in a gambling game, we might ask how much, on the average, we could expect to get in return. Or, if a doctor recommends that we go on a special reducing diet, we might inquire how much weight we could expect to lose. If we stock a stream with 10,000 small trout, how many might we expect to survive to adult size? Each of these questions refers to an average value of a random variable.

Rather than restrict our discussion to simply the average value of a random variable, we can easily refer to the average value of a function of a random variable $H(X)$. This includes as a special case the identity function $H(X) = X$; that is, if we discuss the average of $H(X)$, this discussion includes the average value of $X$ itself. For concreteness, let us assume that the range of $X$ consists of three numbers, $x_1, x_2, x_3$. We can interpret $p_X(x_i)$, $i = 1, 2, 3$, as giving us the relative proportion of the time that we would observe $X = x_i$ if we were to repeatedly perform the experiment for which $X$ is defined. The values $H(X)$ assumes when $X = x_1$ or $x_2$ or $x_3$ are $H(x_i)$, $i = 1, 2, 3$; thus we would expect $H(X)$ to equal $H(x_i)$ the proportion $p_X(x_i)$ of the time. The average value which $H(X)$ would be expected to assume then is

$$H(x_1)p_X(x_1) + H(x_2)p_X(x_2) + H(x_3)p_X(x_3) = \sum_{\substack{\text{range} \\ \text{of } X}} H(x)p_X(x).$$

This is in fact the definition given in Definition 3.3.1 below. If $X$ were a continuous random variable, we would expect to use integration rather than summation to get the average value of $H(X)$, just as we must integrate rather than sum to evaluate probabilities for $X$. Thus it might seem natural to take as the average value of $H(X)$ the value of integral

$$\int_{-\infty}^{\infty} H(x)f_X(x)\,dx.$$

This is in fact what we have in Definition 3.3.1.

DEFINITION 3.3.1. (1) If $X$ is a discrete random variable with probability function $p_X(x)$, the *expected value* of $H(X)$ (written $E[H(X)]$) is defined to be

$$E[H(X)] = \sum_{\substack{\text{range} \\ \text{of } X}} H(x)p_X(x),$$

so long as the sum is absolutely convergent.

(2) If $X$ is a continuous random variable with probability density function $f_X(x)$, the *expected value* of $H(X)$ is defined to be

$$E[H(X)] = \int_{-\infty}^{\infty} H(x)f_X(x)\,dx,$$

so long as the integral is absolutely convergent. If the integral or the sum is not absolutely convergent, we simply say the expected value does not exist.

DEFINITION 3.3.2. The expected value of $X$ itself is called the *mean* or *average value* of $X$ and is denoted by $\mu_X$; that is, $\mu_X = E[X]$.

Graphically, the mean value of the random variable locates the middle of its probability function or density function in a center of gravity sense. That is, if we imagine that the density function of a random variable describes the relative distribution of mass in a metal rod, then $\mu_X$ locates the point at which the rod would balance. Similarly, if we think of the probability distribution of a discrete random variable as giving the location and amount of mass of various tin cans sitting on a board, then $\mu_X$ locates the point at which the board would balance. (See Figure 3.5.)

*Example 3.3.1.* Suppose you bet 1 dollar on the occurrence of a 1 in the game of Chuck-a-Luck (see Example 3.1.7). Let $V$ be the amount you win and compute

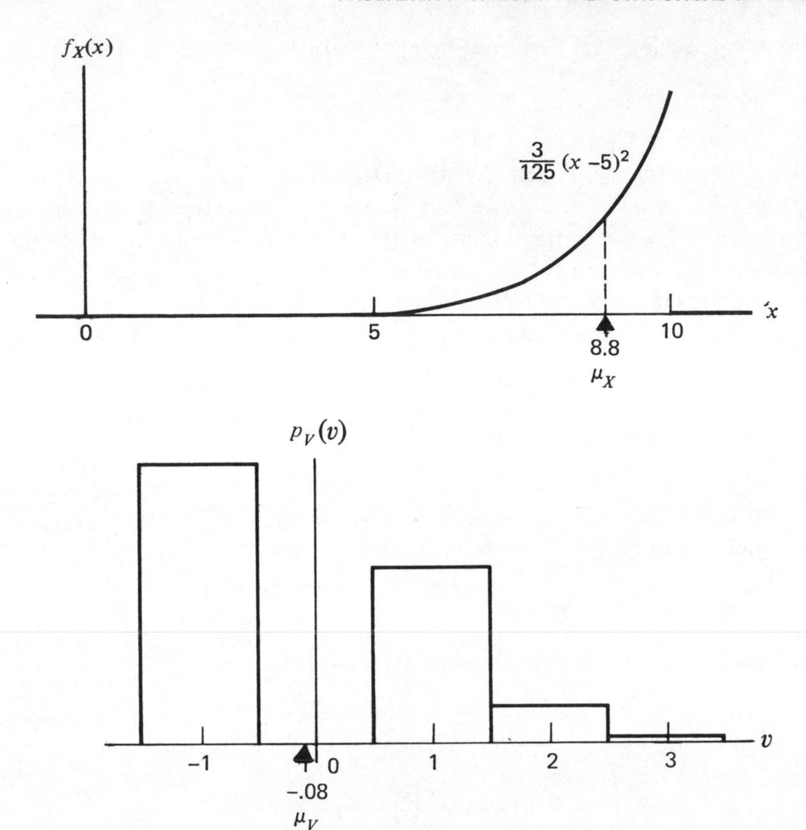


**Figure 3.5.** See Examples 3.3.1 and 3.3.3.

$\mu_V$ and $E[V^2]$. As we saw in Example 3.1.7,

$$\begin{aligned} p_V(v) &= \tfrac{125}{216}, & v &= -1 \\ &= \tfrac{75}{216}, & v &= 1 \\ &= \tfrac{15}{216}, & v &= 2 \\ &= \tfrac{1}{216}, & v &= 3. \end{aligned}$$

Then

$$\begin{aligned} \mu_V = E[V] &= \sum_{\substack{\text{range} \\ \text{of } V}} v p_V(v) \\ &= (-1)\tfrac{125}{216} + (1)\tfrac{75}{216} + (2)\tfrac{15}{216} + (3)\tfrac{1}{216} \\ &= -\tfrac{17}{216} = -.079, \end{aligned}$$

and

$$\begin{aligned} E[V^2] &= \sum_{\substack{\text{range} \\ \text{of } V}} v^2 p_V(v) \\ &= (-1)^2 \tfrac{125}{216} + (1)^2 \tfrac{75}{216} + (2)^2 \tfrac{15}{216} + (3)^2 \tfrac{1}{216} \\ &= \tfrac{269}{216} = 1.245. \end{aligned}$$

Since $\mu_V = -.079$, we could, on the average, expect to lose about 8 cents every time we bet 1 dollar on this game. A game where $\mu_V = 0$ is called fair; thus this is not a fair game. Of course, everything the bettor loses goes to the house, so the casino playing this game can expect to win about 8 cents from every dollar bet on this game, every play. This is part of the reason the game is attractive to a casino.

*Example 3.3.2.* Suppose your doctor recommends that you go a particular diet for 2 weeks. Considering your build and bone structure, he assumes that the amount of weight you will lose is equally likely to lie between 5 and 10 pounds. What is the average amount you might expect to lose on such a diet? If we let $X$ be the number of pounds you will lose, then

$$\begin{aligned} f_X(x) &= \tfrac{1}{5}, \qquad 5 < x < 10 \\ &= 0, \qquad \text{otherwise.} \end{aligned}$$

The amount you might expect to lose is $\mu_X$ where

$$\mu_X = E[X] = \int_5^{10} x \cdot \tfrac{1}{5}\, dx = 7\tfrac{1}{2} \text{ pounds.}$$

*Example 3.3.3.* If, in the previous example, the doctor said that the amount you would lose has probability density function

$$\begin{aligned} f_X(x) &= \tfrac{3}{125}(x-5)^2, \qquad 5 < x < 10 \\ &= 0, \qquad \text{otherwise,} \end{aligned}$$

the expected amount of weight you would lose is

$$\mu_X = E[X] = \int_5^{10} x \cdot \tfrac{3}{125}(x-5)^2\, dx = 8.798 \text{ pounds.}$$

The mean of a random variable gives a measure of the middle of the probability distribution of the random variable. As we shall see, it is desirable in essentially all applications also to have a measure of the variability of the probability distribution (or of the random variable itself). The most commonly used measure of variability of a random variable is the average of the square of the distance between the random variable and its mean $\mu_X$. If $X$ varies relatively little, it will always be close to its mean value and this average squared distance will be relatively small; if $X$ varies a great deal, then it will differ from its mean value by a relatively large amount some of the time and the average squared distance will be relatively large. This average squared deviation is called the variance of the random variable. Its positive square root is called the standard deviation of the random variable. Both are defined below.

DEFINITION 3.3.3. The *variance* of a random variable $X$ (denoted by $\sigma_X^2$) is defined to be

$$\sigma_X^2 = E[(X - \mu_X)^2];$$

its positive square root is denoted by $\sigma_X$ and is called the *standard deviation* of $X$. Thus, $\sigma_X = \sqrt{\sigma_X^2}$.

If there are units attached to the random variable $X$, the variance would be measured in the squares of those units. The standard deviation is of interest since it is measured in the same units as the original random variable.

*Example 3.3.4.* Compute the variance and standard deviation of $V$, defined in Example 3.3.1. By definition,

$$\begin{aligned}\sigma_V^2 &= \sum_{\substack{\text{range}\\ \text{of } V}} [v - (-\tfrac{17}{216})]^2 p_V(v) \\ &= (-\tfrac{199}{216})^2 \tfrac{125}{216} + (\tfrac{233}{216})^2 \tfrac{75}{216} + (\tfrac{449}{216})^2 \tfrac{15}{216} + (\tfrac{665}{216})^2 \tfrac{1}{216} \\ &= 1.239,\end{aligned}$$

and

$$\sigma_V = \sqrt{1.239} = 1.113.$$

*Example 3.3.5.* Compute the variance and standard deviation of the random variable $X$, defined in Example 3.3.2. By definition,

$$\sigma_X^2 = \int_5^{10} (x - 7\tfrac{1}{2})^2 \cdot \tfrac{1}{5}\, dx = \tfrac{25}{12} = 2.083.$$

and

$$\sigma_X = \sqrt{2.083} = 1.443.$$

The operation of taking expected values of random variables has several convenient properties, as is proved in Theorem 3.3.1.

***Theorem 3.3.1.*** If $X$ is a random variable, then

1. $E[c] = c$, where $c$ is a constant
2. $E[cH(X)] = cE[H(X)]$
3. $E[H(X) + G(X)] = E[H(X)] + E[G(X)]$,

so long as the expected values involved exist.

*Proof:* We shall only give the proof for the case of $X$ being continuous. The theorem is true for $X$ being discrete as well, a few details in the following proof needing to be changed appropriately.

1. $E[c] = \int_{-\infty}^{\infty} cf_X(x)\,dx = c\int_{-\infty}^{\infty} f_X(x)\,dx = c$

2. $E[cH(X)] = \int_{-\infty}^{\infty} cH(x)f_X(x)\,dx = c\int_{-\infty}^{\infty} H(x)f_X(x)\,dx = cE[H(X)]$

3. $$\begin{aligned} E[H(X) + G(X)] &= \int_{-\infty}^{\infty} [H(x) + G(x)]f_X(x)\,dx \\ &= \int_{-\infty}^{\infty} H(x)f_X(x)\,dx + \int_{-\infty}^{\infty} G(x)f_X(x)\,dx \\ &= E[H(X)] + E[G(X)]. \end{aligned}$$ ◀

Let us immediately use Theorem 3.3.1 to derive an alternative formula for the variance of a random variable, one that is frequently more easily evaluated than is the definition for $\sigma_X^2$. We proceed as follows:

$$\begin{aligned} \sigma_X^2 = E[(X - \mu_X)^2] &= E[X^2 - 2X\mu_X + \mu_X^2] \\ &= E[X^2] - 2\mu_X E[X] + \mu_X^2 \\ &= E[X^2] - \mu_X^2. \end{aligned} \tag{3.11}$$

Thus, $\sigma_X^2$ can always be calculated as the average of the squares of $X$ minus the square of the average of $X$.

*Example 3.3.6.* Let us recompute the variances of $V$ and of $X$ as given in Examples 3.3.5 and 3.3.6 to verify that we do in fact get the same answers. We saw in Example 3.3.1 that

$$\mu_V = -\tfrac{17}{216}, \qquad E[V^2] = \tfrac{269}{216}.$$

Then

$$\sigma_V^2 = \frac{269}{216} - \left(-\frac{17}{216}\right)^2 = \frac{58{,}104 - 289}{46{,}656} = \frac{57{,}815}{46{,}696} = 1.239,$$

as we found in Example 3.3.5. In Example 3.3.2, we found that $\mu_X = 7\tfrac{1}{2}$. Then

$$E[X^2] = \int_5^{10} x^2\,\tfrac{1}{5}\,dx = \tfrac{175}{3}$$

and

$$\sigma_X^2 = E[X^2] - \mu_X^2 = \frac{175}{3} - \left(\frac{15}{2}\right)^2 = \frac{700 - 675}{12} = \frac{25}{12} = 2.083,$$

the same result we had in Example 3.3.5.

Let us discuss an additional measure of the middle of the probability law for a random variable $X$ (in a different sense) and an additional measure of

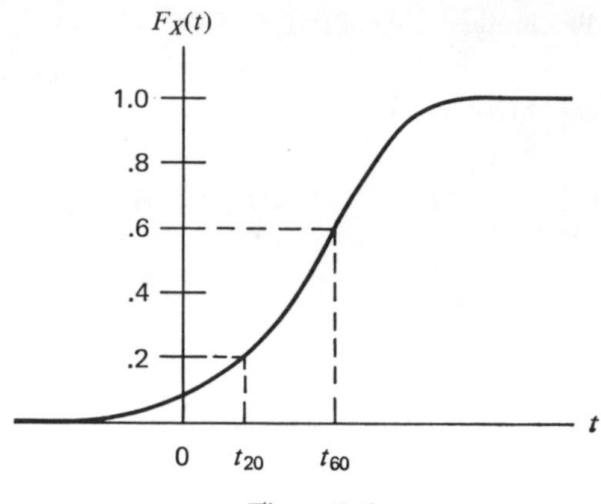


**Figure 3.6.**

the variability of $X$ (also in a different sense). Both will be special functions of the percentiles of $X$, which are defined below.

DEFINITION 3.3.4. The *k-th percentile* of $X$ (denoted by $t_k$) is the smallest possible number such that the probability of not exceeding it is at least $k/100$ where $0 < k < 100$; i.e.,

$$F_X\left(\frac{t_k}{100}\right) \geq \frac{k}{100}.$$

Note, as in Figure 3.6, that the 20-th percentile, $t_{20}$, is the number such that $F_X(t_{20}) = .2$; the 60-th percentile, $t_{60}$, satisfies $F_X(t_{60}) = .60$. Thus, $t_k$ is the smallest number such that the probability that $X \leq t_k$ is $k/100$. Several of the percentiles are given special names; $t_{25}$, $t_{50}$, and $t_{75}$ are called the quartiles of the distribution, because they cut the probability function or density function into four equal probabilities. The 50-th percentile, $t_{50}$, is also called the median of the random variable; it has the property that the random variable is equally likely to lie on either side of it. This is occasionally used as a measure of the middle of the probability law. The difference, $t_{75} - t_{25}$, is called the interquartile range and is sometimes used as a measure of variability. Note that the interquartile range is the length of an interval which brackets the "middle" 50% of the probability distribution for $X$.

The inequality is used in Definition 3.3.4 to define the $k$-th percentile to make sure that the percentiles are defined for all random variables. In particular, if $X$ is a discrete random variable, then its distribution function has jumps in it and there may be no number such that $F_X(t) = k/100$. By using the given definition, the $k$-th percentile is defined to be the number $t_k$ at which the distribution function jumped beyond $k/100$ (see Figure 3.7).

*Example 3.3.7.* Suppose that $X$ is a random variable with distribution function

$$\begin{aligned} F_X(t) &= 0, & t &< 0 \\ &= \sqrt{t}, & 0 &\leq t \leq 1 \\ &= 1, & t &> 1. \end{aligned}$$

Let us find the median of $X$ and the interquartile range for $X$. Since

$$F_X(t) = \sqrt{t},$$

the median is defined by

$$\sqrt{t_{50}} = .5$$

and thus $t_{50} = (.5)^2 = .25$. Similarly,

$$\begin{aligned} t_{25} &= (.25)^2 = .0625 \\ t_{75} &= (.75)^2 = .5625. \end{aligned}$$

The interquartile range is

$$t_{75} - t_{25} = .5.$$

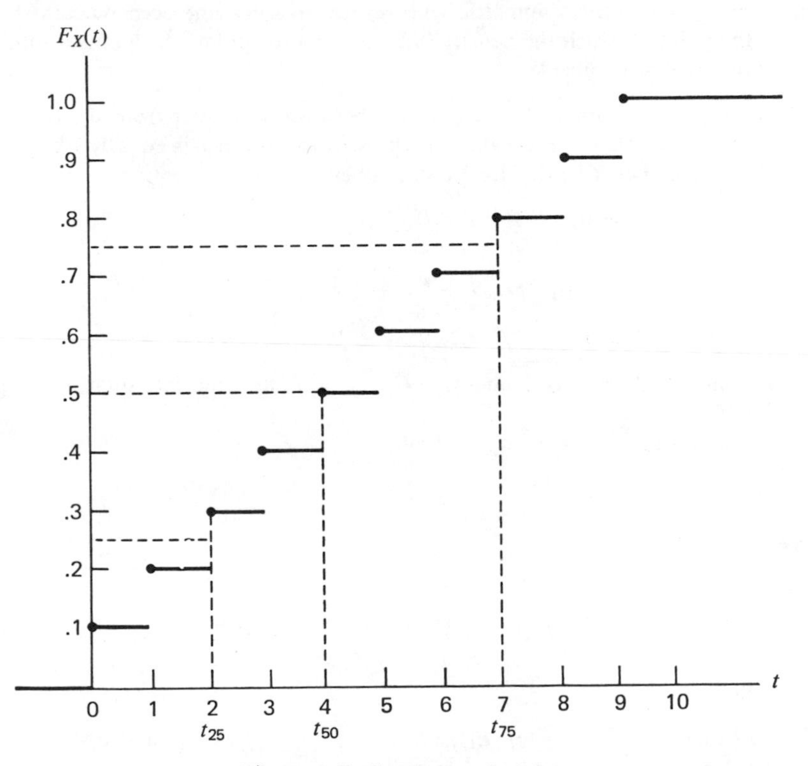


**Figure 3.7.** See Example 3.3.8.

Let us also compute $\mu_X$ and $\sigma_X$. We find that

$$f_X(t) = \frac{1}{2\sqrt{t}}, \qquad 0 < t < 1$$
$$= 0, \qquad \text{otherwise.}$$

Then

$$\mu_X = \int_0^1 x \frac{1}{2\sqrt{x}}\, dx = \tfrac{1}{3}$$

$$E[X^2] = \int_0^1 x^2 \frac{1}{2\sqrt{x}}\, dx = \tfrac{1}{5}$$

$$\sigma_X{}^2 = E[X^2] - \mu_X{}^2 = \tfrac{1}{5} - \tfrac{1}{9} = \tfrac{4}{45},$$

and

$$\sigma_X = \sqrt{\frac{4}{45}} = \frac{2}{3\sqrt{5}}.$$

Note that the median and $\mu_X$ are not equal in this example. Unless the probability density function is symmetric with respect to some line perpendicular to the $x$-axis, the point at which the density balances and the point which cuts it into two equal areas will not coincide.

*Example 3.3.8.* Suppose that we randomly select a number from the telephone book and assume that the last digit in the selected number is equally likely to be 0, 1, 2, . . . , 9. Let $X$ be this final digit. Then

$$F_X(t) = 0, \qquad t < 0$$
$$= \frac{j+1}{10}, \qquad j \le t < j+1, \qquad j = 0, 1, \ldots, 8$$
$$= 1, \qquad t \ge 9.$$

The median for $X$ then is 4 and $t_{25} = 2$, $t_{75} = 7$, and the interquartile range is $7 - 2 = 5$.

We also find that $\mu_X = 4$, $\sigma_X = 3.54$.

## EXERCISE 3.3.

**1.** If

$$p_X(x) = \tfrac{1}{4}, \qquad x = 2, 4, 8, 16$$
$$= 0, \qquad \text{otherwise}$$

compute:

(a) $E[X]$
(b) $E[X^2]$
(c) $E[1/X]$
(d) $E[2^{X/2}]$
(e) $\sigma_X{}^2$ and $\sigma_X$.

**2.** Suppose that $f_X(x) = \frac{1}{2}$, $-1 < x < 1$, compute:

(a) $E[X]$ (c) $E[X + 2]$ (e) $\sigma_X^2$

(b) $E[X^2]$ (d) $E[X/4 + 7]$ (f) $\sigma_X$.

**3.** Given

$$f_X(x) = 2(1 - x), \quad 0 < x < 1$$
$$= 0, \quad \text{otherwise,}$$

compute:

(a) $E[X]$ (c) $E[(X + 10)^2]$ (e) $\sigma_X^2$

(b) $E[X^2]$ (d) $E[1/(1 - X)]$ (f) $\sigma_X$.

**4.** Show that $E[X - \mu_X] = 0$.

**5.** Given

$$F_X(t) = 0, \quad t < 0$$
$$= \frac{t}{100}, \quad 0 \leq t \leq 100$$
$$= 1, \quad t > 0,$$

find: (a) $t_{10}$, (b) $t_{20}$, (c) $t_{55}$, (d) $t_{99}$, (e) $t_{47}$, (f) $t_{80}$, and (g) the interquartile range.

**6.** Suppose that

$$F_U(t) = 0, \quad t < 1$$
$$= \log_e t, \quad 1 \leq t \leq e$$
$$= 1, \quad t > e,$$

find the median, the interquartile range for $U$, $\mu_U$, and $\sigma_U^2$.

**7.** If

$$F_Z(t) = 0, \quad t < 0$$
$$= 2^t - 1, \quad 0 \leq t \leq 1$$
$$= 1, \quad t > 1,$$

find the median, the interquartile range for $Z$, $\mu_Z$, and $\sigma_Z$.

**8.** If

$$F_X(t) = 0, \quad t < 0$$
$$= \tfrac{1}{6}, \quad 0 \leq t < 1$$
$$= \tfrac{1}{3}, \quad 1 \leq t < 2$$
$$= \tfrac{1}{2}, \quad 2 \leq t < 3$$
$$= 1, \quad t \geq 3,$$

find: (a) $t_{50}$, (b) $t_{34}$, (c) $t_{51}$, (d) $t_{16}$ (e) $t_{17}$ and (f) the interquartile range.

**9.** Show that $E[(X - a)^2]$ is minimized if $a = \mu_X$. (**Hint:** Write out analogous to Formula 3.1.)

**10.** The adult height of a 3-year-old boy is equally likely to fall in the interval from 5 feet 6 inches to 5 feet 11 inches. What is his expected height?

**11.** A church lottery is going to give away a 3000-dollar car. They sell 10,000 tickets at 1 dollar apiece. If you buy 1 ticket, what is your expected gain? What is your expected gain if you buy 100 tickets? Compute the variance of your gain in these two instances.

### 3.4. Moments and Generating Functions

The expected values of the integral powers of a random variable $X$ are called the moments of the random variable. It can be shown that complete knowledge of all of the moments of a random variable characterizes its distribution function. That is, the moments are unique for a given distribution function; if two random variables have the same moments they must have the same probability law and vice versa. Thus we shall find that a familiarity with the moments of a random variable provides us with a powerful theoretical tool. The formal definition of the moments of a random variable follows.

DEFINITION 3.4.1. The $k$-th *moment* of a random variable $X$ (denoted by $m_k$) is the expected value of $X$ to the $k$-th power, $k = 1, 2, 3, \ldots$; that is, $m_k = E[X^k]$.

When discussing the mean and variance of a random variable $X$, we were actually using functions of the first two moments of $X$. The first moment $m_1$ is defined to be

$$m_1 = E[X^1] = E[X]$$

and thus $m_1 = \mu_X$. The second moment $m_2$ is

$$m_2 = E[X^2]$$

and thus $m_2 = \sigma_X{}^2 + \mu_X{}^2$. Thus knowledge of $m_1$ and $m_2$ for a random variable would immediately allow us to calculate the mean and the variance of the random variable. These two moments give information about the middle of the probability law and the relative variability about that middle value. The higher order moments $m_3, m_4, m_5, \ldots$, can be shown to give information about other facets of the probability law: the relative peakedness, how similar the two tails are on either side of the mean value, etc. Let us look at some particular examples.

*Example 3.4.1.* Suppose that we throw a dart at a target and have probability .9 of hitting the bull's-eye. Let $X$ be 1 if we hit the bull's-eye and 0 if we miss it. Then,

$$\begin{aligned} p_X(x) &= .1, && \text{at } x = 0 \\ &= .9, && \text{at } x = 1 \\ &= 0, && \text{otherwise.} \end{aligned}$$

The $k$-th moment of $X$ is

$$m_k = E[X^k] = 0^k(.1) + 1^k(.9) = .9 \qquad \text{for } k = 1, 2, 3, \ldots.$$

Thus, this random variable $X$ has constant moments of .9 for all $k$.

*Example 3.4.2.* Suppose that we are going to drive a car to an athletic event. From past experience in making this trip we are willing to assume that our driving time is equally likely to be anywhere from 20 to 30 minutes. If we let $X$ be the number of minutes it will take us to get there, then we have

$$\begin{aligned} f_X(x) &= \tfrac{1}{10}, \qquad 20 < x < 30 \\ &= 0, \qquad \text{otherwise.} \end{aligned}$$

The $k$-th moment of this random variable is

$$m_k = E[X^k] = \int_{20}^{30} \frac{x^k}{10}\, dx = \frac{30^{k+1} - 20^{k+1}}{10(k+1)}, \qquad k = 1, 2, 3, \ldots.$$

Thus,

$$m_1 = \frac{(30)^2 - (20)^2}{10(2)} = 25$$

$$m_2 = \frac{(30)^3 - (20)^3}{10(3)} = 633\tfrac{1}{3},$$

etc., and we have $\mu_X = 25$, $\sigma_X{}^2 = 8\frac{1}{3}$.

A particular function, which can be used to generate the moments of a random variable $X$ by differentiation, is the moment generating function. This is defined as follows.

DEFINITION 3.4.2. The *moment-generating function* (denoted by $m_X(t)$) for a random variable $X$ is defined to be $m_X(t) = E[e^{tX}]$.

To see why $m_X(t)$ is called a moment generating function, let us look at its derivatives with respect to $t$, evaluated at $t = 0$. We define

$$m_X^{(k)}(0) = \frac{d^k}{dt^k}\, m_X(t)\bigg|_{t=0}, \qquad k = 1, 2, 3, \ldots.$$

Then

$$m_X^{(1)}(t) = \frac{d}{dt}\, E[e^{tX}].$$

Assuming we can interchange the operations of differentiation and expectation, we have

$$m_X^{(1)}(t) = E\left[\frac{d}{dt}\, e^{tX}\right] = E[Xe^{tX}]$$

and thus

$$m_X^{(1)}(0) = E[Xe^0] = E[X] = m_1.$$

(The interchange of the two operations is permissible if $m_X(t)$ exists and, if the integral or sum defining $m_X(t)$ is improper, if the integral or sum converges uniformly.) Similarly, we have

$$m_X^{(2)}(t) = \frac{d^2}{dt^2} E[e^{tX}]$$

$$= E\left[\frac{d^2}{dt^2} e^{tX}\right]$$

$$= E[X^2 e^{tX}],$$

so that $m_X^{(2)}(0) = E[X^2] = m_2$. Similarly, we see that $m_X^{(k)}(0) = E[X^k] = m_k$. Thus the successive derivatives of $m_X(t)$ evaluated at the origin generate the moments of $X$. For this reason, $m_X(t)$ is called the moment generating function for $X$.

*Example 3.4.3.* Suppose that the length of time a transistor will work (in a given circuit) is a random variable $Y$ with density function

$$f_Y(y) = 1000e^{-1000y}, \qquad y > 0$$

$$= 0, \qquad \text{otherwise.}$$

The moment generating function for $Y$ is

$$m_Y(t) = E(e^{tY})$$

$$= \int_0^\infty e^{ty} 1000e^{-1000y}\, dy$$

$$= \int_0^\infty 1000e^{-y(1000-t)}\, dy$$

$$= \frac{1000}{1000 - t}, \qquad \text{for } t < 1000.$$

From this we find that

$$\mu_Y = m_1 = m^{(1)}(0) = \frac{1}{1000}$$

$$m_2 = m^{(2)}(0) = \frac{2}{(1000)^2},$$

so

$$\sigma_Y^2 = m_2 - \mu_Y^2 = \frac{1}{(1000)^2}$$

and

$$\sigma_Y = \frac{1}{1000}.$$

*Example 3.4.4.* A fair coin is flipped twice. Let $Z$ be the number of heads that occur. Then

$$\begin{aligned} p_Z(z) &= \tfrac{1}{4}, \qquad \text{at } z = 0 \\ &= \tfrac{1}{2}, \qquad \text{at } z = 1 \\ &= \tfrac{1}{4}, \qquad \text{at } z = 2 \\ &= 0, \qquad \text{otherwise.} \end{aligned}$$

The moment generating function for $Z$ is

$$\begin{aligned} m_Z(t) &= E[e^{tZ}] \\ &= \tfrac{1}{4} + \tfrac{1}{2}e^t + \tfrac{1}{4}e^{2t} \\ &= \tfrac{1}{4}(1 + e^t)^2, \end{aligned}$$

from which we find $\mu_Z = m_1 = 1$, $\sigma_Z^2 = m_2 - \mu_Z^2 = \frac{1}{2}$.

The fact that the moments characterize the distribution function of a random variable is stated in the following theorem, which we shall not prove but shall make use of as the need arises.

***Theorem 3.4.1.*** Assume that $X$ and $Y$ are random variables with moment generating functions $m_X(t)$ and $m_Y(t)$, respectively. Then $m_X(t) = m_Y(t)$ if and only if $F_X(t) = F_Y(t)$ for all real $t$.

We shall delay until later an example illustrating this theorem.

For some uses we shall want to be able to compute what are called moments about the mean of the random variable. These are defined as follows.

DEFINITION 3.4.3. The quantity $E[(X - \mu_X)^k] = \mu_k$ is called the $k$-th *moment about the mean* of $X$ where $k = 1, 2, 3, \ldots$.

As we saw in problem 4 of the last section, $\mu_1 = E[(X - \mu_X)] = 0$ for any random variable $X$. Furthermore,

$$\mu_2 = E[(X - \mu_X)^2]$$

is the variance $\sigma_X^2$ of the random variable. By using the binomial theorem (see Appendix 2), we can derive the following relationship between the moments of the random variable and the moments about the mean of the random variable.

$$\begin{aligned} \mu_k &= E[(X - \mu_X)^k] \\ &= E\left[\sum_{i=0}^{k} \binom{k}{i} X^i(-\mu_X)^{k-i}\right] \\ &= \sum_{i=0}^{k} \binom{k}{i} m_i(-\mu_X)^{k-i}. \end{aligned}$$

Thus,

$$\mu_1 = m_1 - \mu_X = 0,$$

as already noted, and

$$\begin{aligned}\mu_2 &= (-\mu_X)^2 + 2m_1(-\mu_X) + m_2 \\ &= m_2 - \mu_X^2 \\ \mu_3 &= (-\mu_X)^3 + 3m_1(-\mu_X)^2 + 3m_2(-\mu_X) + m_3 \\ &= m_3 - 3m_2\mu_X + 2\mu_X^3,\end{aligned}$$

etc. Similarly, we can write

$$\begin{aligned}m_k &= E[X^k] = E[(\{X - \mu_X\} + \mu_X)^k] \\ &= E\left[\sum_{i=0}^{k} \binom{k}{i} \{X - \mu_X\}^i \mu_X^{k-i}\right] \\ &= \sum_{i=0}^{k} \binom{k}{i} \mu_i \mu_X^{k-i},\end{aligned}$$

and thus

$$\begin{aligned}m_1 &= \mu_X + \mu_1 = \mu_X \\ m_2 &= \mu_X^2 + 2\mu_1\mu_X + \mu_2 \\ &= \mu_X^2 + \mu_2 \\ m_3 &= \mu_X^3 + 3\mu_1\mu_X^2 + 3\mu_2\mu_X + \mu_3 \\ &= \mu_3 + 3\mu_2\mu_X + \mu_X^3,\end{aligned}$$

etc., since $\mu_1 = 0$. Thus, as is not surprising, knowledge of the moments of $X$ is equivalent to knowledge of the moments of $X$ about its mean and vice versa.

We can also easily find a generating function for moments about the mean. We define

$$m_{X-\mu_X}(t) = E[e^{t(X-\mu_X)}]$$

and find that $\mu_k = m_{X-\mu_X}^{(k)}(0)$. Furthermore,

$$\begin{aligned}E[e^{t(X-\mu_X)}] &= E[e^{-t\mu_X}e^{tX}] \\ &= e^{-t\mu_X}E[e^{tX}],\end{aligned}$$

so that

$$m_{X-\mu_X}(t) = e^{-t\mu_X}m_X(t). \tag{4.1}$$

Thus, if we know the moment generating function, we simply multiply it by $e^{-t\mu_X}$ to derive the function that generates moments about $\mu_X$.

*Example 3.4.5.* For the random variable $Y$ defined in Example 3.4.3, we found

$$m_Y(t) = \frac{1000}{1000 - t}, \qquad m_1 = \mu_Y = \frac{1}{1000}.$$

Then

$$m_{Y-1/1000}(t) = e^{-t/1000}\frac{1000}{1000 - t}$$

will generate moments about the mean of $Y$ and it is easily verified that

$$\mu_1 = m^{(1)}_{Y-1/1000}(0) = 0$$
$$\sigma_Y{}^2 = m^{(2)}_{Y-1/1000}(0) = \frac{1}{(1000)^2}\,.$$

For the random variable $Z$ defined in Example 3.4.4, we found

$$m_Z(t) = \tfrac{1}{4}(1 + e^t)^2, \qquad m_1 = \mu_Z = 1,$$

and thus

$$m_{Z-1}(t) = \frac{e^{-t}}{4}(1 + e^t)^2$$

will generate moments of $Z$ about its mean. We can also verify for this random variable that

$$\mu_1 = 0, \qquad \mu_2 = \tfrac{1}{2} = \sigma_Z{}^2.$$

Before concluding this section let us study briefly one additional generating function, one that is frequently called either the factorial moment generating function or the probability generating function. Although we shall use the former name, it is commonly referred to as the probability generating function, especially in stochastic process applications of probability. It is defined as follows.

DEFINITION 3.4.4. The *factorial moment generating function* for a random variable $X$ is defined to be

$$\psi_X(t) = E[t^X].$$

To see first of all why this might be called the factorial moment generating function, we note that:

$$\psi^{(1)}_X(t) = E[Xt^{X-1}]$$
$$\psi^{(2)}_X(t) = E[X(X-1)t^{X-2}]$$
$$\vdots$$
$$\psi^{(k)}_X(t) = E[X(X-1)\cdots(X-k+1)t^{X-k}],$$

and thus

$$\xi_1 = \psi^{(1)}_X(1) = E[X] = \mu_X$$
$$\xi_2 = \psi^{(2)}_X(1) = E[X(X-1)]$$
$$\vdots$$
$$\xi_k = \psi^{(k)}_X(1) = E[X(X-1)\cdots(X-k+1)].$$

The quantities $\xi_k$ are called the factorial moments of $X$ (because of their factorial-like structure). The factorial moments are especially useful for discrete random variables, as we shall see. It can be shown that knowledge of $\xi_k$, $k = 1, 2, \ldots,$ is equivalent to knowledge of the moments of $X$ and vice versa. Thus, if we are easily able to get the factorial moments, then we are also able to evaluate the moments of the random variable, should we desire to do so. In fact, since

$$t^X = e^{X \log_e t},$$

we can see that

$$\psi_X(t) = m_X(\log_e t)$$

and that

$$\psi_X(e^t) = m_X(t).$$

To see why $\psi_X(t)$ is also called a probability generating function, let us assume that $X$ is a discrete random variable and that the range of $X$ is the set of integers $1, 2, \ldots, n$. Then

$$\begin{aligned}\psi_X(t) &= E[t^X] \\ &= \sum_{i=1}^{n} t^i p_X(i) \\ &= tp_X(1) + t^2 p_X(2) + \cdots + t^n p_X(n).\end{aligned}$$

Then

$$\begin{aligned}\psi_X^{(1)}(t) &= p_X(1) + 2tp_X(2) + \cdots + nt^{n-1}p_X(n) \\ \psi_X^{(2)}(t) &= 2p_X(2) + 3 \cdot 2tp_X(3) + \cdots + n(n-1)t^{n-2}p_X(n) \\ &\vdots \\ \psi_X^{(n-1)}(t) &= (n-1)!\, p_X(n-1) + n!\, tp_X(n) \\ \psi_X^{(n)}(t) &= n!\, p_X(n),\end{aligned}$$

and we see that

$$\psi_X^{(k)}(0) = k!\, p_X(k), \qquad k = 1, 2, \ldots, n.$$

That is, the derivatives of $\psi_X(t)$ evaluated at $t = 0$ gives a known constant times the values of the probability function for $X$. (Note as well that $\psi_X(0) = p_X(0)$.)

*Example 3.4.6.* For the discrete random variable $Z$ defined in Example 3.4.4, we have

$$\begin{aligned}\psi_Z(t) &= E[t^Z] \\ &= \tfrac{1}{4} + \tfrac{1}{2}t + \tfrac{1}{4}t^2 = \tfrac{1}{4}(1 + t)^2.\end{aligned}$$

Note that

$$\begin{aligned}
\psi_Z(0) &= \tfrac{1}{4} = p_Z(0) \\
\psi_Z^{(1)}(0) &= \tfrac{1}{2} = p_Z(1) \\
\psi_Z^{(2)}(0) &= \tfrac{1}{2} = 2!\, p_Z(2) \\
\psi_Z^{(k)}(t) &= 0, \qquad k = 3, 4, 5, \ldots,
\end{aligned}$$

and that

$$\begin{aligned}
\psi_Z^{(1)}(1) &= 1 = \mu_Z \\
\psi_Z^{(2)}(1) &= \tfrac{1}{2} = E[Z^2] - \mu_Z.
\end{aligned}$$

*Example 3.4.7.* Suppose that we flip a fair coin until we get a head (see Example 2.8.1) and let $W$ be the number of flips required to conclude the experiment. Then

$$p_W(k) = \frac{1}{2^k}, \qquad k = 1, 2, 3, \ldots,$$

and

$$\begin{aligned}
\psi_W(t) &= E[t^W] \\
&= \sum_{i=1}^{\infty} t^i \frac{1}{2^i} \\
&= \sum_{i=1}^{\infty} \left(\frac{t}{2}\right)^i = \frac{1}{1 - t/2} - 1 = \frac{t}{2 - t}, \qquad \text{if } |t| < 2.
\end{aligned}$$

We find that

$$\begin{aligned}
\psi_W^{(1)}(t) &= \frac{2}{(2 - t)^2} \\
\psi_W^{(2)}(t) &= \frac{4}{(2 - t)^3} \\
\psi_W^{(k)}(t) &= \frac{2 \cdot k!}{(2 - t)^{k+1}}, \qquad k = 1, 2, 3, \ldots.
\end{aligned}$$

Thus

$$\frac{1}{k!}\,\psi_W^{(k)}(0) = \frac{1}{2^k} = p_W(k), \qquad k = 1, 2, 3, \ldots$$

and

$$\begin{aligned}
\psi_W^{(1)}(1) &= 2 = \mu_W \\
\psi_W^{(2)}(1) &= 4 = E[W^2] - \mu_W,
\end{aligned}$$

etc.

**EXERCISE 3.4.**

**1.** A single die is rolled 1 time. Let $X$ be the number of 6's that occur. Compute $m_X(t)$ and use it to evaluate the first 3 moments of $X$.

**2.** Assume that $Y$, the number of minutes it takes you to eat lunch on an average day, is equally likely to lie in the interval from 30 to 40 minutes. Compute $m_Y(t)$ and $m_{Y-\mu_Y}(t)$.

**3.** The number of hours of satisfactory operation that a certain brand of TV set will give (without repair) is a random variable $Z$ whose probability density function is

$$\begin{aligned} f_Z(z) &= 500e^{-500z}, \qquad z > 0 \\ &= 0, \qquad z \leq 0. \end{aligned}$$

Derive $m_Z(t)$ and use it to compute $\mu_Z$ and $\sigma_Z^2$.

**4.** A fair coin is flipped 3 times. Let $U$ be the total number of heads that occur. Derive $\psi_U(t)$ and use it to compute $\mu_U$ and $\sigma_U^2$.

**5.** One integer is selected at random from the set $\{1, 2, 3, \ldots, n\}$. Let $V$ be the selected integer and derive $\psi_V(t)$.

**6.** Express $\xi_1$, $\xi_2$, and $\xi_3$ in terms of $m_1$, $m_2$, and $m_3$.

**7.** Express $m_1$, $m_2$, and $m_3$ in terms of $\xi_1$, $\xi_2$, and $\xi_3$.

**8.** Rachel is going to throw darts at a target until she hits the bull's-eye. Assume that she has constant probability .9 of hitting the bull's-eye each toss and that the tosses are independent. Let $W$ be the number of tosses required. Derive $\psi_W(t)$ and use it to compute $\mu_W$ and $\sigma_W^2$.

**9.** Suppose that $X$ is a random variable whose range is $\{-1, -2, \ldots, -n\}$. Show that $E[t^{-X}]$ would be a probability generating function for $X$.

**10.** Show that $m_X(0) = m_{X-\mu_X}(0) = 1$, for any random variable $X$.

### 3.5 Functions of Random Variables

In many problems, we may be interested in making statements about functions of random variables. For example, a chemist may want to make statements about the temperature at which a certain reaction will occur and also to have the ability to make such statements for temperatures measured either on the absolute scale or on the Fahrenheit scale. When computing probability statements about random variables measured in time, we might want to use seconds, minutes, or hours as units. When computing probabilities about random variables measured in distance, we might want to use inches, feet, yards, etc. as our units. In each of these cases, a simple linear transformation of the random variable concerned enables us to use any units we feel are appropriate. The following theorem shows how we can derive the distribution function for $Y$ from the distribution function for $X$ if $Y$ is a linear function of $X$.

***Theorem 3.5.1.*** $X$ is a random variable with distribution function $F_X(t)$.

Define $Y = a + bX$ where $b > 0$. Then:

$$F_Y(t) = F_X\left(\frac{t-a}{b}\right)$$

$$F_Y(a + bt) = F_X(t), \qquad \text{for all } t.$$

*Proof:* By definition,

$$\begin{aligned} F_Y(t) &= P(Y \leq t) \\ &= P(a + bX \leq t) \\ &= P\left(X \leq \frac{t-a}{b}\right) \\ &= F_X\left(\frac{t-a}{b}\right). \end{aligned}$$

The second conclusion follows by replacing $t$ by $a + bt$. ◀

*Example 3.5.1.* Suppose that the length of time it takes Earl to drive to work every day is a random variable $X$ (measured in hours) with distribution function

$$\begin{aligned} F_X(t) &= 0, & t &< \tfrac{1}{4} \\ &= 4(t - \tfrac{1}{4})^2, & \tfrac{1}{4} &\leq t \leq \tfrac{3}{4} \\ &= 1, & t &> \tfrac{3}{4}. \end{aligned}$$

Let $Y$ be the time it takes Earl to make the drive measured in minutes, and let $Z$ be the length of time it takes him measured in seconds. Then, $Y = 60X$ and

$$\begin{aligned} F_Y(t) &= F_X\left(\frac{t}{60}\right) \\ &= 0, \qquad \frac{t}{60} < \frac{1}{4}, \qquad \text{i.e., } t < 15 \\ &= 4\left(\frac{t}{60} - \frac{1}{4}\right)^2 = \frac{1}{900}(t - 15)^2, \qquad \text{if } \frac{1}{4} \leq \frac{t}{60} \leq \frac{3}{4}, \qquad \text{i.e., } 15 \leq t \leq 45 \\ &= 1, \qquad \text{if } \frac{t}{60} > \frac{3}{4}, \qquad \text{i.e., } t > 45. \end{aligned}$$

We also have $Z = 3600X$ and

$$\begin{aligned} F_Z(t) &= F_X\left(\frac{t}{3600}\right) \\ &= 0, \qquad \text{if } \frac{t}{3600} < \frac{1}{4}, \qquad \text{i.e., } t < 900 \\ &= 4\left(\frac{t}{3600} - \frac{1}{4}\right)^2 = \frac{1}{3{,}240{,}000}(t - 900)^2, \qquad \text{if } \frac{1}{4} \leq \frac{t}{3600} \leq \frac{3}{4}, \\ &\qquad \text{i.e., } 900 \leq t \leq 2700 \\ &= 1, \qquad \text{if } \frac{t}{3600} > \frac{3}{4}, \qquad \text{i.e., } t > 2700. \end{aligned}$$

*Example 3.5.2.* Suppose that the temperature at which a particular reaction will occur is a random variable $U$ with distribution function

$$\begin{aligned} f_U(t) &= 0, & t &< 170 \\ &= \tfrac{1}{2}\sqrt{t - 170}, & 170 &\le t \le 174 \\ &= 1, & t &> 174, \end{aligned}$$

where $U$ is measured in degrees Fahrenheit. Let $V$ be the temperature at which the reaction will occur, if $V$ is now measured on the Kelvin (absolute) scale. Then $V = \frac{5}{9}U + 256\frac{7}{9}$ and

$$\begin{aligned} F_V(t) &= F_U\left(\frac{t - 256\frac{7}{9}}{\frac{5}{9}}\right) \\ &= F_U\left(\frac{9t - 2311}{5}\right) \\ &= 0, \qquad \text{if } t < 351\tfrac{2}{9} \\ &= \frac{1}{2}\left[\frac{9t - 2311}{5} - 170\right]^{1/2} = \frac{1}{2}\left[\frac{9t - 3161}{5}\right]^{1/2} \qquad \text{if } 351\tfrac{2}{9} \le t \le 353\tfrac{4}{9}, \\ &= 1, \qquad \text{if } t > 353\tfrac{4}{9}. \end{aligned}$$

If $X$ is a continuous random variable and $Y = a + bX$, then $Y$ is also a continuous random variable and its density function may be derived from that of $X$. As was proved in Theorem 3.5.1,

$$F_Y(t) = F_X\left(\frac{t - a}{b}\right),$$

and thus

$$\begin{aligned} f_Y(t) &= \frac{d}{dt} F_Y(t) \\ &= \frac{d}{dt} F_X\left(\frac{t - a}{b}\right) \\ &= \frac{1}{b} f_X\left(\frac{t - a}{b}\right). \end{aligned}$$

*Example 3.5.3.* The random variable $X$ defined in Example 3.5.1 has density function

$$\begin{aligned} f_X(t) &= 8t - 2, & \tfrac{1}{4} &< t < \tfrac{3}{4} \\ &= 0 & &\text{otherwise,} \end{aligned}$$

and thus $Y = 60X$ and $Z = 3600X$ have density functions

$$\begin{aligned} f_Y(t) &= \frac{1}{60} f_X\left(\frac{t}{60}\right) = \frac{1}{60}\left(\frac{8t}{60} - 2\right), & 15 &< t < 45 \\ &= 0, & &\text{otherwise,} \end{aligned}$$

and

$$f_Z(t) = \frac{1}{3600} f_X\left(\frac{t}{3600}\right) = \frac{1}{3600}\left(\frac{8t}{3600} - 2\right), \qquad 900 < t < 2700$$
$$= 0, \qquad \text{otherwise.}$$

The random variable $U$ defined in Example 3.5.2 has density function

$$f_U(t) = \frac{1}{4\sqrt{t - 170}} \qquad 170 < t < 174$$
$$= 0, \qquad \text{otherwise.}$$

Thus, $V = \frac{5}{9}U + 256\frac{7}{9}$ has density function

$$f_V(t) = \frac{1}{4\sqrt{(t - 256\frac{7}{9})/\frac{5}{9}) - 170}}$$
$$= \frac{\sqrt{5}}{4\sqrt{9t - 3161}}, \qquad 351\tfrac{2}{9} < t < 353\tfrac{4}{9}$$
$$= 0, \qquad \text{otherwise.}$$

Since the distribution functions of $X$ and $Y$, a linear function of $X$, are related in a simple way we would expect the same to be true for the percentiles and the moments of the random variables. In fact, let $Y = a + bX$, let $t_k$ be the $k$-th percentile of $X$, and let $s_k$ be the $k$-th percentile of $Y$. Then, by Definition 3.3.4,

$$F_Y(s_k) = \frac{k}{100} = F_X\left(\frac{s_k - a}{b}\right);$$

but $F_X(t_k) = k/100$ and thus $(s_k - a)/b = t_k$. That is, $s_k = a + bt_k$ for any $k$, and thus the $k$-th percentile of $Y$ is the same linear function of $t_k$. If $X$ is a random variable with $t_{50} = 4$ and $t_{90} = 7$, then $Y = 10 + 20X$ has $s_{50} = 10 + 20(4) = 90$ and $s_{90} = 10 + 20(7) = 150$.

Again, suppose that $Y = a + bX$. Then,

$$m_Y(t) = E[e^{tY}] = E[e^{t(a+bX)}]$$
$$= E[e^{ta}e^{tbX}] = e^{ta}E[e^{tbX}]$$
$$= e^{ta}m_X(tb).$$

Differentiating with respect to $t$, we find that

$$m_Y^{(1)}(t) = ae^{ta}m_X(tb) + be^{ta}m_X^{(1)}(tb)$$
$$m_Y^{(2)}(t) = a^2e^{ta}m_X(tb) + 2abe^{ta}m_X^{(1)}(tb) + b^2e^{ta}m_X^{(2)}(tb),$$

and thus

$$m_Y^{(1)}(0) = a + bm_X^{(1)}(0)$$
$$m_Y^{(2)}(0) = a^2 + 2abm_X^{(1)}(0) + b^2m_X^{(2)}(0).$$

That is,

$$\mu_Y = a + b\mu_X$$
$$E[Y^2] = a^2 + 2ab\mu_X + b^2E[X^2],$$

and thus

$$\begin{aligned}\sigma_Y^2 &= E[Y^2] - \mu_X^2 \\ &= a^2 + 2ab\mu_X + b^2E[X^2] - (a + b\mu_X)^2 \\ &= b^2E[X^2] - b^2\mu_X^2 \\ &= b^2\sigma_X^2.\end{aligned}$$

So we see that if $Y = a + bX$, then the mean of $Y$ is the same linear function of the mean of $X$ and the variance of $Y$ is $b^2\sigma_X^2$, regardless of the value of $a$. If, for example, $X$ is a random variable with $\mu_X = 10$ and $\sigma_X^2 = 1$, then $Y = 5 + 10X$ has mean $\mu_Y = 105$ and variance $\sigma_Y^2 = 10^2\sigma_X^2 = 100$.

An important linear function of a random variable $X$ is called the standard form of $X$. It is defined as follows.

DEFINITION 3.5.1. The *standard form* of a random variable $X$ is the random variable $Z = (X - \mu_X)/\sigma_X$.

Note immediately then that $\mu_Z = -\mu_X/\sigma_X + (1/\sigma_X)\,\mu_X = 0$ and $\sigma_Z^2 = (1/\sigma_X^2)\,\sigma_X^2 = 1$. Thus the standard form of any random variable will always have mean 0 and variance 1.

We shall occasionally have use for the distribution of the square of a continuous random variable $X$. This is given in Theorem 3.5.2.

***Theorem 3.5.2.*** Suppose that $X$ is a continuous random variable with distribution function $F_X(t)$. If $Y = aX^2$, $a > 0$, then the distribution function of $Y$ is

$$\begin{aligned}F_Y(t) &= 0, && t < 0 \\ &= F_X\left(\sqrt{\frac{t}{a}}\right) - F_X\left(-\sqrt{\frac{t}{a}}\right), && t \geq 0\end{aligned}$$

and the density function of $Y$ is

$$\begin{aligned}f_Y(t) &= \frac{1}{2\sqrt{at}}\left[f_X\left(\sqrt{\frac{t}{a}}\right) + f_X\left(-\sqrt{\frac{t}{a}}\right)\right], && t > 0 \\ &= 0, && \text{otherwise.}\end{aligned}$$

*Proof:* If $Y = aX^2$, then the number which $Y$ assigns to any experimental outcome (in the sample space $S$) is $a$ times the square of the number assigned that outcome by $X$; thus the range of $Y$ consists of only nonnegative numbers and $F_Y(t) = 0$, $t < 0$. For any $t \geq 0$

$$\begin{aligned} F_Y(t) &= P(Y \leq t) \\ &= P(aX^2 \leq t) \\ &= P\left(-\sqrt{\frac{t}{a}} \leq X \leq \sqrt{\frac{t}{a}}\right) \\ &= F_X\left(\sqrt{\frac{t}{a}}\right) - F_X\left(-\sqrt{\frac{t}{a}}\right), \end{aligned}$$

since $X$ is continuous. Then, to derive $f_Y(t)$, we differentiate $F_Y(t)$ which yields

$$\begin{aligned} \frac{d}{dt} F_Y(t) &= f_X\left(\sqrt{\frac{t}{a}}\right)\left(\frac{1}{2\sqrt{at}}\right) - f_X\left(-\sqrt{\frac{t}{a}}\right)\left(-\frac{1}{2\sqrt{at}}\right) \\ &= \frac{1}{2\sqrt{at}}\left[f_X\left(\sqrt{\frac{t}{a}}\right) + f_X\left(-\sqrt{\frac{t}{a}}\right)\right], \qquad t > 0, \end{aligned}$$

and we thus find

$$\begin{aligned} f_Y(t) &= \frac{1}{2\sqrt{at}}\left[f_X\left(\sqrt{\frac{t}{a}}\right) + f_X\left(-\sqrt{\frac{t}{a}}\right)\right], \qquad t > 0 \\ &= 0, \qquad t \leq 0, \end{aligned}$$

as was to be proved. ◀

*Example 3.5.4.* A nursery specializes in installation of circular flower beds. For each such bed the workman lays out the circle in the desired area, removes six inches of dirt, and replaces the dirt with a topsoil-peat mixture. When laying out the circle, a workman puts down a peg in the center, cuts a length of rope (already tied to the stake) equal to the radius of the desired circle, and then uses this to mark out the bed on the ground. Assume that the desired radius is $r$ (in feet). Also assume that the workman is a little sloppy and is equally likely to cut the rope to any length in the interval $(r - .1, r + .1)$. Then, if we let $X$ be the length of the cut piece,

$$\begin{aligned} f_X(x) &= 5, \qquad r - .1 < x < r + .1 \\ &= 0, \qquad \text{otherwise}, \end{aligned}$$

the actual surface area of the circle produced then is

$$Y = \pi X^2$$

with density

$$\begin{aligned} f_Y(t) &= \frac{1}{2\sqrt{\pi t}}\left[f_X\left(\sqrt{\frac{t}{\pi}}\right) + f_X\left(-\sqrt{\frac{t}{\pi}}\right)\right] \\ &= \frac{5}{2\sqrt{\pi t}} \qquad \text{for } \pi(r - .1)^2 \leq t \leq \pi(r + .1)^2 \\ &= 0, \qquad \text{otherwise}. \end{aligned}$$

Note then that the expected surface area of the flower bed is $\pi r^2 + (.001)(5/3)\pi$; that is, the average surface area exceeds $\pi r^2$ even though the average radius is $r$. The probability that the surface area exceeds $\pi r^2$ is

$$P(Y > \pi r^2) = \int_{\pi r^2}^{\pi(r+.1)^2} \frac{5}{2\sqrt{\pi t}}\,dt$$

$$= .5.$$

The following theorem gives the probability law of the square root of a continuous random variable. We require $X$ to be a positive random variable so that its square root is real. The proof of Theorem 3.5.3 is asked for in the problems that follow.

***Theorem 3.5.3.*** Assume that $X$ is a continuous random variable and $F_X(0) = 0$. If $Y = \sqrt{X}$, then

$$\begin{aligned} F_Y(t) &= 0, && \text{for } t < 0 \\ &= F_X(t^2), && \text{for } t \geq 0 \end{aligned}$$

and

$$\begin{aligned} f_Y(t) &= 0, && \text{for } t < 0 \\ &= 2tf_X(t), && \text{for } t \geq 0. \end{aligned}$$

**EXERCISE 3.5.**

**1.** Let $X$ be a random variable with distribution function $F_X(t)$ and let $Y = a + bX$ where $b < 0$. Derive the distribution function for $Y$.

**2.** Suppose that $b = 0$ in problem 1 above. Derive the distribution function for $Y$, defined as in that problem.

**3.** Given

$$\begin{aligned} F_X(t) &= 0, && t < -1 \\ &= \frac{t+1}{2}, && -1 \leq t \leq 1 \\ &= 1, && t > 1, \end{aligned}$$

find the distribution function for $Y = 15 + 2X$ and the density function for $Y$.

**4.** Suppose that

$$\begin{aligned} F_W(t) &= 0, && t < 0 \\ &= t^3, && 0 \leq t \leq 1 \\ &= 1, && t > 1 \end{aligned}$$

and let $Z = W - 1$. Find $F_Z(t)$ and $f_Z(t)$.

**5.** If

$$\begin{aligned} F_X(t) &= 0, & t &< -10 \\ &= \tfrac{1}{4}, & -10 &\le t < 0 \\ &= \tfrac{3}{4}, & 0 &\le t < 10 \\ &= 1, & t &\ge 10, \end{aligned}$$

find the distribution function for

$$U = 7X - 50 \text{ and } p_U(u).$$

**6.** If

$$\begin{aligned} F_Y(t) &= 1 - e^{-t}, & t &\ge 0 \\ &= 0, & t &< 0, \end{aligned}$$

find $F_X(t)$ and $f_X(t)$ where $X = 2Y - 7$.

**7.** If

$$\begin{aligned} F_X(t) &= 0, & t &< 0 \\ &= t, & 0 &\le t \le 1 \\ &= 1, & t &> 1, \end{aligned}$$

find the distribution function and the density function for the standard form of $X$.

**8.** If

$$\begin{aligned} F_Z(t) &= 1 - e^{-t}, & t &> 0 \\ &= 0, & t &\le 0, \end{aligned}$$

find the distribution function and the density function for the standard form of $Z$.

**9.** Given

$$\begin{aligned} F_U(t) &= 0, & t &< -1 \\ &= \tfrac{1}{3}, & -1 &\le t < 0 \\ &= \tfrac{2}{3}, & 0 &\le t < 1 \\ &= 1, & t &> 1, \end{aligned}$$

find the distribution function for the standard form of $U$.

**10.** Let

$$\begin{aligned} F_X(t) &= 0, & t &< 0 \\ &= t, & 0 &\le t \le 1 \\ &= 1, & t &> 1, \end{aligned}$$

and find $F_Z(t)$ and $f_Z(t)$ where $Z = X^2$.

**11.** If

$$\begin{aligned} F_X(t) &= 0, & t &< -1 \\ &= \frac{t+1}{2}, & -1 &\le t \le 1 \\ &= 1, & t &> 1, \end{aligned}$$

find the distribution function and the density function for $Z = X^2$.

**12.** Prove Theorem 3.5.3.

# 4

# Some Standard Distributions

Some random variables occur frequently in many different contexts in applied problems. Because they do occur often in practice, these random variables are given special names; we shall study several of these random variables and their properties in this chapter.

## 4.1. Binomial and Bernoulli Random Variables

Before looking at the definitions of these two random variables we shall first define a Bernoulli trial.

DEFINITION 4.1.1. A *Bernoulli trial* is an experiment which has two possible outcomes, generally called success and failure.

The sample space for a Bernoulli trial will in general be written $S = \{s, f\}$. Many different examples of Bernoulli trials can be quoted: a flip of a single coin (head or tail), the flight of a missile (if we call it simply a success or not), performance of a student in a particular course (pass or fail), or performance of an athletic team (win or not win). Any chance mechanism whose outcomes are grouped into two classes can be looked at as being a Bernoulli trial.

A frequently used notation is

$$P(\{s\}) = p$$
$$P(\{f\}) = q = 1 - p,$$

which we shall also use; the quantity $p$ is of course free to take on any value in the interval from 0 to 1, inclusive, for various types of Bernoulli trials. In Section 2.7 we briefly discussed the idea of independent trials (see Definition 2.7.4). We shall apply some of that discussion now. Suppose that we have an experiment which consists of $n$ (a positive integer) repeated independent Bernoulli trials. The sample space for this experiment then is the Cartesian product of the sample spaces of the individual trials. Thus $S = S_1 \times S_2 \times \cdots \times S_n$ where $S_i = \{s, f\}, i = 1, 2, \ldots, n$, and $P_i(\{s\}) = p$ for all $i$. (See Definition 2.7.4.) The binomial random variable is defined as follows for this sample space.

DEFINITION 4.1.2. Let $X$ be the total number of successes in $n$ repeated independent Bernoulli trials with probability $p$ of success on a given trial. $X$ is called the *binomial* random variable with parameters $n$ and $p$.

The range of the random variable $X$ is the integers $0, 1, 2, \ldots, n$; thus $X$ is a discrete random variable and as such must have a probability function. The statement above that $X$ has parameters $n$ and $p$ means that the probability function for $X$ is completely specified if the values of $n$ and $p$ are known. This probability function is derived as follows.

***Theorem 4.1.1.*** If $X$ is binomial with parameters $n$ and $p$, then

$$p_X(x) = \binom{n}{x} p^x q^{n-x}, \qquad x = 0, 1, 2, \ldots, n$$

$$= 0, \qquad \text{otherwise.}$$

*Proof:* As mentioned earlier, the sample space for the experiment (of $n$ trials) is $S = S_1 \times S_2 \times \cdots \times S_n$ where $S_i = \{s, f\}$ and $P_i(\{s\}) = p$, $i = 1, 2, \ldots, n$. Then, since the trials are independent, we assign probabilities to the single-element events of $S$ by multiplying the values of the probabilities for the individual trials. Clearly, the random variable $X$ assigns, to each element $\omega$ of $S$, the number of $s$'s contained in $\omega$. Then

$$p_X(k) = P[\{(x_1, x_2, \ldots, x_n)\colon \text{exactly } k\ x_i\text{'s equal } s\}].$$

The probability on the right is evaluated by summing together the probabilities of the single-element events making it up. But every one of these single-element events must consist of a single $n$-tuple which contains exactly $k$ $s$'s. The probability assigned to every such single-element event is $p^k q^{n-k}$, using the product rule mentioned earlier. Thus, if we can merely count the number of $n$-tuples in $S$ having exactly $k$ $s$'s, the product of this number with $p^k q^{n-k}$ will give us $p_X(k)$. The number of $n$-tuples having exactly $k$ $s$'s is

$\binom{n}{k}$. Thus,

$$\begin{aligned} p_X(k) &= \binom{n}{k} p^k q^{n-k}, \qquad k = 0, 1, 2, \ldots, n \\ &= 0, \qquad \text{otherwise.} \end{aligned}$$

◄

We should first of all check to see that $p_X(k)$ is a probability function (see Equations 3.1 and 3.2). Obviously,

$$p_X(k) \geq 0, \qquad \text{for all } k.$$

The sum of $p_X(k)$ over the range of $X$ is

$$\begin{aligned} \sum_{k=0}^{n} p_X(k) &= \sum_{k=0}^{n} \binom{n}{k} p^k q^{n-k} \\ &= (p + q)^n \\ &= (p + 1 - p)^n = 1, \end{aligned}$$

and thus $p_X(k)$ is a probability function.

Figure 4.1 presents graphs of four different binomial distributions (that is, four different pairs of values for $n$ and $p$). These graphs are histograms or bar charts of the probability distributions; centered above each value in the range of the random variable is a bar whose base is one unit and whose height is proportional to the probability that the random variable equals that value. Thus the areas of the bars are proportional to the probabilities of occurrence of the various values in the range of the random variable. Note that if $p = .5$, then when $n$ equals either 10 or 20 the bar chart is symmetric about a line passing through the mean of the random variable. Increasing $p$ from .5 to .9 causes the majority of the area (probability) to be shifted to the right; but when $n = 20$, notice that the histogram is again close to being symmetric about a line through the mean. As we shall see in Chapter 5, the larger that $n$ is, the closer the histogram is to this symmetric situation, no matter what the value of $p$.

*Example 4.1.1.* Suppose that a basketball player has probability $\frac{3}{4}$ of sinking a free throw and that his shots are independent. If he gets 5 free throws in a particular game, define $X$ to be the number that he makes. Then $X$ is binomial with parameters $n = 5, p = \frac{3}{4}$, and

$$\begin{aligned} p_X(k) &= \binom{5}{k} \left(\frac{3}{4}\right)^k \left(\frac{1}{4}\right)^{5-k}, \qquad k = 0, 1, 2, 3, 4, 5 \\ &= 0, \qquad \text{otherwise.} \end{aligned}$$

Thus, the probability that he makes them all is

$$p_X(5) = (\tfrac{3}{4})^5 = .237,$$

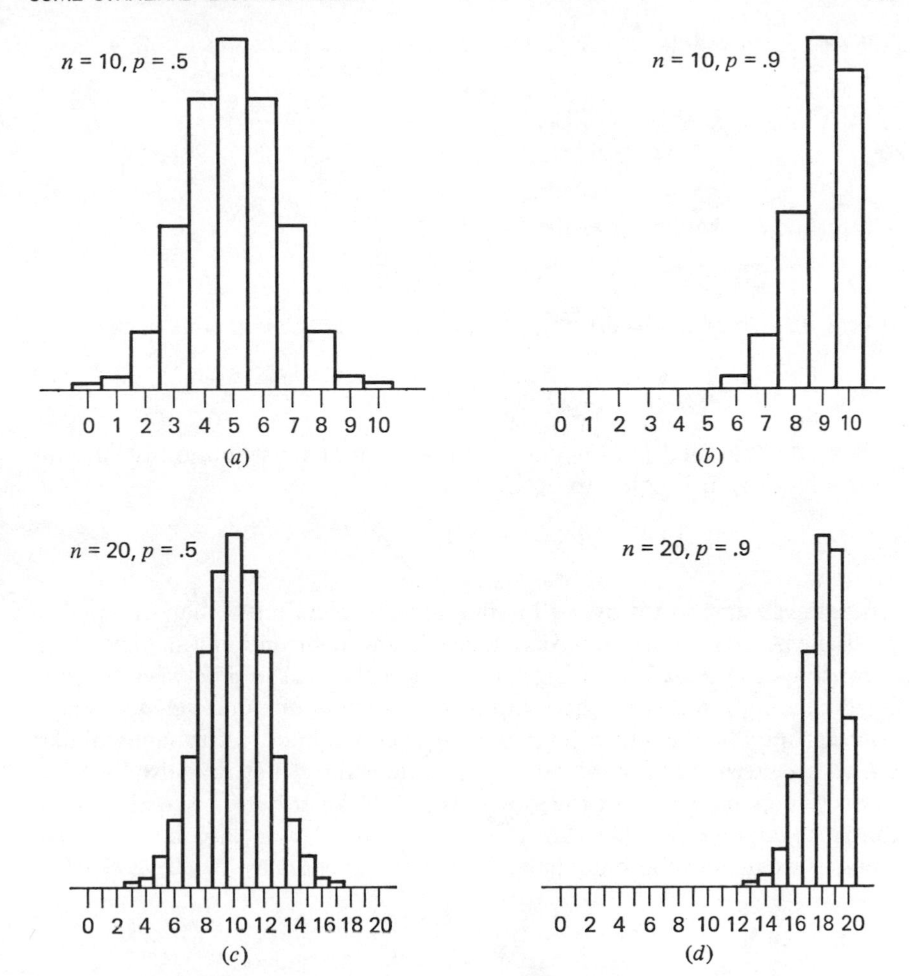


**Figure 4.1.** Binomial random variable.

that he misses them all is

$$p_X(0) = (\tfrac{1}{4})^5 = .001,$$

and that he makes at least half of them is

$$P(X \geq 3) = \sum_{k=3}^{5} \binom{n}{k} \left(\frac{3}{4}\right)^k \left(\frac{1}{4}\right)^{5-k} = .8965.$$

*Example 4.1.2.* A student takes a multiple-choice exam which contains 8 questions, each with 3 alternative answers. Assume that he is guessing when answering each question. Then the probability that he answers a question correctly is $\frac{1}{3}$ for each question. Let $Y$ be the total number he gets correct, and we have

$$p_Y(k) = \binom{8}{k}\left(\frac{1}{3}\right)^k\left(\frac{2}{3}\right)^{8-k}, \qquad k = 0, 1, 2, \ldots, 8$$
$$= 0, \qquad \text{otherwise.}$$

The probability that he misses them all is

$$p_Y(0) = (\tfrac{2}{3})^8 = .039.$$

If he must get 6 or more correct to pass, then the probability that he passes is

$$\sum_{k=6}^{8}\binom{8}{k}\left(\frac{1}{3}\right)^k\left(\frac{2}{3}\right)^{8-k} = .020.$$

Since the binomial random variable is discrete, its distribution function is discontinuous; it is equal to

$$F_X(t) = \sum_{k\leq t}\binom{n}{k}p^kq^{n-k}$$

where the summation is over all nonnegative integers $k$ less than or equal to $t$. Table A, Appendix 5, gives values of the distribution function for $n$ increasing in steps of 1 from 2 to 20 and selected values of $p$ between 0 and $\frac{1}{2}$. By evaluating the size of the jump in the distribution function, we can of course compute the individual binomial probabilities. Although tabular entries are given only for values of $p$ between 0 and $\frac{1}{2}$, it should also be noted that by interchanging the roles of $p$ and $q$ the distribution function for a binomial random variable with $p$ between $\frac{1}{2}$ and 1 may also be evaluated. Thus, if we denote the binomial distribution function by $F_X(k; n, p)$, then

$$F_X(k; n, p) = \sum_{x=0}^{k}\binom{n}{x}p^xq^{n-x},$$

where $k$ is any integer between 0 and $n$ inclusive. Now

$$\sum_{x=0}^{k}\binom{n}{x}p^xq^{n-x} = 1 - \sum_{x=k+1}^{n}\binom{n}{x}p^xq^{n-x}$$
$$= 1 - \sum_{n-x=0}^{n-k-1}\binom{n}{n-x}q^{n-x}p^{n-(n-x)}$$
$$= 1 - F_X(n-k-1; n, q).$$

Thus, for example,

$$F_X(15; 20, .8) = 1 - F_X(4; 20, .2)$$
$$F_X(7; 14, .9) = 1 - F_X(6; 14, .1).$$

Let us derive the moment generating function for the binomial random variable and use it to evaluate the mean and variance of $X$. By definition

$$\begin{aligned} m_X(t) &= E(e^{tX}) \\ &= \sum_{k=0}^{n} e^{tk}\binom{n}{k} p^k q^{n-k} \\ &= \sum_{k=0}^{n} \binom{n}{k} (pe^t)^k q^{n-k} \\ &= (q + pe^t)^n. \end{aligned}$$

Then

$$\begin{aligned} m_X^{(1)}(t) &= n(q + pe^t)^{n-1}pe^t \\ m_X^{(2)}(t) &= n(q + pe^t)^{n-1}pe^t + n(n-1)(q + pe^t)^{n-2}(pe^t)^2 \end{aligned}$$

and

$$\begin{aligned} E[X] &= m_X^{(1)}(0) = np \\ E[X^2] &= m_X^{(2)}(0) = np + n(n-1)p^2 \end{aligned}$$

so that

$$\begin{aligned} \mu_X &= np, \\ \sigma_X^2 &= E[X^2] - \mu_X^2 \\ &= npq. \end{aligned}$$

*Example 4.1.3.* The binomial random variable $X$ discussed in Example 4.1.1 has mean value $\mu_X = 5 \cdot \frac{3}{4} = 3\frac{3}{4}$ and variance $\sigma_X^2 = 5 \cdot \frac{3}{4} \cdot \frac{1}{4} = \frac{15}{16}$. Thus, on the average, this player would make $3\frac{3}{4}$ shots out of 5, with a standard deviation of $\sqrt{15}/4$. The binomial random variable $Y$ discussed in Example 4.1.2 has mean value $\mu_Y = 8 \cdot \frac{1}{3} = 2\frac{2}{3}$ and variance $8 \cdot \frac{1}{3} \cdot \frac{2}{3} = \frac{16}{9}$. Thus this student would, by guessing, answer $2\frac{2}{3}$ questions correctly, on the average, with a standard deviation of $\frac{4}{3}$ about that mean value.

The Bernoulli random variable is actually a special case of the binomial, as the following definition shows.

DEFINITION 4.1.3. Suppose that $X$ is a binomial random variable with parameters $n = 1$ and $p$. Then $X$ is called the *Bernoulli* random variable with parameter $p$.

The Bernoulli random variable then is simply the number of successes we observe in a single Bernoulli trial and has probability function

$$\begin{aligned} p_X(x) &= p, && \text{for } x = 1 \\ &= q, && \text{for } x = 0 \\ &= 0, && \text{otherwise.} \end{aligned}$$

Its moment generating function is

$$m_X(t) = (q + pe^t)^1 = q + pe^t$$

and it has $\mu_X = p$, $\sigma_X^2 = pq$.

**EXERCISE 4.1.**

**1.** Five fair dice are rolled 1 time. Let $X$ be the number of 1's that occur. Compute the mean of $X$, the variance of $X$, $P(1 \leq X < 4)$, and $P(X \geq 2)$.

**2.** An urn contains 8 red and 2 black balls. Twenty balls are drawn with replacement. Let $Y$ be the number of red balls that occur; compute $\mu_Y$, $\sigma_Y$, $P(Y = 16)$, $P(Y < 14)$, and $P(Y > 18)$.

**3.** Ten coins are tossed onto a table. Let $Z$ be the number of coins that land head up. Compute $P(Z = 5)$ and $\mu_Z$.

**4.** Suppose that it is known that 1% of the glasses made by a certain glass-blowing machine will be defective in some way. If we randomly select 10 glasses made by this machine, what is the probability that none of them is defective? How many would we expect to be defective?

**5.** Given that $X$ is a binomial random variable with parameters $n$ and $p$, derive directly the fact that $\mu_X = np$ (see Formula A2.2 in Appendix 2.)

**6.** Derive the variance of the binomial random variable by evaluating $E[X(X - 1)]$ and then add $\mu_X - \mu_X^2$ (see Formula A2.3 in Appendix 2).

**7.** $Y$ is known to be a binomial random variable with mean $\mu_Y = 6$ and variance $\sigma_Y^2 = 4$. Find the probability distribution for $Y$ (that is, evaluate $n$ and $p$).

**8.** The probability that an individual seed of a certain type will germinate is known to be .9. A nursery man wants to sell flats of this type of plant and will claim that each flat contains 100 plants. If he plants 110 seeds in a flat (which we assume will sprout independently), how many plants should we expect an "average" flat to contain? Is there any number of seeds he can plant in a flat in order to be certain that the flat will contain 100 plants?

**9.** Derive the factorial moment generating function for a binomial random variable with parameters $n$ and $p$.

**10.** Derive the moment generating function for the standard form of the binomial random variable.

## 4.2. Geometric and Hypergeometric Random Variables

Suppose that we are able to perform Bernoulli trials and know the probability of success is $p$ for each trial. If we perform independent trials until we get a success, we define the geometric random variable as follows.

DEFINITION 4.2.1. Independent Bernoulli trials are performed until we get a success. The probability of success on each trial is $p$ where $0 < p \leq 1$. Let $Y$ be the number of trials necessary (to get the first success). $Y$ is called the *geometric* random variable with parameter $p$.

We see immediately that the range of $Y$ is the set of positive integers, $1, 2, 3, \ldots$, since the number of trials necessary may equal any of these values. (The reader might like to refer to Example 2.8.1 which is an example of this sort of experiment.) Thus $Y$ is a discrete random variable. The probability function for the geometric random variable is given in the following theorem.

***Theorem 4.2.1.*** If $Y$ is a geometric random variable with parameter $p$, then

$$\begin{aligned} p_Y(k) &= q^{k-1}p, \qquad k = 1, 2, 3, \ldots \\ &= 0, \qquad \text{otherwise.} \end{aligned}$$

*Proof:* The probability that $Y = 1$ is clearly $p$, because that is specified as the probability of a success on the first trial (as well as all others). Next, $Y = 2$ if and only if we get a failure on the first trial followed by a success on the second trial; since the trials are independent the probability of this occurring is $qp$. Thus, $p_Y(2) = qp$. $Y = 3$ if and only if we get failures on each of the first two trials followed by a success on the third; the probability of this occurring is $q \cdot q \cdot p$, so $p_Y(3) = q^2p$.

In general, $Y = k$ if and only if we get failures on each of the first $k - 1$ trials followed by a success on the $k$-th trial; the probability of this happening is $q^{k-1}p$ and thus

$$\begin{aligned} p_Y(k) &= q^{k-1}p, \qquad k = 1, 2, 3, \ldots \\ &= 0, \qquad \text{otherwise.} \end{aligned}$$ ◀

The geometric random variable derives its name from the fact that its probability function forms a geometric progression. To check, first of all, that $p_Y(k)$ is a probability function, note that

1. $p_Y(k) \geq 0, \qquad \text{for all } k.$

2. $$\begin{aligned} \sum_{k=1}^{\infty} p_Y(k) &= \sum_{k=1}^{\infty} q^{k-1}p \\ &= p\sum_{k=1}^{\infty} q^{k-1} \\ &= p\,\frac{1}{1-q} = 1 \end{aligned}$$

since $q = 1 - p$. (See discussion of geometric progressions in Appendix 3.)

*Example 4.2.1.* Suppose that Roger is going to position himself at the free-throw line on a basketball court and shoot until he makes a basket. If we assume that his shots are independent, and that he has constant probability .8 of making each shot, and if we let $Y$ be the number of shots he takes until he makes a basket, then $Y$ is a geometric random variable with parameter .8; its probability function is

$$\begin{aligned} p_Y(k) &= (.2)^{k-1}(.8), \qquad k = 1, 2, 3, \ldots \\ &= 0, \qquad \text{otherwise.} \end{aligned}$$

Then, the probability that it takes him less than 5 shots is

$$P(Y < 5) = \sum_{k=1}^{4} (.2)^{k-1}(.8) = .9984.$$

The probability that it takes him an even number of shots is

$$\begin{aligned} \sum_{k=1}^{\infty} (.2)^{2k-1}(.8) &= (.2)(.8) \sum_{k=1}^{\infty} (.2)^{2k-2} \\ &= (.2)(.8) \frac{1}{1 - (.2)^2} = \frac{1}{6}. \end{aligned}$$

Let us derive the factorial moment generating function for the geometric random variable $Y$ and use it to evaluate the mean and the variance of $Y$. By definition,

$$\begin{aligned} \psi_Y(t) &= E(t^Y) \\ &= \sum_{k=1}^{\infty} t^k q^{k-1} p \\ &= pt \sum_{k=1}^{\infty} (qt)^{k-1} \\ &= \frac{pt}{1 - qt}, \qquad \text{for } |qt| < 1, \qquad \text{i.e., } |t| < \frac{1}{q}. \end{aligned}$$

Then

$$\psi_Y^{(1)}(t) = \frac{p}{1 - qt} + \frac{qpt}{(1 - qt)^2} = \frac{p}{(1 - qt)^2}$$

$$\psi_Y^{(2)}(t) = \frac{2qp}{(1 - qt)^3}$$

and thus

$$\mu_Y = \psi_Y^{(1)}(1) = \frac{p}{(1 - q)^2} = \frac{p}{p^2} = \frac{1}{p}$$

$$E[X(X - 1)] = E[X^2] - E[X] = \psi_Y^{(2)}(1) = \frac{2qp}{(1 - q)^3} = \frac{2q}{p^2}$$

so that

$$\sigma_Y{}^2 = \frac{2q}{p^2} + \frac{1}{p} - \frac{1}{p^2} = \frac{q}{p^2}.$$

Thus the average number of shots that Roger would have to make (in Example 4.2.1) is $\mu_Y = \frac{1}{.8} = 1\frac{1}{4}$ and the variance of his number of shots is $\sigma_Y{}^2 = \frac{.2}{(.8)^2} = \frac{5}{16}$. This relatively small variance implies that the probability of his number of shots being close to the average number of shots ($\mu_Y = 1\frac{1}{4}$) is quite high, as is verified by realizing that $p_Y(1) = .8$, $p_Y(2) = .16$.

*Example 4.2.2.* Suppose that 1 copy in 10 of the *San Francisco Chronicle-Examiner* daily newspaper bears a special prize-winning number. Let $Z$ be the number of papers you must buy to get 1 prize. Then $Z$ is a geometric random variable with probability function

$$\begin{aligned} p_Z(z) &= (.9)^{k-1}(.1), \qquad && k = 1, 2, 3, \ldots \\ &= 0, && \text{otherwise,} \end{aligned}$$

and $\mu_Z = \frac{1}{.1} = 10$, $\sigma_Z{}^2 = \frac{.9}{(.1)^2} = 90$. Note that this geometric random variable has a much larger variance than does the one discussed in Example 4.2.1 (since it has a small value of $p$); thus we would expect much more variability in $Z$ than we would in $Y$ from one repetition of the experiment to another.

Figure 4.2 gives the histograms of the probability functions of two geometric random variables, one with $p = .1$ and the other with $p = .9$. Note

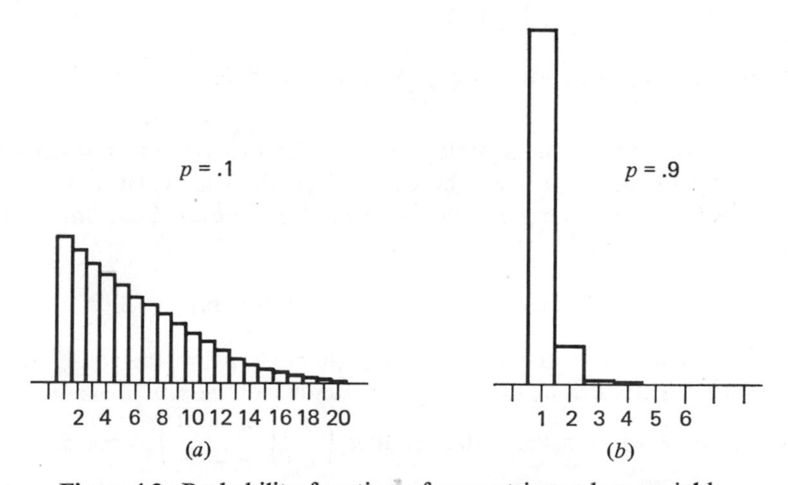


**Figure 4.2.** Probability function of geometric random variable.

that in both cases (as in all others) the probabilities of occurrence of succeedingly larger integers are decreasing geometrically. The difference between the binomial and the geometric random variables should be mentioned once more: in the binomial, the total number of trials is fixed and the random variable is the number of successes to occur; in the geometric, the number of successes to be observed (one) is fixed and the random variable is the number of trials required.

Let us now define a random variable which is similar to the binomial in a sense.

DEFINITION 4.2.2. An urn contains $M$ balls of which $W$ are white. Define $Z$ to be the number of white balls that occur in a sample of $n$ balls drawn at random from the urn without replacement. $Z$ is called the *hypergeometric* random variable.

Since the number of white balls drawn could take on only integer values, $Z$ is another example of a discrete random variable. If the sampling from the urn had been done with replacement rather than without, $Z$ would be a binomial random variable with parameters $n$ and $p = W/M$; this is the similarity between the binomial and hypergeometric random variables mentioned above. The probability function for $Z$ is derived in Theorem 4.2.2.

***Theorem 4.2.2.*** If $Z$ is the random variable given in Definition 4.2.2., then

$$p_Z(k) = \frac{\binom{W}{k}\binom{M-W}{n-k}}{\binom{M}{n}}, \qquad k = 0, 1, 2, \ldots, n,$$

where we are using the convention $\binom{b}{a} = 0$ for $a > b$.

*Proof:* The elements of the sample space for the experiment of selecting $n$ balls without replacement from the urn will be the subsets of $n$ balls that could occur; since the sampling is done at random, these subsets are equally likely to occur. Define:

$A_k$: exactly $k$ white balls are drawn.

Then $P(A_k)$ is the ratio of the number of elements belonging to $A_k$ to the number of elements belonging to $S$. The number of subsets of size $n$, each having exactly $k$ white balls, is the product $\binom{W}{k}\binom{M-W}{n-k}$, since any $k$ of the white balls could occur in conjunction with any $n - k$ nonwhite balls;

the number of elements belonging to $S$ is $\binom{M}{n}$. Thus,

$$P(A_k) = \frac{\binom{W}{k}\binom{M-W}{n-k}}{\binom{M}{n}}.$$

Since $Z = k$ if and only if $A_k$ occurs, $P(Z = k) = P(A_k)$; i.e.,

$$p_Z(k) = \frac{\binom{W}{k}\binom{M-W}{n-k}}{\binom{M}{n}}, \qquad k = 0, 1, 2, \ldots, n$$

$$p_Z(k) = 0, \qquad \text{otherwise.}$$ ◀

Obviously, $p_Z(k) \geq 0$ for all $k$ since the combinatorial coefficients are all nonnegative. It can be shown that $\sum_{k=0}^{n} p_Z(k) = 1$; however, this uses some special summation results which we shall not derive here.

*Example 4.2.3.* A milk case contains 5 quarts of milk, only 4 of which are fresh. If we randomly select 2 of these (without replacement) and let $Z$ be the number of nonfresh quarts that we select, then $Z$ is a hypergeometric random variable with

$$p_Z(k) = \frac{\binom{1}{k}\binom{4}{2-k}}{\binom{5}{2}}, \qquad k = 0, 1, 2$$

$$= 0, \qquad \text{otherwise.}$$

(Note we have made an analogy between balls in an urn and milk quarts in the case; $M = 5$, $W = 1$, $n = 2$.)

The probability that we get 2 fresh quarts is

$$p_Z(0) = \frac{\binom{1}{0}\binom{4}{2}}{\binom{5}{2}} = \frac{3}{5};$$

the probability that the unfresh quart is one of those we selected is

$$p_Z(1) = \frac{\binom{1}{1}\binom{4}{1}}{\binom{5}{2}} = \frac{2}{5}.$$

*Example 4.2.4.* Assume that $M$ people are eligible to vote in a certain election and that $W$ of them favor proposition $A$ while the remainder do not. If we select $n$ of these voters at random without replacement and let $X$ be the number of people in the sample that favor proposition $A$, then $X$ is a hypergeometric random variable. The probability that a majority of people in the sample favor proposition $A$ then is gotten by summing $p_X(k)$ from the smallest integer larger than $n/2$ on up. In such sample-survey problems, $M$ and $n$ usually are known, as is the observed value of $X$ after the sample is taken, but $W$ is generally unknown. A problem in statistical estimation, which we shall study in Chapter 7, is to use the observed value of $X$ from the sample to guess the unknown value of $W$.

The mean value of a hypergeometric random variable $Z$ is

$$\mu_Z = \sum_{k=0}^{n} k \frac{\binom{W}{k}\binom{M-W}{n-k}}{\binom{M}{n}};$$

this sum can be shown to equal $\mu_Z = n\, W/M$.

Similarly, the variance of $Z$ can be shown to be

$$\sigma_Z^2 = n \frac{W}{M} \cdot \frac{(M-W)}{M} \cdot \frac{(M-n)}{M-1}.$$

Thus, for the random variable $Z$ defined in Example 4.2.3, we find

$$\mu_Z = 2 \cdot \tfrac{1}{5} = \tfrac{2}{5}, \qquad \sigma_Z^2 = \tfrac{2}{5} \cdot \tfrac{4}{5} \cdot \tfrac{3}{4} = \tfrac{6}{25}.$$

Notice that if we set $n = M$, that is we sample all of the balls in the urn, and let $Z$ be the number of white balls in the sample, then

$$\mu_Z = M \frac{W}{M} = W, \qquad \sigma_Z^2 = \frac{W}{M} \frac{(M-W)}{M} \frac{(M-M)}{M} = 0.$$

Thus the average number of white balls we get is $W$, as it should be, and the variance of $Z$ about its mean ($W$) is zero (it never varies).

It was mentioned earlier that if the sampling were done with replacement in Definition 4.2.2, then $Z$ would be a binomial random variable with parameters $n$ and $p = W/M$. As such, $Z$ would then have mean $np = n\; W/M$ and variance $npq = n(W/M)(1 - W/M) = n(W/M)(M - W)/M$. We note then that the expected number of white balls in the sample is the same whether the sampling is done with or without replacement. The variance, though, is reduced by a factor of $(M - n)/(M - 1)$ if the sampling is done without replacement rather than with replacement. This is intuitively reasonable since, if the same ball cannot occur in the sample more than once, the number of balls to be sampled from is constantly decreasing. Thus the

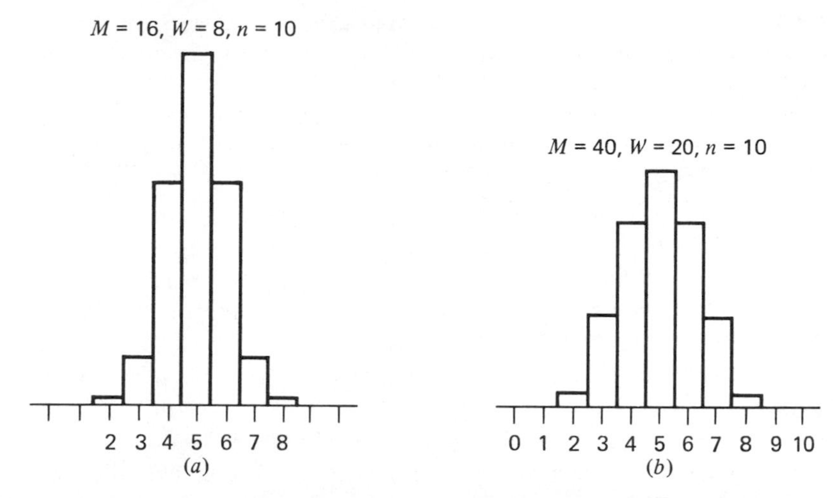


**Figure 4.3.** Probability function of hypergeometric random variable.

variability of the balls to be sampled from is decreased, as is reflected in the factor $(M - n)/(M - 1)$ by which the variance is reduced. It can be seen that as $M$ increases (with $n$ held constant) the variance of the hypergeometric gets closer and closer to the variance of the corresponding binomial random variable because $(M - n)/(M - 1)$ gets closer and closer to 1. In fact, it can be shown that if $M \to \infty$ and $W \to \infty$ in such a way that $W/M$ remains constant at $p$ ($n$ is held constant), the distribution function of the hypergeometric random variable converges to the distribution function of the binomial random variable with parameters $n$ and $p$. This is also intuitively reasonable, since the larger that $M$ (the number of balls in the urn) becomes, the less difference it makes whether the balls are replaced or not before subsequent draws.

Figure 4.3 shows the histograms for two hypergeometric random variables; in part *a* we have $M = 16$, $W = 8$, $n = 10$; in part *b* $M = 40$, $W = 20$, $n = 10$. Note how similar the histogram in Figure 4.3*b* is to the histogram in Figure 4.1*a* which gives the probability distribution for a binomial random variable with $n = 10$, $p = .5$.

**EXERCISE 4.2.**

**1.** A fair coin is flipped until a head occurs. What is the probability that less than 3 flips are required? That less than 4 flips are required?

**2.** An American roulette wheel commonly has 38 spots on it of which 18 are black, 18 are red, and 2 are green. Let $X$ be the number of spins necessary to get the first red number. Give the probability function for $X$ and the mean for $X$.

**3.** Let $Y$ be the number of spins necessary to observe the first green number for the roulette wheel mentioned in problem 2. What is the probability function, mean, and variance for $Y$?

**4.** Suppose an urn contains 10 balls of which 1 is black. Let $Z$ be the number of draws, with replacement, necessary to observe the black ball. What is the probability function for $Z$? The mean of $Z$?

**5.** Suppose that the drawing in problem 4 is done without replacement. Find the probability function for $Z$ and the mean of $Z$.

**6.** A box contains 5 marbles of which 3 are chipped. Two marbles are chosen randomly without replacement from the box. What is the probability function for the number of chipped marbles in the sample?

**7.** Thirteen cards are drawn randomly without replacement from a regular 52-card deck. What is the probability function for the number of red cards in the sample? What are the mean and the variance of the number of red cards?

**8.** A bag contains 10 flashbulbs, 8 of which are good. If 5 flashbulbs are chosen from the bag at random, what is the probability function for the number of good flashbulbs? For the number of bad flashbulbs?

**9.** An ice cream company makes chocolate-covered ice cream bars on sticks which sell for 10 cents. Suppose that they put a star on every 50-th stick; anyone who buys a bar with a starred stick gets a free ice cream bar. If you decide to buy ice cream bars until you get a free one, how much would you expect to spend before getting a free bar?

**10.** A panel of 7 judges is to decide which of 2 final contestants in a beauty contest will be declared the winner; a simple majority of the judges will determine the winner. Assume that 4 of the judges will vote for Marie and that the other 3 will vote for Sue. If we randomly select 3 of the judges and ask them who they are going to vote for, what is the probability that a majority of the judges in the sample will favor Marie?

**11.** Repeated independent Bernoulli trials, each with probability $p$ of success, are performed until $r \geq 1$ successes are obtained. Let $X$ be the number of trials required and derive the probability function for $X$. ($X$ is called the negative binomial or Pascal random variable.)

**12.** Assume that every time you drive your car the probability is .001 that you will get a ticket for speeding. Also assume that you will lose your license once you have received three such tickets. Let $X$ be the number of times you will drive your car until you get the third such ticket and derive the probability function for $X$ (assume that you have the same probability, .001, of getting a ticket each time you drive and that the occurrences of such tickets are independent).

### 4.3. The Poisson Random Variable

As we shall see, the Poisson random variable has many important practical applications. Before defining the random variable, let us first discuss what is known as a Poisson process. Many problems consist of observing what could be called discrete events in a continuous interval. For example, we might observe the arrival of cars at a supermarket between the hours of 10 a.m. and 11 a.m. on a specified day. The arrival of a car at the parking lot is a discrete event because its arrival time is a single point in our continuous 1-hour period. Or, we might observe the number of calls coming into the switchboard of a large corporation from 1 p.m. to 3 p.m. The particular calls that come in are the discrete events, since again the time of arrival of any single call is a single point in the 2-hour period. Or, if we were the manufacturers of insulated wire, we could count the number of defects in insulation that occurred in a particular 100-foot segment. In this case the continuous interval is the 100 feet of wire and the discrete events are the defects in insulation that occur, assuming that each defect could be thought of as occurring at a single point along the length of wire.

Many similar examples could be given. Each of these could be looked at as being a Poisson process if we are willing to assume that the occurrences of the discrete events are generated in such a manner that Definition 4.3.1 is satisfied.

DEFINITION 4.3.1. In a Poisson process with parameter $\lambda$, discrete events are generated in a continuous interval (time, length, etc.) in the following manner. (1) We can take a sufficiently short interval, of length $h$, such that: (i) the probability of exactly 1 occurrence in the interval is approximately $\lambda h$, and (ii) the probability of 2 or more occurrences in the interval is approximately 0. (2) The occurrence of an event in 1 interval of length $h$ has no effect on the occurrence or nonoccurrence in another nonoverlapping interval of length $h$. (The occurrences are statistically independent of one another.)

It would seem reasonable when observing the arrival of cars at a supermarket parking lot to assume that no more than 1 car would arrive in a millionth of a second, for example, and that the arrival or nonarrival of a car in the first millionth of a second would have no effect on whether or not a car will arrive in the next or any succeeding millionth of a second. Granted these assumptions, the arrivals would constitute a Poisson process. Similarly, it seems reasonable that no more than 1 call would arrive at the industrial switchboard during a very short period of time and that what happens in 1 short period of time would have no effect on succeeding short periods of time; this would also be an example of a Poisson process. Similarly, the occurrences of defects along the 100 foot length of insulated wire might well satisfy the requirements for a Poisson process.

If we observe a Poisson process for a unit length of time, the number of events that occur is a random variable. Assuming that they occur at a constant rate $\lambda$ does not imply that exactly $\lambda$ events occur in a unit interval but rather that the average number that occurs per unit is $\lambda$. The Poisson random variable assigns to any unit interval of length the number of events to occur in that interval, as given in the following definition.

DEFINITION 4.3.2. A Poisson process with parameter $\lambda$ is observed for $s$ units of time. Let $X$ be the number of events to occur. Then $X$ is called the *Poisson* random variable with parameter $\lambda s$.

Notice that the parameter of the Poisson process ($\lambda$) is multiplied by the length of time the process is observed ($s$) to give the parameter of $X$. If we expect $\lambda$ events to occur on the average in an interval of unit length, then we would expect $\lambda s$ events to occur on the average in an interval of length $s$. Thus we might expect that the Poisson random variable $X$ will have mean $\lambda s$. The probability law for $X$ is given as Theorem 4.3.1.

***Theorem 4.3.1.*** If $X$ is a Poisson random variable with parameter $\lambda s$, then

$$p_X(k) = \frac{(\lambda s)^k}{k!} e^{-\lambda s}, \qquad k = 0, 1, 2, \ldots$$

$$= 0, \qquad \text{otherwise.}$$

*Proof:* A Poisson process with parameter $\lambda$ is observed for $s$ units of time and $X$ is the number of events that occur. To derive the probability function for $X$, we assume that the interval of time is divided into $n = s/h$ nonoverlapping intervals, each of length $h$ ($h$ is very small). Then from Definition 4.3.1, it is reasonable to assume that either 0 or 1 event will occur in each of these short intervals and that the probability that exactly 1 event will occur is $\lambda h$ for each. Furthermore, the occurrence or nonoccurrence of an event in one of these short intervals has no effect on the occurrence or nonoccurrence of an event in another interval (thus they might be called independent). Thus the interval of time of length $s$ has been subdivided into $n$ repeated independent Bernoulli trials, each an interval of length $h$, with probability of success equal to $\lambda h$. Then, if we let $X$ be the total number of events occurring in the interval, $X$ is approximately a binomial random variable with parameters $n = s/h$ and $p = \lambda h = \lambda s/n$; then

$$p_X(k) \doteq \binom{n}{k}\left(\frac{\lambda s}{n}\right)^k\left(1 - \frac{\lambda s}{n}\right)^{n-k}, \qquad k = 0, 1, 2, \ldots, n$$

$$= 0, \qquad \text{otherwise.}$$

If we take the limit of this expression as $n \to \infty$ (thus the individual pieces are each shrinking to zero length), we get the exact probability function for the Poisson random variable. Now

$$\binom{n}{k}\left(\frac{\lambda s}{n}\right)^k\left(1-\frac{\lambda s}{n}\right)^{n-k}$$

$$= \frac{(\lambda s)^k}{k!}\left(1-\frac{\lambda s}{n}\right)^{-k}\left(1-\frac{\lambda s}{n}\right)^n \frac{n(n-1)\cdots(n-k+1)}{n^k}$$

$$= \frac{(\lambda s)^k}{k!}\left(1-\frac{\lambda s}{n}\right)^{-k}\left(1-\frac{\lambda s}{n}\right)^n \cdot 1\cdot\left(1-\frac{1}{n}\right)\left(1-\frac{2}{n}\right)\cdots\left(1-\frac{k-1}{n}\right).$$

Since

$$\lim_{n\to\infty}\frac{(\lambda s)^k}{k!} = \frac{(\lambda s)^k}{k!}$$

$$\lim_{n\to\infty}\left(1-\frac{\lambda s}{n}\right)^{-k} = 1$$

$$\lim_{n\to\infty}\left(1-\frac{\lambda s}{n}\right)^{n} = e^{-\lambda s}$$

$$\lim_{n\to\infty} 1\left(1-\frac{1}{n}\right)\left(1-\frac{2}{n}\right)\cdots\left(1-\frac{k-1}{n}\right) = 1,$$

we get

$$\lim_{n\to\infty}\binom{n}{k}\left(\frac{\lambda s}{n}\right)^k\left(1-\frac{\lambda s}{n}\right)^{n-k} = \frac{(\lambda s)^k}{k!}e^{-\lambda s}.$$

Thus, the probability function for $X$ is

$$p_X(k) = \frac{(\lambda s)^k}{k!}e^{-\lambda s}, \qquad k = 0, 1, 2, \ldots$$

$$= 0, \qquad \text{otherwise.}$$ ◄

Before looking at some particular examples of the Poisson random variable, let us check to make sure that $p_X(k)$ is a probability function. First

$$\frac{(\lambda s)^k}{k!}e^{-\lambda s} > 0$$

for all values of $k$ so $p_X(k) \geq 0$. Secondly,

$$\sum_{k=0}^{\infty} p_X(k) = \sum_{k=0}^{\infty}\frac{(\lambda s)^k}{k!}e^{-\lambda s}$$

$$= e^{-\lambda s}\sum_{k=0}^{\infty}\frac{(\lambda s)^k}{k!}$$

$$= e^{-\lambda s}e^{\lambda s} = 1.$$

(See Formula 3.5, Appendix 3.) Thus $p_X(k)$ does satisfy the requirements for a probability function.

*Example 4.3.1.* Assume that the number of deaths by suicide in Manhattan is a Poisson process with parameter $\lambda = 3$ per month. Let $X$ be the number of deaths by suicide that will occur between January 1 and March 31, 1975, inclusive. Then $X$ is a Poisson random variable with parameter

$$\lambda s = 3 \cdot 3 = 9$$

and

$$p_X(k) = \frac{(9)^k}{k!} e^{-9}, \qquad k = 0, 1, 2, \ldots .$$

The probability that there will be at least 10 deaths by suicide in this period then is

$$P(X \geq 10) = \sum_{k=10}^{\infty} \frac{(9)^k}{k!} e^{-9} = .4216.$$

(See Table B, Appendix 5, and the discussion below.) The probability of exactly 9 deaths is

$$P(X = 9) = F_X(10; 9) - F_X(9; 9) = .1317.$$

The distribution function for the Poisson random variable is discontinuous since the Poisson is a discrete random variable. In fact, denoting the distribution function for a Poisson random variable $X$ with parameter $\lambda s$ by $F_X(t; \lambda s)$, we have

$$\begin{aligned} F_X(t; \lambda s) &= 0, & t < 0 \\ &= \sum_{k \leq t} p_X(k), & t \geq 0 \end{aligned}$$

where the sum is over all nonnegative integers less than or equal to $t$. Table B, Appendix 5, gives a table of $F_X$ for selected values of $\lambda s$. The number tabulated is the value of $F_X(t; \lambda s)$; to derive the value of the probability function at some integer $k$, we must compute the size of the jump in $F_X$ at $t = k$. Thus,

$$p_X(k) = F_X(k; \lambda s) - F_X(k - 1; \lambda s).$$

For example,

$$F_X(9; 9) = .5874$$

$$F_X(8; 9) = .4557$$

and thus

$$P(X = 9) = .5874 - .4557 = .1317 \qquad (\text{with } \lambda s = 9).$$

Furthermore

$$F_X(2; 1) = .9197$$

$$F_X(1; 1) = .7358$$

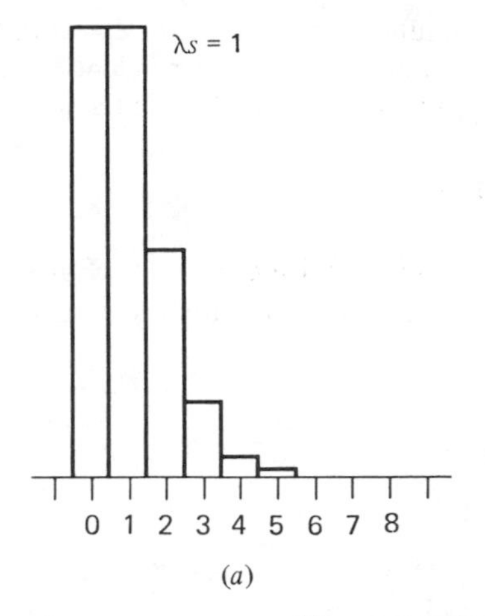

(a)

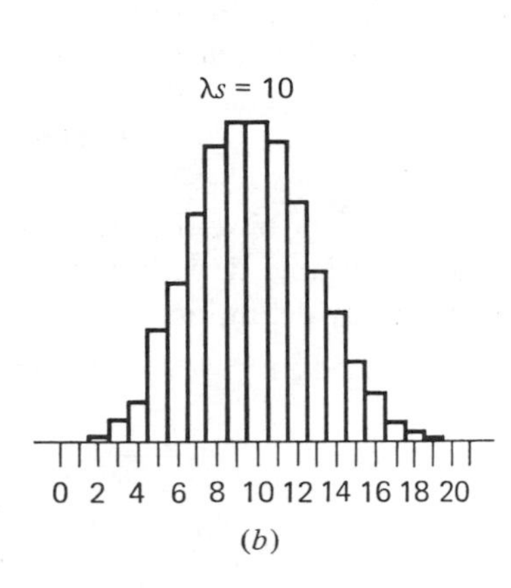

(b)

**Figure 4.4.** Poisson random variable.

so that

$$P(X = 2) = .9197 - .7358 = .1839 \qquad (\text{with } \lambda s = 1)$$

and

$$P(X \geq 2) = 1 - F_X(1; 1) = .2642$$
$$P(X \geq 3) = 1 - F_X(2; 1) = .0803.$$

Figure 4.4 shows the graphs of the probability functions for two Poisson random variables, one with $\lambda s = 1$, the other with $\lambda s = 10$. Note how much more symmetric the histogram is with $\lambda s = 10$ than with $\lambda s = 1$. As we shall see later, this increased symmetry with increasing $\lambda s$ continues and gives a basis for approximations of the Poisson probability function.

*Example 4.3.2.* Suppose that the number of calls arriving at an industrial switchboard is a Poisson process with parameter $\lambda = 120$ calls per hour. Let $X$ be the number of calls that arrive in a 1-minute period. Then $X$ is a Poisson random variable with parameter $\lambda s = 2$ and

$$p_X(k) = \frac{2^k}{k!} e^{-2}, \qquad k = 0, 1, 2, \ldots.$$

The probability of no calls in this 1-minute interval is

$$p_X(0) = e^{-2} = .1353;$$

the probability of between 1 and 5 calls, inclusive, arriving is

$$P(1 \leq X \leq 5) = F_X(5; 2) - F_X(0; 2)$$
$$= .9834 - .1353 = .8481.$$

If we look at 3 successive, nonoverlapping 1-minute periods, the probability of between 1 and 5 calls, inclusive, in every 1-minute interval is $(.8481)^3 = .6100$, since the numbers of calls in these 1-minute intervals are independent. The probability of between 1 and 5 calls, inclusive, in exactly 2 of the 3 intervals is

$$\binom{3}{2}(.8481)^2(.1519) = .3278$$

Let us now derive the factorial moment generating function for a Poisson random variable $X$ with parameter $\lambda s$ and use it to evaluate $\mu_X$ and $\sigma_X{}^2$ We have, by definition,

$$\begin{aligned}\psi_X(t) &= E[t^X] \\ &= \sum_{k=0}^{\infty} t^k \frac{(\lambda s)^k}{k!} e^{-\lambda s} \\ &= e^{-\lambda s} \sum_{k=0}^{\infty} \frac{(\lambda st)^k}{k!} \\ &= e^{-\lambda s} e^{\lambda st}.\end{aligned}$$

Then

$$\begin{aligned}\psi_X^{(1)}(t) &= \lambda s e^{-\lambda s} e^{\lambda st} \\ \psi_X^{(2)}(t) &= (\lambda s)^2 e^{-\lambda s} e^{\lambda st},\end{aligned}$$

so that

$$\begin{aligned}\psi_X^{(1)}(1) &= \lambda s \\ \psi_X^{(2)}(1) &= (\lambda s)^2\end{aligned}$$

and thus

$$\begin{aligned}\mu_X &= \psi_X^{(1)}(1) = \lambda s \\ \sigma_X{}^2 &= \psi_X^{(2)}(1) + \mu_X - \mu_X{}^2 = \lambda s.\end{aligned}$$

Thus, in Example 4.3.1, the expected number of suicides in the 3-month period is 9, as is the variance of the number of suicides. In Example 4.3.2, the expected number of calls and the variance of the number of calls in a 1-minute period are 2.

*Example 4.3.3.* It is not necessary that the events observed be occurrences distributed in time for the Poisson random variable to be useful. For example, suppose that 2000 micro-organisms of some kind are distributed uniformly and independently throughout a gallon of water. Samples of size $1/a$ of a gallon can then be thought of as being a Poisson process with parameter $\lambda = 2000/a$, because this is the rate at which micro-organisms would occur in the sample material (independently). Suppose, for example, that $a = 4000$ (thus we are selecting $\frac{1}{4000}$ of a gallon, about .95 cubic centimeter) and let $X$ be the number of micro-organisms to occur in the sample. Then $X$ is a Poisson random variable with parameter

$\lambda = = \frac{2000}{4000} = \frac{1}{2}$, and the probability function for $X$ is

$$p_X(k) = \left(\frac{1}{2}\right)^k \frac{1}{k!} e^{-1/2}, \qquad k = 0, 1, 2, \ldots$$
$$= 0, \qquad \text{otherwise.}$$

Thus the probability that we would have no micro-organisms in the sample is

$$p_X(0) = .6065$$

and the probability of getting exactly 1 is

$$p_X(1) = .3033.$$

The mean number we would get is

$$\mu_X = \tfrac{1}{2}$$

and the variance of the number of organisms in the sample is

$$\sigma_X{}^2 = \tfrac{1}{2}.$$

If we select a sample of $\frac{1}{1000}$ of a gallon (about 3.8 cubic centimeters) and let $Y$ be the number of organisms in this size sample, then $Y$ is a Poisson random variable with parameter 2. We find that $P(Y = 0) = .1353$, $P(Y \geq 2) = .5940$, $P(Y \leq 6) = .9955$.

We might also inquire how large a fraction, $1/a$, of the total volume we should take in order that we have probability .9 or more of getting at least 1 micro-organism in our sample. Then, as above, $X$ is a Poisson random variable with parameter $2000/a$ and

$$P(X = 0) = e^{-2000/a}.$$

Then we require that

$$P(X = 0) \leq .1;$$

that is,

$$e^{-2000/a} \leq .1$$

and

$$-\frac{2000}{a} \leq \ln (.1),$$

$$\frac{1}{a} \geq \frac{\ln (.1)}{-2000} = \frac{\ln (10)}{2000} = .00115.$$

Thus, if we take at least .00115 of the total volume as our sample, we have probability of at least .9 of getting 1 or more micro-organisms in the sample.

In certain circumstances the Poisson probability function gives a good approximation to the binomial probability function. As we saw in the proof of Theorem 4.3.1, the limit of the binomial probability function, where $p = \lambda s/n$, is actually given by the Poisson probability function (for the same number of successes, $k$). Since $p$ is inversely related to $n$, then, as $n \to \infty$, the parameter $p \to 0$. Thus, we might expect that if $n$ is large and $p$ is small, the Poisson probability function with $\lambda s = np$ should give a good approximation to the binomial probability function. This is in fact the case. The following example illustrates this idea.

*Example 4.3.4.* A college professor, based on his past experience, feels that there is a probability of .001 that he will be late to any given class and that his being late or not for any class has no effect on whether he is late or not for any other class. Then the number of times $X$ that he will be late to his next 100 classes is a binomial random variable with parameters $n = 100$ and $p = .001$, and the exact probabilities of his being late exactly 0 times and exactly 1 time are, respectively,

$$P(X = 0) = (.99)^{100} = .9057$$

$$P(X = 1) = \binom{100}{1}(.999)^{999}(.001) = .0897.$$

Note that in this case $n$ is fairly large and $p$ is rather small so that we would expect the above probabilities to be well approximated by those for a Poisson random variable with $\lambda s = np = 100(.001) = .1$. The Poisson values are

$$\frac{(.1)^0}{0!}e^{-.1} = .9048$$

$$\frac{(.1)^1}{1!}e^{-.1} = .0905$$

and the approximation is quite good.

A number of rules-of-thumb have been given which describe when this Poisson approximation is accurate; many texts say that if $n \geq 20$ and $p \leq .05$, it is quite accurate. It is very good if $n \geq 100$ and $np \leq 10$.

## EXERCISE 4.3.

**1.** It has been observed that cars pass a certain point on a rural road at the average rate of 3 per hour. Assume that the instants at which the cars pass are independent and let $X$ be the number that pass this point in a 30-minute interval. Compute $P(X = 0)$, $P(X \geq 2)$.

**2.** It has been observed empirically that deaths per hour, due to traffic accidents, occur at a rate of 8 per hour on long holiday weekends in the United States. Assuming that these deaths occur independently, compute the probability that a 1-hour period would pass with no deaths; that a 15-minute period would pass with no deaths; that 4 consecutive, nonoverlapping 15-minute periods would pass with no deaths.

**3.** It has been observed that packages of Hamm's beer are removed from the shelf of a particular supermarket at a rate of 10 per hour during rush periods. What is the probability that at least 1 package is removed during the first 10 minutes of a rush period? What is the probability that at least 1 is removed from the shelf during each of 3 consecutive, nonoverlapping 10-minute intervals?

**4.** At a certain manufacturing plant, accidents have been occurring at the rate of 1 every 2 months. Assuming that the accidents occur independently, what is the expected number of accidents per year? What is the standard deviation of the number of accidents per year? What is the probability of there being no accidents in a given month?

**5.** If $X$ is a Poisson random variable with parameter $\lambda s$ (a positive integer), show that $p_X(\lambda s) = p_X(\lambda s - 1)$.

**6.** If $X$ is a Poisson random variable with parameter $\lambda s$ (a positive integer), show that the largest value which $p_X(k)$ takes on is $p_X(\lambda s)$. (**Hint:** Look at $p_X(k)/p_X(k - 1)$, $k = 1, 2, \ldots$.)

**7.** Derive the moment generating function for the Poisson random variable.

**8.** Assume that 1 baby in 10,000 is born blind. If a large city hospital has 5000 births in 1970, approximate the probability that none of the babies born that year was blind at birth. Also approximate the probabilities that exactly 1 is born blind and that at least 2 are born blind. (Use the Poisson approximation.)

**9.** Assume that 1 new tire in a 1000 has a weak spot in its side wall. If you buy 4 new cars, approximate the probability that none of your cars has a tire with a weak spot and compare this with the exact binomial value.

**10.** A bakery makes chocolate chip cookies; a batch consists of 1000 cookies. A total of 3000 chocolate chips is added to the batter for each batch and the batter is well mixed. If we select 1 cookie at random from a batch, what is the probability that it contains no chocolate chips? That it contains exactly 3 chips? How many cookies with exactly 1 chocolate chip each would you expect in a batch?

### 4.4. Uniform and Exponential Random Variables

In this section we shall discuss two commonly occurring continuous random variables. The simplest possible continuous random variable is called the uniform random variable; it is defined as follows.

DEFINITION 4.4.1. $X$ is a *uniform* random variable on the interval $(a, b)$ if

1. The range of $X$ is the interval $(a, b)$.
2. All points in the interval are equally likely to occur as the value for $X$.

($X$ is uniformly distributed on $(a, b)$.)

We have, in fact, already seen some examples of a uniform random variable (see Examples 2.8.3 and 2.8.4 and also problems 5 through 8 in Exercise 2.8). In these cases we did not formally define a random variable but could easily have done so by defining $X(\omega) = \omega$. Then the fact that the sample outcome is equally likely to take on any value in a certain interval implies that $X$ is equally likely to take on values in that same interval. (See also Examples 3.1.3, 3.2.5, and 3.2.7.) If $X$ takes on only values in the interval $(a, b)$, then its density function must be zero outside this interval (in order that the probability of $X$ lying outside the interval be zero). The density function of a uniform random variable is given as Theorem 4.4.1.

***Theorem 4.4.1.*** If $X$ is uniformly distributed on $(a, b)$, then

$$\begin{aligned} f_X(x) &= \frac{1}{b-a}, \qquad a < x < b \\ &= 0, \qquad \text{otherwise.} \end{aligned}$$

*Proof:* As was mentioned in Section 2.8, if $X$ is equally likely to take on values in $(a, b)$, then the probability of $X$ lying in any subinterval of $(a, b)$ must be proportional to the length of the subinterval. Since the probability of $X$ lying in any interval is given by the integral of its density function over that interval, the density function of $X$ must be constant on the interval $(a, b)$. As noted above, the density of $X$ must be zero outside this interval; thus

$$\begin{aligned} f_X(x) &= c, \qquad a < x < b \\ &= 0, \qquad \text{otherwise.} \end{aligned}$$

The condition

$$\int_{-\infty}^{\infty} f_X(x)\, dx = 1$$

then implies

$$\int_a^b c\, dx = 1;$$

that is

$$c\int_a^b dx = c(b - a) = 1.$$

Thus $c = 1/(b - a)$ and the theorem is proved. ◀

The uniform random variable gets its name from the fact that its density function is uniform (constant) on the interval $(a, b)$. The distribution function for a uniform random variable is

$$\begin{aligned} F_X(t) &= \int_{-\infty}^{t} f_X(x)\, dx \\ &= 0, \qquad t < a \\ &= \frac{t-a}{b-a}, \qquad a \le t \le b \\ &= 1, \qquad t > b; \end{aligned}$$

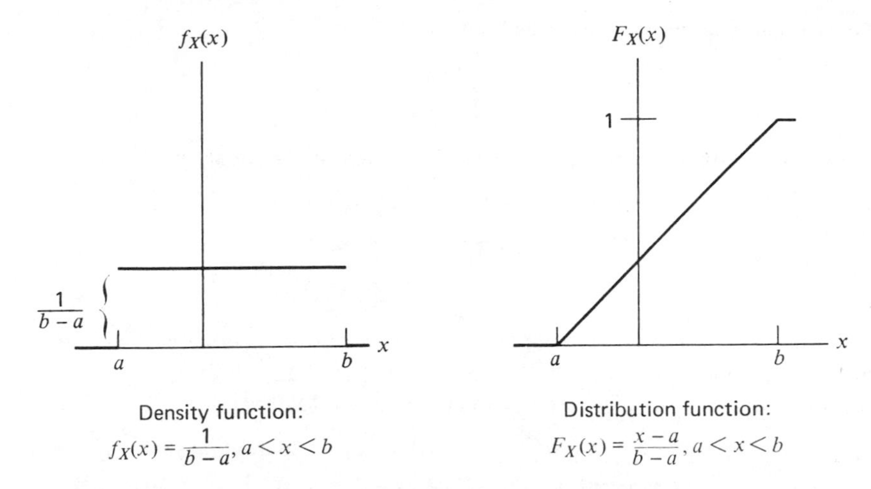


**Figure 4.5.** Uniform random variable.

thus $F_X$ is increasing linearly on the interval $(a, b)$. Figure 4.5 shows graphs of the density function and the distribution function of a uniform random variable on the interval $(a, b)$.

Suppose that $X$ is uniformly distributed on $(a, b)$. Then

$$\mu_X = E[X] = \int_a^b \frac{x}{b-a}\,dx = \frac{b+a}{2}$$

$$E[X^2] = \int_a^b \frac{x^2}{b-a}\,dx = \frac{b^2 + ab + a^2}{3},$$

so that

$$\sigma_X{}^2 = E[X^2] - \mu_X{}^2 = \frac{(b-a)^2}{12}.$$

*Example 4.4.1.* Suppose that $X$ is uniformly distributed on the interval $(0, 1)$ and we construct a square having side length $X$. Then

$$\begin{aligned} f_X(x) &= 1, \qquad 0 < x < 1 \\ &= 0, \qquad \text{otherwise.} \end{aligned}$$

The average side length of our square then is $\mu_X = \frac{1}{2}$ and the variance of our side length is $\frac{1}{12}$. If we let $Y$ be the area of the square so constructed, then (see Theorem 3.5.2)

$$\begin{aligned} f_Y(y) &= \frac{1}{2\sqrt{y}} \qquad 0 < y < 1 \\ &= 0, \qquad \text{otherwise.} \end{aligned}$$

The average area of the constructed square is

$$\mu_Y = \int_0^1 \frac{y}{2\sqrt{y}}\, dy = \frac{1}{3}$$

(note that $\mu_Y \neq \mu_X{}^2$) and the variance of the area of the square is

$$\begin{aligned}\sigma_Y{}^2 &= E[Y^2] - \mu_Y{}^2 \\ &= \int_0^1 \frac{y^2}{2\sqrt{y}}\, dy - \frac{1}{9} = \frac{4}{45}.\end{aligned}$$

Another frequently occurring continuous random variable is the exponential random variable. Although it may occur in many different contexts, we shall introduce it through a Poisson process (see Definition 4.3.1) as follows.

DEFINITION 4.4.2. Given a Poisson process with parameter $\lambda$, we designate by zero the time at which we begin observing the process. Let $T$ be the time that passes until the first event occurs. $T$ is called the *exponential* random variable with parameter $\lambda$.

Since time is measured continuously (and positively), it is immediately apparent that $T$ is a continuous random variable with its range being positive numbers. The form of the distribution function and density function for $T$ is given as Theorem 4.4.2.

***Theorem 4.4.2.*** Suppose that $T$ is an exponential random variable with parameter $\lambda$. Then

$$\begin{aligned}F_T(s) &= 0, && s < 0 \\ &= 1 - e^{-\lambda s}, && s \geq 0\end{aligned}$$

and

$$\begin{aligned}f_T(s) &= \lambda e^{-\lambda s}, && s > 0 \\ &= 0, && \text{otherwise.}\end{aligned}$$

*Proof:* Since time is measured positively, it is immediately apparent that $P(T \leq s) = 0$ if $s < 0$; thus $F_T(s) = 0$, $s < 0$. Consider some time $s \geq 0$. The time to the first event will *exceed* $s$ if and only if there are no events in the interval $(0, s)$. But the probability of no events in $(0, s)$ we have seen is $e^{-\lambda s}$ (see Theorem 4.3.1, set $k = 0$); thus $P(T > s) = e^{-\lambda s}$. But $P(T \leq s) = 1 - P(T > s)$ and therefore $F_T(s) = 1 - e^{-\lambda s}$ for $s \geq 0$. The density function for $T$ is given by the derivative of $F_T$; since

$$\begin{aligned}\frac{d}{ds} F_T(s) &= \lambda e^{-\lambda s}, && s > 0 \\ &= 0, && s < 0,\end{aligned}$$

we have

$$f_T(s) = \lambda e^{-\lambda s}, \qquad s > 0$$
$$= 0, \qquad s \leq 0,$$

which completes the proof. ◀

It is easily shown that $F_T$ is, in fact, a distribution function and that $f_T$ is a density function. The moment-generating function for $T$ is

$$m_T(t) = E[e^{tT}]$$
$$= \int_0^\infty e^{ts}\lambda e^{-\lambda s}\,ds$$
$$= \int_0^\infty \lambda e^{-s(\lambda - t)}\,ds$$
$$= \frac{\lambda}{\lambda - t}, \qquad \text{for } t < \lambda.$$

Then

$$m_T^{(1)}(t) = \frac{\lambda}{(\lambda - t)^2}$$
$$m_T^{(2)}(t) = \frac{2\lambda}{(\lambda - t)^3}$$

so that

$$\mu_T = m_T^{(1)}(0) = \frac{1}{\lambda}$$
$$\sigma_T^{\,2} = m_T^{(2)}(0) - \mu_T^{\,2} = \frac{1}{\lambda^2}.$$

Figure 4.6 shows graphs of the density function and the distribution function for an exponential random variable $X$ with parameter $\lambda$.

*Example 4.4.2.* Breakdowns in equipment at a large industrial plant have been observed to be approximately a Poisson process with parameter $\lambda = \frac{1}{2}$ per hour (1 every 2 hours). If we arrive at this plant at 9 a.m. on a Monday morning and let $T$ be the time (from our arrival) until the first breakdown, then $T$ has density function

$$f_T(s) = \tfrac{1}{2}e^{-s/2}, \qquad s > 0$$
$$= 0, \qquad s \leq 0.$$

The probability that it is at least 1 hour until the first breakdown is

$$P(T > 1) = \int_1^\infty \tfrac{1}{2}e^{-s/2}\,ds = e^{-1/2} = .6065;$$

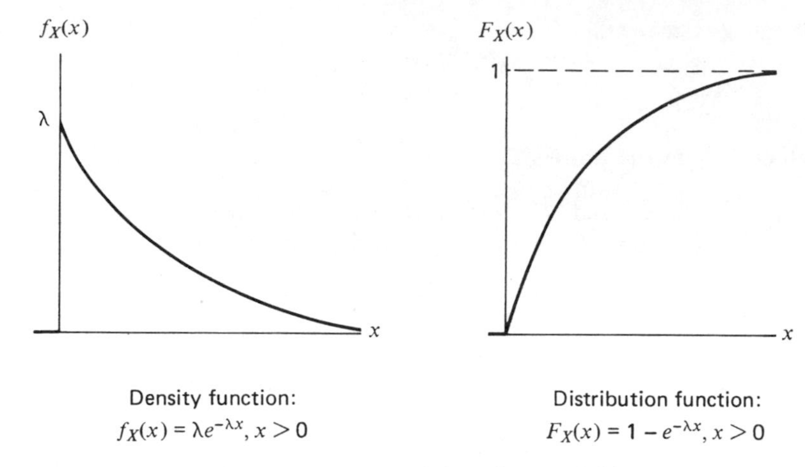


Density function:
$f_X(x) = \lambda e^{-\lambda x}, x > 0$

Distribution function:
$F_X(x) = 1 - e^{-\lambda x}, x > 0$

**Figure 4.6.** Exponential random variable.

the probability that it is no more than 4 hours to the first breakdown is

$$P(T \leq 4) = \int_0^4 \tfrac{1}{2} e^{-s/2}\, ds = 1 - e^{-2} = .8647.$$

The average time to the first breakdown is $\mu_T = 1/\lambda = 2$ hours; the probability that the time to the next breakdown is greater than the average (2 hours) is

$$P(T > 2) = \int_2^\infty \tfrac{1}{2} e^{-s/2}\, ds = e^{-1} = .3679.$$

*Example 4.4.3.* It is commonly assumed that failures of electron tubes of various sorts occur like events in a Poisson process; specifically, if a tube is put into operation, the time $T$ until it fails is an exponential random variable with parameter $\lambda$. Let us suppose that a particular device contains 5 electron tubes; the time (in hours) until failure for each of the 5 is assumed to be an exponential random variable with parameter $\lambda = \frac{1}{1000}$. Suppose further that the device works only so long as all 5 tubes are working and that the tubes operate (and fail) independently. We might then ask what is the probability that the device will operate at least $a$ hours. Each tube has density function

$$\begin{aligned} f_T(s) &= \tfrac{1}{1000} e^{-s/1000}, \qquad & s > 0 \\ &= 0, & s \leq 0. \end{aligned}$$

The probability that a particular tube will last at least $a$ hours is

$$P(T > a) = \int_a^\infty \tfrac{1}{1000} e^{-s/1000}\, ds = e^{-a/1000}.$$

Then, since the tubes are operating independently and the device continues working only so long as all the tubes do, we find that the probability the device will work at least $a$ hours is $(e^{-a/1000})^5 = e^{-a/200}$. Thus, the probability that it will operate at least 100 hours is $e^{-1/2} = .6065$; the probability that it will operate at least 1000 hours is $e^{-5} = .0067$.

There are a number of similarities between the discrete geometric random variable and the continuous exponential random variable. It will be recalled that the geometric random variable is the number of trials $X$ until the first success in repeated independent Bernoulli trials with parameter $p$. As we have seen, the exponential random variable is the length of time $T$ until the first event (which could be called a success) in a Poisson process with parameter $\lambda$. The mean value of $X$ was seen to be $1/p$; the mean value of $T$ is $1/\lambda$. Theorem 4.4.3 gives a particular property of the exponential random variable which is also shared by the geometric (as you are asked to prove in problem 12 below).

***Theorem 4.4.3.*** If $T$ is an exponential random variable, then

$$P(T > a + b \mid T > a) = P(T > b).$$

*Proof:* Let us first get a clear understanding of the notation used. Let the event that $T > a$ be denoted by $A$, the event that $T > b$ be denoted by $B$, and the event that $T > a + b$ be denoted by $C$. Then we are to prove that $P(C \mid A) = P(B)$. By definition, $P(C \mid A) = P(C \cap A)/P(A)$. The event $C$, that $T > a + b$, is a subset of the event $A$, that $T > a$. Thus, $C \cap A = C$ and we have $P(C \mid A) = P(C)/P(A)$. Now

$$P(A) = P(T > a) = \int_a^\infty \lambda e^{-\lambda s}\, ds = e^{-\lambda a}$$

$$P(B) = P(T > b) = \int_b^\infty \lambda e^{-\lambda s}\, ds = e^{-\lambda b}$$

$$P(C) = P(T > a + b) = \int_{a+b}^\infty \lambda e^{-\lambda s}\, ds = e^{-\lambda(a+b)}$$

and we see that

$$P(C \mid A) = \frac{P(C)}{P(A)} = \frac{e^{-\lambda(a+b)}}{e^{-\lambda a}} = e^{-\lambda b} = P(B),$$

as was to be proved. ◀

Suppose, for example, that the time to failure of an electron tube is an exponential random variable. Then Theorem 4.4.3 shows that if the tube has already been used for 100 hours without failure, the probability that it will last another 50 hours is the same as the probability of its lasting 50 hours when it was new. Since this property is also shared by the geometric random variable, we can say that if we have already had 10 trials without a success, the probability of 5 more trials without a success is identical with the original probability of having 5 trials with no success when we first started. Thus, both the exponential random variable and the geometric random variable can be said to have no memory; the fact that they have already achieved a certain value has no effect on the probability of achieving a larger value (relative to the original probability of equalling the incremental amount). This property is a direct consequence of the independence properties assumed: that the Bernoulli trials are independent and that the events in the Poisson process occur independently.

*Example 4.4.4.* Studies have shown that the occurrence of accidents to individuals (broken legs, bumped heads, car accidents, etc.) appear to be well described by Poisson processes; that is, the occurrences of accidents to individuals seem to occur in a manner consistent with the Poisson process assumptions. People who have a relatively large value of $\lambda$ (the rate of occurrence) are called accident prone. Suppose you are a person to whom accidents occur at a rate of 1 per year in a Poisson manner. Then Theorem 4.4.3 says that even if it has been 10 years since your last accident, your probability of having an accident within the next year is no higher than it was during any preceding year. Small comfort, perhaps, but comfort nonetheless.

## EXERCISE 4.4.

**1.** Suppose that $X$ is uniformly distributed on the interval $(1, 2)$ and we construct a square having sides of length $X$. Derive the probability density function of $Y = X^2$, the area of the square, and compute $P(Y > 2)$.

**2.** If $X$ is uniformly distributed on the interval $(1, 4)$, derive the density function of $Z = X^{1/2}$.

**3.** Derive the moment generating function for a random variable $X$ which is uniformly distributed on the interval $(a, b)$ and use it to evaluate $\mu_X$ and $\sigma_X{}^2$.

**4.** Derive the factorial moment generating function for a random variable $X$ uniformly distributed on $(a, b)$.

**5.** Suppose that quarter-pound bars of butter are cut from larger slabs by a machine. We assume that the larger slabs are quite uniform in density; if the length of the bar is exactly $3\frac{3}{8}$ inches, then the bar will weigh $\frac{1}{4}$ pound. Suppose that the true length $X$ of a bar cut by this machine is equally likely to lie in the interval from 3.35 inches to 3.45 inches. Assuming that the lengths of bars cut by this machine are independent, what is the probability that all 4 bars in a particular pound package of butter will weigh at least $\frac{1}{4}$ pound? That exactly 3 will weigh at least $\frac{1}{4}$ pound?

**6.** $X$ is uniformly distributed on $(0, 2)$ and $Y$ is exponential with parameter $\lambda$. Find the value of $\lambda$ such that $P(X < 1) = P(Y < 1)$.

**7.** Calls arrive at a switchboard according to a Poisson process with parameter $\lambda = 5$ per hour. If we are at the switchboard, what is the probability that it is at least 15 minutes until the next call? That it is no more than 10 minutes? That it is exactly 5 minutes?

**8.** A newsboy is selling papers on a busy street. The papers he sells are events in a Poisson process with parameter $\lambda = 50$ per hour. If we have just purchased a paper from him, what is the probability that it will be at least 2 minutes until he sells another? If it is already 5 minutes since his last sale, what is the probability it will be at least 2 more minutes until his next sale?

**9.** $X$ is uniform on $(-1, 3)$ and $Y$ is exponential with parameter $\lambda$. Find $\lambda$ such that ${\sigma_X}^2 = {\sigma_Y}^2$.

**10.** $X$ is geometric with parameter $p$ and $Y$ is exponential with parameter $\lambda$. Find $\lambda$ such that $P(X > 1) = P(Y > 1)$.

**11.** We are given a Poisson process with parameter $\lambda$. We begin observing the process at time zero; let $S$ be the time until the second event occurs. Derive the probability density function for $S$.

**12.** Prove Theorem 4.4.3 for a geometric random variable with parameter $p$.

### 4.5. Normal Random Variable

The normal probability density function is undoubtedly the most frequently used of all probability laws, both because the normal random variable does frequently occur in practical problems and, as we shall see in the next chapter, because it provides an accurate approximation to a large number of other probability laws. As can be seen from Definition 4.5.1, the density function for a normal random variable is a symmetric, bell-shaped curve, familiar to many people as the curve that is used when tests are "graded-on-the-curve." Due to the bell-shape of the density function, a normally distributed random variable has the biggest probability of taking on a value close to $\mu$ (in a sense) and correspondingly less of taking on values further from $\mu$ (on either side). The formal definition follows.

DEFINITION 4.5.1. A random variable $X$ is *normally* distributed if and only if its probability density function is

$$f_X(x) = \frac{1}{\sigma\sqrt{2\pi}}\, e^{-(x-\mu)^2/2\sigma^2}$$

for all real $x$. The parameter $\mu$ can equal any real number while the parameter $\sigma$ must be positive.

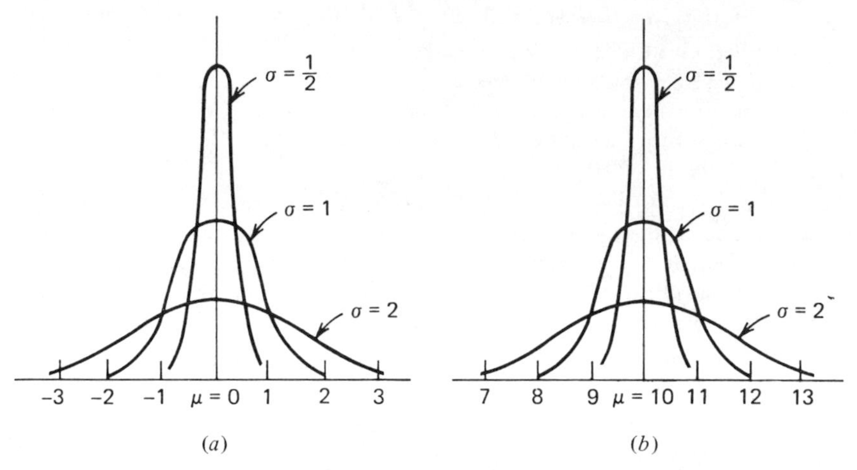


**Figure 4.7.** Normal random variable.

Since the exponential function is nonnegative, we obviously have $f_X(x) \geq 0$ for all $x$. You are asked to show in problem 7 below that $\int_{-\infty}^{\infty} f_X(x)\, dx = 1$; granting that this is the case, $f_X(x)$ is a density function.

Figure 4.7 shows the graphs of some typical density functions for various values of the parameters $\mu$ and $\sigma^2$. Notice that in any event the density function is a symmetric, bell-shaped curve centered at $\mu$ (thus we would guess that $\mu$ is, in fact, the mean of the random variable; this is proved below). The parameter $\sigma$ controls the relative flatness of the bell. Keeping $\mu$ constant and decreasing $\sigma$ causes the density function to become more sharply peaked, thus giving higher probabilities of $X$ being close to $\mu$. Increasing $\sigma$ causes the density function to flatten, thus giving lower probabilities of $X$ being close to $\mu$. If $\sigma$ is held constant and $\mu$ is varied, the density function's shape is held constant with its midpoint moving to the location of $\mu$.

Let us derive the moment generating function for a normal random variable.

***Theorem 4.5.1.*** If $X$ is a normal random variable with probability density function given in Definition 4.5.1, then the moment generating function for $X$ is

$$m_X(t) = e^{t\mu + t^2\sigma^2/2}.$$

*Proof:* By definition,

$$\begin{aligned} m_X(t) = E[e^{tX}] &= \int_{-\infty}^{\infty} e^{tx} \frac{1}{\sigma\sqrt{2\pi}} e^{-(x-\mu)^2/2\sigma^2}\, dx \\ &= \frac{1}{\sigma\sqrt{2\pi}} \int_{-\infty}^{\infty} e^{tx-(x-\mu)^2/2\sigma^2}\, dx. \end{aligned}$$

Completing the square in the exponent, we have

$$tx - \frac{(x-\mu)^2}{2\sigma^2} = -\frac{1}{2\sigma^2}[x^2 - 2\mu x + \mu^2 - 2\sigma^2 tx]$$

$$= -\frac{1}{2\sigma^2}[(x-\mu-\sigma^2 t)^2 - 2\sigma^2 t\mu - \sigma^4 t^2]$$

$$= -\frac{1}{2\sigma^2}[(x-\mu-\sigma^2 t)^2] + t\mu + \frac{\sigma^2 t^2}{2}.$$

Thus

$$m_X(t) = e^{t\mu+\sigma^2t^2/2}\int_{-\infty}^{\infty}\frac{1}{\sigma\sqrt{2\pi}}e^{-[x-\mu-\sigma^2t]^2/2\sigma^2}\,dx$$

$$= e^{t\mu+\sigma^2t^2/2},$$

since the integral has value 1 (see problem 7 below). ◀

Let us use $m_X$ to find the mean and variance of the normal random variable;

$$m_X^{(1)}(t) = (\mu + t\sigma^2)e^{t\mu+\sigma^2t^2/2}$$

$$m_X^{(2)}(t) = [\sigma^2 + (\mu + t\sigma^2)^2]e^{t\mu+\sigma^2t^2/2}.$$

Thus,

$$\mu_X = m_X^{(1)}(0) = \mu$$

$$\sigma_X{}^2 = m_X^{(2)}(0) - \mu^2 = \sigma^2;$$

that is, the parameter $\mu$ in the density function is the mean of the random variable $X$ and the parameter $\sigma^2$ in the density is the variance of $X$. The distribution function for a normal random variable $X$ is

$$F_X(t) = \int_{-\infty}^{t}\frac{1}{\sigma\sqrt{2\pi}}e^{-(x-\mu)^2/2\sigma^2}\,dx.$$

Unfortunately, this integration cannot be carried out in closed form. Numerical techniques could be used, of course, to evaluate the integral for specific values of $\mu$ and $\sigma^2$. Luckily, it is necessary only to use numerical integration to tabulate the distribution function of the standard normal random variable ($\mu = 0$, $\sigma^2 = 1$; see Definition 3.5.1). Once such a table is available for the standard normal distribution function, it can be used to evaluate the distribution function for a normal random variable with any mean $\mu$ and any variance $\sigma^2$. Since the standard normal distribution function is of such importance, we shall use a special symbol for it, as given in Definition 4.5.2.

DEFINITION 4.5.2. If $Z$ is a normal random variable with $\mu = 0$, $\sigma^2 = 1$, then $Z$ is called the *standard normal* random variable. Its density function is

$$n_Z(z) = \frac{1}{\sqrt{2\pi}} e^{-z^2/2}\, dz,$$

and its distribution function is

$$N_Z(t) = \int_{-\infty}^{t} \frac{1}{\sqrt{2\pi}} e^{-z^2/2}\, dz.$$

Table C, Appendix 5, gives a table of values of $N_Z(t)$. Let us consider some examples of its use.

*Example 4.5.1.* If $Z$ is a standard normal random variable, let us compute the following probabilities.

$$P(Z < 1) = N_Z(1) = .8413$$

$$P(1 < Z < 2) = N_Z(2) - N_Z(1) = .9773 - .8413 = .1360$$

$$P(|Z| < \tfrac{1}{2}) = P(-\tfrac{1}{2} < Z < \tfrac{1}{2}) = N_Z(\tfrac{1}{2}) - N_Z(-\tfrac{1}{2})$$
$$= .5915 - .3085 = .3830.$$

*Example 4.5.2.* We can also use Table C to answer questions such as these. Find the number $b$ such that $P(Z > b) = .25$. Since $P(Z > b) = 1 - N_Z(b)$, we are asked to find $b$ such that $N_Z(b) = .75$; by scanning the values of $N_Z$, we find that $b = .674$. We note that for a standard normal random variable, since its density function is symmetric about zero, we have $P(Z > a) = P(Z < -a)$ for any $a > 0$. That is $1 - N_Z(a) = N_Z(-a)$. Then suppose that we want to find the number $c$ such that $P(|Z| < c) = .9$. That is, since

$$P(|Z| < c) = P(-c < Z < c) = N_Z(c) - N_Z(-c)$$
$$= 2N_Z(c) - 1,$$

we want to find the point $c$ such that $N_Z(c) = .95$. By scanning the values of $N_Z$, we find that $c = 1.645$.

It was mentioned above that the standard normal distribution function could be used to evaluate the distribution function for a normal random variable with arbitrary $\mu$ and $\sigma$. The actual technique by which this is done is given as Theorem 4.5.2.

***Theorem 4.5.2.*** If $X$ is a normal random variable with mean $\mu$ and variance $\sigma^2$, then

$$F_X(t) = N_Z\left(\frac{t-\mu}{\sigma}\right).$$

*Proof:* By definition

$$F_X(t) = \int_{-\infty}^{t} \frac{1}{\sigma\sqrt{2\pi}} e^{-(x-\mu)^2/2\sigma^2}\, dx;$$

in this integral we change to $z$ as the variable of integration where

$$z = \frac{x - \mu}{\sigma}, \qquad dz = \frac{dx}{\sigma}.$$

The limits for $x$ were $-\infty < x < t$ which implies

$$-\infty < \frac{x - \mu}{\sigma} = z < \frac{t - \mu}{\sigma}.$$

Then we have

$$F_X(t) = \int_{-\infty}^{t-\mu/\sigma} \frac{1}{\sqrt{2\pi}} e^{-z^2/2}\,dz = N_Z\left(\frac{t - \mu}{\sigma}\right),$$

as was to be proved. ◀

*Example 4.5.3.* Suppose that the true weight $X$ of a pound of butter cut from a larger slab by a machine is a normal random variable with $\mu = 1.02$ pounds, $\sigma = .01$ pound. (We shall say that the weights of packages made by this machine are normally distributed with $\mu = 1.02$ and $\sigma = .01$; see Chapter 6). Find the probability that the weight of this package is less than 1 pound; also that the weight is more than 1.05 pounds.

$$P(X < 1) = F_X(1) = N_Z\left(\frac{1 - 1.02}{.01}\right) = N_Z(-2) = .0227$$

$$\begin{aligned} P(X > 1.05) &= 1 - F_X(1.05) = 1 - N_Z\left(\frac{1.05 - 1.02}{.01}\right) \\ &= 1 - N_Z(3) = .0013. \end{aligned}$$

*Example 4.5.4.* Assume that the distance $X$ that a particular athlete will be able to put a shot (on his first try) is a normal random variable with parameters $\mu = 50$ feet, $\sigma = 5$ feet. Compute the probability that he tosses it no less than 55 feet and the probability that his toss travels between 50 feet and 60 feet.

$$\begin{aligned} P(X > 55) &= 1 - F_X(55) \\ &= 1 - N_Z\left(\frac{55 - 50}{5}\right) \\ &= 1 - N_Z(1) \\ &= .1587. \\ P(50 < X < 60) &= F_X(60) - F_X(50) \\ &= N_Z(2) - N_Z(0) \\ &= .4773. \end{aligned}$$

When we get into statistical inference problems, we shall derive the distribution of what is called a $\chi^2$ (chi-square) random variable (see Chapter 6). The $\chi^2$ random variable has a single parameter called its degrees of freedom; let us note now that the square of a standard normal random variable $Z$ is a $\chi^2$ random variable with 1 degree of freedom. That is, if we let $W = Z^2$, then $W$ is a $\chi^2$ random variable with 1 degree of freedom. We can in fact evaluate the distribution function of $W$ from the distribution function for $Z$. As was derived in Example 4.5.2, if $Z$ is standard normal, then

$$N_Z(-a) = 1 - N_Z(a).$$

We saw in Theorem 3.5.2 that if $W = Z^2$, then

$$F_W(t) = N_Z(\sqrt{t}) - N_Z(-\sqrt{t}), \qquad \text{for } t > 0;$$

thus, if $W$ is a $\chi^2$ random variable with one degree of freedom,

$$\begin{aligned} F_W(t) &= N_Z(\sqrt{t}) - [1 - N_Z(\sqrt{t})] \\ &= 2N_Z(\sqrt{t}) - 1. \end{aligned}$$

Our table of $N_Z$ then enables us to evaluate probabilities for $W$ as well.

*Example 4.5.5.* If $X$ is a normal random variable with parameters $\mu$ and $\sigma$, the distribution function of the $\chi^2$ random variable just discussed can be used to evaluate probabilities that $X$ would differ from $\mu$ by more than $k\sigma$. Now

$$\begin{aligned} P(|X - \mu| > k\sigma) &= P\left(\frac{|X - \mu|}{\sigma} > k\right) \\ &= P\left(\frac{|X - \mu|^2}{\sigma^2} > k^2\right) \\ &= P(W > k^2), \end{aligned}$$

where $W$ is a $\chi^2$ random variable with 1 degree of freedom since $(X - \mu)/\sigma$ is a standard normal random variable. Then, for example,

$$\begin{aligned} P(|X - \mu| > \sigma) = P(W > 1) &= 1 - F_W(1) \\ &= 2 - 2N_Z(1) = .3174 \\ P(|X - \mu| > 2\sigma) = P(W > 4) &= 1 - F_W(4) \\ &= 2 - 2N_Z(2) = .0454 \\ P(|X - \mu| < 3\sigma) = P(W < 9) &= F_W(9) \\ &= 2N_Z(3) - 1 = .9974 \end{aligned}$$

## EXERCISE 4.5.

**1.** Assume that the time $X$ required for a distance runner to run a mile is a normal random variable with parameters $\mu = 4$ minutes, 1 second and $\sigma = 2$ seconds. What is the probability that this athlete will run the mile in less than 4 minutes? In more than 3 minutes, 55 seconds?

**2.** The length $X$ of an adult rock cod caught in Monterey Bay is a normal random variable with parameters $\mu = 16$ inches and $\sigma = 1$ inch. If you catch one of these fish, what is the probability that it will be at least 14 inches long? That it will be no more than 17 inches long? That its length will be between 12 inches and 15 inches?

**3.** If $Z$ is a standard normal random variable and we define $U = |Z|$, then $U$ is called the folded standard normal variable. Express $F_U(t)$ in terms of $N_Z(t)$.

**4.** Suppose that we are given a target with a vertical straight line drawn through its center. Let us assume that if we throw a dart at this target and measure the distance $Z$ between the point we hit and the center line, then $Z$ is a standard normal random variable (if the dart lands right of the center line the measurement is positive, if it lands to the left of the center line the measurement is negative). Then, the distance from the point we hit to the center line is $|Z| = U$, the folded normal random variable defined in problem 3. Compute $P(U > 1)$ and $P(U < \frac{1}{2})$.

**5.** Find the median and the interquartile range for a standard normal random variable $Z$.

**6.** Find the median and the interquartile range for a normal random variable $X$ with parameters $\mu$ and $\sigma$.

**7.** Show that

$$\int_{-\infty}^{\infty} \frac{1}{\sigma\sqrt{2\pi}} e^{-(x-\mu)^2/2\sigma^2}\, dx = 1$$

for any $\mu$ and for $\sigma > 0$. (**Hint:** If

$$A = \int_{-\infty}^{\infty} \frac{1}{\sigma\sqrt{2\pi}} e^{-(x-\mu)^2/2\sigma^2}\, dx,$$

then

$$A^2 = \int_{-\infty}^{\infty}\int_{-\infty}^{\infty} \frac{1}{2\pi\sigma^2} e^{-(x-\mu)^2/2\sigma^2-(y-\mu)^2/2\sigma^2}\, dx\, dy;$$

let $u = (x - \mu)/\sigma$, $v = (y - \mu)/\sigma$, and transform to polar coordinates to show $A^2 = 1$ which implies $A = 1$.)

**8.** If $W$ is a $\chi^2$ random variable with 1 degree of freedom, find the median and the interquartile range for $W$.

**9.** Evaluate $f_W(t)$ where $W$ is a $\chi^2$ random variable with 1 degree of freedom. (See Theorem 3.5.3.)

**10.** What is the moment generating function for a standard normal random variable?

**11.** Derive the moment generating function for a $\chi^2$ random variable with 1 degree of freedom.

# 5

# Jointly Distributed Random Variables

In many cases we shall see experiments for which more than one random variable are simultaneously defined. We shall then want to study the joint behavior of such random variables. This chapter is devoted to the definitions of a number of new terms and to the derivation of techniques appropriate to the study of several variables. Some important topics are discussed; although this is not the place for a complete discussion of all of them, hopefully enough material is presented to give the central ideas.

## 5.1. Two-Dimensional Random Variables

The simplest possible example of jointly distributed random variables is given by the two-dimensional case. The extension from a single random variable to two is the biggest step however; the extension from two to an arbitrary $n$ is relatively straightforward.

DEFINITION 5.1.1. Given an experiment, the pair $(X, Y)$ is called a *two-dimensional* random variable if each of $X$ and $Y$ associates a real number with every element of $S$.

Thus, $(X, Y)$ is a pair of real-valued element functions defined on $S$; as such each (of $X$ and $Y$) is then a random variable in its own right.

*Example 5.1.1.* Suppose a fair die is rolled 2 times. Let $X$ be the number that occurs on the first roll and let $Y$ be the number that occurs on the second roll. Then $(X, Y)$ is a two-dimensional random variable.

*Example 5.1.2.* One student is selected at random from the student body at San Francisco State. Let $U$ be the weight of the selected student and let $V$ be the height of the selected student. Then $(U, V)$ is a two-dimensional random variable.

As we might guess on the basis of our previous probability theory, if we have a two-dimensional random variable, it is possible that both are discrete, both are continuous, or one is discrete and the other continuous. We shall, in the main, be concerned only with the first two cases; it will be quite easy to formulate results for the third case based on a good understanding of the first two.

In the case of a single random variable $X$, we could think of the values $X$ takes on as being points on a line. Thus we defined the probability function for $X$ or the density function for $X$, both of which are functions of a real variable, depending on whether $X$ was discrete or continuous. We derived these one-dimensional functions (at least in theory) for computing probabilities about $X$ from the probability measure defined on the subsets of the sample space $S$. In the case of a two-dimensional random variable $(X, Y)$, it is convenient to think of the values the (two-dimensional) random variable takes on as being points in a Cartesian plane. We formally think of deriving the (two-dimensional) probability function or density function for computing probabilities about $(X, Y)$ again from the probability measure defined on the elements of $S$. This derivation is illustrated in the next two examples.

*Example 5.1.3.* An urn contains 3 balls numbered 1, 2, 3, respectively. We draw 2 balls at random, with replacement, from the urn. Let $X$ be the number on the first ball drawn and let $Y$ be the number on the second ball drawn. The collection of 2-tuples (pairs of numbers) that could be drawn would be a reasonable sample space for the experiment. Moreover, since the balls are drawn at random, the single-element events (2-tuples) are equally likely to occur. The range of $X$ is $\{1, 2, 3\}$, as is the range of $Y$. Table 5.1 has as entries the probabilities of occurrence

**Table 5.1**

| $y$ | $x$ | | |
|---|---|---|---|
| | 1 | 2 | 3 |
| 1 | $\frac{1}{9}$ | $\frac{1}{9}$ | $\frac{1}{9}$ |
| 2 | $\frac{1}{9}$ | $\frac{1}{9}$ | $\frac{1}{9}$ |
| 3 | $\frac{1}{9}$ | $\frac{1}{9}$ | $\frac{1}{9}$ |

of the possible pairs of values for $(X, Y)$. The probability of occurrence of any other pair of values $(x, y)$ is zero.

*Example 5.1.4.* Suppose that we draw 2 balls without replacement from the same urn discussed in Example 5.1.3 and again let $X$ be the number on the first ball drawn and let $Y$ be the number on the second ball. Again the ranges of the two random variables are each $\{1, 2, 3\}$, except now it is not possible for both $X$ and $Y$ to equal the same number simultaneously. The entries in Table 5.2 are the probabilities of occurrence of the possible pairs of values for $(X, Y)$, and again the probability of occurrence of any other pair $(x, y)$ is 0.

**Table 5.2**

| $y$ | $x$ = 1 | 2 | 3 |
|---|---|---|---|
| 1 | 0 | $\frac{1}{6}$ | $\frac{1}{6}$ |
| 2 | $\frac{1}{6}$ | 0 | $\frac{1}{6}$ |
| 3 | $\frac{1}{6}$ | $\frac{1}{6}$ | 0 |

The function which we have tabled in each of the two preceding examples is actually the two-dimensional probability function for $(X, Y)$; the definition of this function follows.

DEFINITION 5.1.2. $(X, Y)$ is a two-dimensional *discrete* random variable if the range of each is a discrete set. (We shall also say that $X$ and $Y$ are jointly discrete.) The probability function for a discrete two-dimensional random variable (written $p_{X,Y}(x, y)$) is defined to be

$$p_{X,Y}(x, y) = P(X = x,\ Y = y)$$

for all real pairs $(x, y)$.

Two-dimensional probability functions follow the same rules as do their one-dimensional analogs, namely

$$p_{X,Y}(x, y) \geq 0, \qquad \text{for all } (x, y)$$

$$\sum_{\substack{\text{range} \\ \text{of } X}} \sum_{\substack{\text{range} \\ \text{of } Y}} p_{X,Y}(x, y) = 1.$$

Furthermore, any function of a pair of real variables that satisfies these two equations can be thought of as being the probability function for some two-dimensional random variable.

Given $p_{X,Y}(x, y)$, we can easily evaluate $p_X(x)$ and $p_Y(y)$, the probability functions of $X$ and $Y$ individually; for

$$p_X(x) = p(X = x) = \sum_{\substack{\text{range} \\ \text{of } Y}} P(X = x, Y = y)$$

$$= \sum_{\substack{\text{range} \\ \text{of } Y}} p_{X,Y}(x, y).$$

We must sum over $y$ with $x$ fixed because this will give us the total probability of $X = x$. Similarly,

$$p_Y(y) = \sum_{\substack{\text{range} \\ \text{of } X}} p_{X,Y}(x, y).$$

The two one-dimensional functions, $p_X(x)$ and $p_Y(y)$, are called the marginal probability functions for $X$ and for $Y$, respectively. The word marginal is used because the values of these functions are given by the totals at the margins of the rows and columns of a tabular presentation of $p_{X,Y}(x, y)$, as in Tables 5.1 and 5.2. Notice that in each of these tables if we sum across a row ($y$ fixed) the total is $p_Y(y)$; if we sum down a column ($x$ fixed) the column total is simply $p_X(x)$. It is important to realize that knowledge of the marginals, $p_X(x)$ and $p_Y(y)$, generally is not equivalent to knowledge of $p_{X,Y}(x, y)$. This is immediately apparent by referring again to Tables 5.1 and 5.2; the marginals are the same in both cases but the two-dimensional probability functions are not.

To compute probabilities about two-dimensional continuous random variables, we must integrate two-dimensional density functions. These are defined as follows.

DEFINITION 5.1.3. $(X, Y)$ is a *continuous* two-dimensional random variable if $P(X = x, Y = y) = 0$ for all $(x, y)$ and the ranges of $X$ and $Y$ are both uncountable sets. (We shall say that $X$ and $Y$ are jointly continuous random variables.) The density function (written $f_{X,Y}(x, y)$) of $(X, Y)$ is nonzero in a region $R$ of the plane and is such that

$$P(a_1 < X < b_1, a_2 < Y < b_2) = \int_{x=a_1}^{b_1} \int_{y=a_2}^{b_2} f_{X,Y}(x, y)\, dy\, dx$$

for all $a_1$, $b_1$, $a_2$, $b_2$.

As in the one-dimensional case, it can be shown that a density function must satisfy two requirements:

1. $f_{X,Y}(x, y) \geq 0 \qquad$ for all $(x, y)$
2. $\displaystyle\int_{-\infty}^{\infty} \int_{-\infty}^{\infty} f_{X,Y}(x, y)\, dx\, dy = 1.$

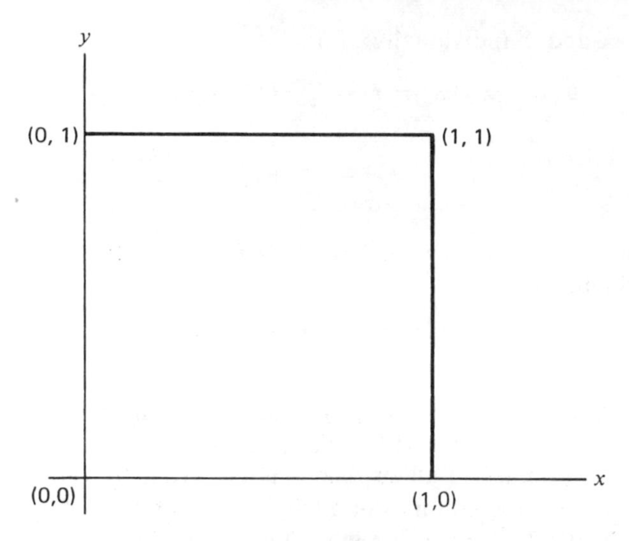


**Figure 5.1.**

Furthermore, any function of two variables that satisfies these two requirements can be thought of as being the density function for some two-dimensional random variable.

*Example 5.1.5.* Suppose we select one point at random from the unit square (see Figure 5.1). Let $X$ be the $x$-coordinate of the selected point and let $Y$ be the $y$-coordinate of the selected point. Then $(X, Y)$ is a two-dimensional continuous random variable. Since we are selecting the single point from this square at random, we must have

$$\begin{aligned} f_{X,Y}(x, y) &= c, \qquad 0 < x < 1, \qquad 0 < y < 1 \\ &= 0, \qquad \text{otherwise} \end{aligned}$$

because, presumably, all points have the same chance of occurring. Furthermore,

$$\int_0^1 \int_0^1 c \, dx \cdot dy = 1$$

which implies that $c = 1$; thus the density function for $(X, Y)$ is

$$\begin{aligned} f_{X,Y}(x, y) &= 1, \qquad 0 < x < 1, \qquad 0 < y < 1 \\ &= 0, \qquad \text{otherwise.} \end{aligned}$$

Knowledge of the two-dimensional density $f_{X,Y}(x, y)$ enables us to find the marginal densities for $X$ and for $Y$ (that is, the one-dimensional density functions for the individual random variables). In the case where $X$ and $Y$ were discrete, we got the marginal probability function for $X$ by summing $p_{X,Y}(x, y)$ over $y$. Analogously, if $X$ and $Y$ are continuous, we get the marginal density function for $X$ by integrating $f_{X,Y}(x, y)$ over $y$; that is

$$f_X(x) = \int_{-\infty}^{\infty} f_{X,Y}(x, y)\, dy.$$

Similarly, the marginal density function for $Y$ is

$$f_Y(y) = \int_{-\infty}^{\infty} f_{X,Y}(x, y)\, dx.$$

That this is the case can be motivated by reasoning that for any $a$ and $b$,

$$\begin{aligned} P(a < X < b) &= P(a < X < b, -\infty < Y < \infty) \\ &= \int_{x=a}^{b} \left\{ \int_{y=-\infty}^{\infty} f_{X,Y}(x, y)\, dy \right\} dx. \end{aligned}$$

The quantity in braces is what must be integrated from $a$ to $b$ to compute $P(a < X < b)$, thus it must be $f_X(x)$.

*Example 5.1.6.* Suppose that $X$ and $Y$ are jointly continuous with density function

$$\begin{aligned} f_{X,Y}(x, y) &= 2(x + y - 3xy^2), \qquad 0 < x < 1, \qquad 0 < y < 1 \\ &= 0, \qquad \text{otherwise.} \end{aligned}$$

The marginal for $X$ is

$$\begin{aligned} f_X(x) &= \int_0^1 2(x + y - 3xy^2)\, dy = 1, \qquad 0 < x < 1 \\ &= 0, \qquad \text{otherwise;} \end{aligned}$$

the marginal for $Y$ is

$$\begin{aligned} f_Y(y) &= \int_0^1 2(x + y - 3xy^2)\, dx = 1 + 2y - 3y^2, \qquad 0 < y < 1 \\ &= 0, \qquad \text{otherwise.} \end{aligned}$$

The distribution function for a two-dimensional random variable is defined quite analogously to its one-dimensional counterpart.

DEFINITION 5.1.4. The *distribution function* (written $F_{X,Y}(t_1, t_2)$) for a two-dimensional random variable $(X, Y)$ is defined to be $F_{X,Y}(t_1, t_2) = P(X \leq t_1, Y \leq t_2)$, for all real pairs $(t_1, t_2)$.

Thus, if $(X, Y)$ is discrete, we can derive the distribution function for $(X, Y)$ by summing values of $p_{X,Y}(x, y)$; if $(X, Y)$ is continuous, we can derive its distribution function by integrating $f_{X,Y}(x, y)$. If $(X, Y)$ is a discrete random variable, then $F_{X,Y}(t_1, t_2)$ is a discontinuous function of $t_1$ and $t_2$; if $(X, Y)$ is a continuous random variable, then $F_{X,Y}(t_1, t_2)$ is a continuous function of $t_1$ and $t_2$.

We shall not have much need for the distribution function of $(X, Y)$ in general. Our biggest concern will be with $F_{X,Y}$ when $X$ and $Y$ are independent random variables (see Section 5.3); in that case $F_{X,Y}$ is easily defined from $F_X$ and $F_Y$.

## EXERCISE 5.1.

**1.** A fair coin has a 1 painted on one side and a 2 painted on the other. The coin is flipped twice. Let $X$ be the sum of the two numbers that occur and let $Y$ be the difference of the two (the first minus the second). Compute $p_{X,Y}(x, y)$, $p_X(x)$, and $p_Y(y)$.

**2.** The fair coin mentioned in problem 1 is flipped 3 times. Let $X$ be the sum of the first two numbers and let $Y$ be the sum of the last two numbers and compute $p_{X,Y}(x, y)$.

**3.** Two cards are drawn at random (without replacement) from a standard 52-card bridge deck. Let $X$ be the number of aces that occur and let $Y$ be the number of spades that occur. Derive $p_{X,Y}(x, y)$ and compute $P(X > Y)$.

**4.** A bag contains 3 black and 7 red pieces of candy. We select 2 pieces without replacement. Let $X$ be the number of black pieces we draw and let $Y$ be the number of red pieces we draw. Derive $p_{X,Y}(x, y)$ and compute $P(X \leq Y)$.

**5.** Suppose that we select a point at random from the interior of the circle centered at the origin, with radius 1. Let $X$ be the $x$-coordinate and $Y$ be the $y$-coordinate of the selected point and derive $f_{X,Y}(x, y)$, $f_X(x)$, and $f_Y(y)$.

**6.** What must $A$ equal if

$$
\begin{aligned}
f_{X,Y}(x, y) &= A\frac{x}{y}, \qquad 0 < x < 1, \qquad 1 < y < 2 \\
&= 0, \qquad \text{otherwise}
\end{aligned}
$$

is to be a density function?

**7.** A family has 2 young boys. Let $X$ be the adult height of the older boy and let $Y$ be the adult height of the younger boy. Suppose $(X, Y)$ is equally likely to fall in the rectangle with corners (66, 68), (66, 72), (71, 68), (71, 72). Compute the probability that the older boy will be taller than the younger as adults.

**8.** Show that

$$P(a < X \leq b,\ Y \leq d) = F_{X,Y}(b, d) - F_{X,Y}(a, d).$$

(**Hint:** Draw a picture.)

**9.** Show that

$$
\begin{aligned}
P(a < X \leq b,\ c < Y \leq d) = F_{X,Y}(b, d) - F_{X,Y}(a, d) - F_{X,Y}(b, c) \\
+ F_{X,Y}(a, c).
\end{aligned}
$$

(**Hint:** Use problem 8 and a new picture.)

## 5.2. Expected Values and Moments

We shall find it useful to be able to compute expected or average values of functions of two random variables; the definition is quite analogous to that for the expected value of a function of a single random variable. (See Definition 3.3.1.)

DEFINITION 5.2.1. Suppose that $(X, Y)$ is a two-dimensional random variable and $H(X, Y)$ is a function of $X$ and $Y$. (1) If $X$ and $Y$ are jointly discrete, the *expected value* of $H(X, Y)$ is defined to be

$$E[H(X, Y)] = \sum_{\substack{\text{range} \\ \text{of } X}} \sum_{\substack{\text{range} \\ \text{of } Y}} H(x, y) p_{X,Y}(x, y),$$

so long as the sum is absolutely convergent. (2) If $X$ and $Y$ are jointly continuous, the *expected value* of $H(X, Y)$ is defined to be

$$E[H(X, Y)] = \int_{-\infty}^{\infty} \int_{-\infty}^{\infty} H(x, y) f_{X,Y}(x, y)\, dx\, dy,$$

so long as the integral is absolutely convergent.

Then, if we know the probability law of $(X, Y)$, we can use it to compute the expected value of such things as $X + Y$, $X - Y$, and $XY$.

*Example 5.2.1.* A pair of fair dice (1 red, 1 green) is rolled 1 time. Let $X$ be the number on the red die and let $Y$ be the number on the green die. Then

$$\begin{aligned} p_{X,Y}(x, y) &= \tfrac{1}{36}, \qquad x = 1, 2, \ldots, 6, \qquad y = 1, 2, \ldots, 6 \\ &= 0, \qquad \text{otherwise.} \end{aligned}$$

The expected value of the sum of the 2 numbers is

$$E[X + Y] = \sum_{x=1}^{6} \sum_{y=1}^{6} (x + y)\tfrac{1}{36} = 7;$$

the expected value of the product of the 2 numbers is

$$E[XY] = \sum_{x=1}^{6} \sum_{y=1}^{6} xy\tfrac{1}{36} = 12\tfrac{1}{4}.$$

*Example 5.2.2.* Suppose that both the husband and the wife in a certain family are professional artists. Let $X$ be the husband's income and $Y$ be the wife's income (both from selling their paintings) in the current year. From their past experience, they feel that the joint density function of their incomes (in thousands of dollars) is well represented by

$$\begin{aligned} f_{X,Y}(x, y) &= \tfrac{1}{15}, \qquad 10 < x < 15, \qquad 8 < y < 11 \\ &= 0, \qquad \text{otherwise.} \end{aligned}$$

The expected value of the family income (the sum of their incomes for this year) is

$$E[X + Y] = \int_{8}^{11} \int_{10}^{15} (x + y)\tfrac{1}{15}\, dx\, dy = 22;$$

the expected value of the difference between the husband's income and his wife's income is

$$E[X - Y] = \int_{8}^{11} \int_{10}^{15} (x - y)\tfrac{1}{15}\, dx\, dy = 3.$$

The following theorem gives a result which frequently makes the computations of some expected values easier.

***Theorem 5.2.1.*** If $(X, Y)$ is a two-dimensional random variable, then $E[G(X) + H(Y)] = E[G(X)] + E[H(Y)]$.

*Proof:* We shall give the proof only for the continuous case. The discrete case follows in exactly the same way, with summation replacing integration. By definition, if $X$ and $Y$ are jointly continuous,

$$\begin{aligned} E[G(X) + H(Y)] &= \int_{-\infty}^{\infty} \int_{-\infty}^{\infty} [G(x) + H(y)] f_{X,Y}(x, y)\, dx\, dy \\ &= \int_{-\infty}^{\infty} G(x) \left\{ \int_{-\infty}^{\infty} f_{X,Y}(x, y)\, dy \right\} dx \\ &\quad + \int_{-\infty}^{\infty} H(y) \left\{ \int_{-\infty}^{\infty} f_{X,Y}(x, y)\, dx \right\} dy \\ &= \int_{-\infty}^{\infty} G(x) f_X(x)\, dx + \int_{-\infty}^{\infty} H(y) f_Y(y)\, dy \\ &= E[G(X)] + E[H(Y)]. \end{aligned}$$

◀

We note immediately then that we can in general compute $E[G(X) + H(Y)]$ from knowledge only of the two marginals $f_X(x)$ and $f_Y(y)$; we do not need to know the joint density function of $(X, Y)$. This theorem pertains only to additive functions, not to multiplicative ones. In general, we find

$$E[G(X)H(Y)] \neq E[G(X)]E[H(Y)].$$

(An exception to this occurs in the case of independent random variables; see Section 5.3.)

In Chapter 3 we discussed the moments of a single random variable and noted that they characterized the distribution function for $X$. In the two-dimensional case, we define the joint moments of $(X, Y)$ as follows.

DEFINITION 5.2.2. The $ij$-th *joint moment* of $(X, Y)$ is defined to be

$$m_{ij} = E[X^i Y^j], \qquad i = 0, 1, 2, \ldots, \qquad j = 0, 1, 2, \ldots$$

($i$ and $j$ are not simultaneously zero).

Then, for example,

$$m_{11} = E[XY], \qquad m_{02} = E[X^0 Y^2] = E[Y^2]$$

$$m_{12} = E[XY^2], \qquad \text{etc.}$$

Note in particular that $m_{i0}$, $i = 1, 2, 3, \ldots$, are simply the moments of $X$ and $m_{0j}$, $j = 1, 2, 3, \ldots$, are the moments of $Y$. Thus the joint moments of $X$ and $Y$ include as special cases the moments of $X$ and the moments of $Y$; these special moments then describe the marginals of $X$ and of $Y$, respectively. The mixed moments $m_{ij}$, for which neither $i$ nor $j$ is zero, measure various aspects of the joint distribution of $X$ and of $Y$. Just as in the one-dimensional case, we can define a function that generates the moments of $(X, Y)$ through differentiation; this is called the moment generating function for $(X, Y)$ and is defined as follows.

DEFINITION 5.2.3. The function

$$m_{X,Y}(t_1, t_2) = E[e^{t_1 X + t_2 Y}]$$

is called the *moment generating* function for $(X, Y)$ (so long as the expectation exists).

That this function does generate moments may be seen by taking partial derivatives of $m_{X,Y}$ with respect to $t_1$ and $t_2$ and then evaluating these derivatives at the origin. Thus,

$$\begin{aligned}\frac{\partial}{\partial t_1} m_{X,Y}(t_1, t_2) &= \frac{\partial}{\partial t_1} E[e^{t_1 X + t_2 Y}] \\ &= E[X e^{t_1 X + t_2 Y}] \\ \frac{\partial}{\partial t_2} m_{X,Y}(t_1, t_2) &= E[Y e^{t_1 X + t_2 Y}]\end{aligned}$$

and, in general,

$$\frac{\partial^{i+j}}{\partial t_1^{\,i} \partial t_2^{\,j}} m_{X,Y}(t_1, t_2) = E[X^i Y^j e^{t_1 X + t_2 Y}].$$

And by setting $(t_1, t_2) = (0, 0)$, we see that we do generate the moments of $(X, Y)$.

*Example 5.2.3.* Suppose that $(X, Y)$ is continuous with density function

$$\begin{aligned} f_{X,Y}(x, y) &= e^{-x-y}, \qquad x > 0, \qquad y > 0 \\ &= 0, \qquad \text{otherwise.}\end{aligned}$$

The moment generating function for $(X, Y)$ is

$$\begin{aligned}
m_{X,Y}(t_1, t_2) &= E[e^{t_1X+t_2Y}] \\
&= \int_0^\infty \int_0^\infty e^{t_1x+t_2y}e^{-x-y}\,dx\,dy \\
&= \int_0^\infty e^{-x(1-t_1)}\,dx \int_0^\infty e^{-y(1-t_2)}\,dy \\
&= \frac{1}{(1-t_1)(1-t_2)}.
\end{aligned}$$

Then, for example,

$$\begin{aligned}
\frac{\partial}{\partial t_1} m_{X,Y}(t_1, t_2) &= \frac{1}{(1-t_1)^2(1-t_2)} \\
\frac{\partial}{\partial t_2} m_{X,Y}(t_1, t_2) &= \frac{1}{(1-t_1)(1-t_2)^2} \\
\frac{\partial^2}{\partial t_1\,\partial t_2} m_{X,Y}(t_1, t_2) &= \frac{1}{(1-t_1)^2(1-t_2)^2},
\end{aligned}$$

so that $m_{10} = 1$, $m_{01} = 1$, $m_{11} = 1$.

*Example 5.2.4.* Suppose that $(X, Y)$ is continuous with density

$$\begin{aligned}
f_{X,Y}(x, y) &= x + y, \qquad 0 < x < 1, \qquad 0 < y < 1 \\
&= 0, \qquad \text{otherwise.}
\end{aligned}$$

Let us find the moments of $(X, Y)$ directly, without using the moment generating function. By definition

$$\begin{aligned}
m_{ij} &= E[X^iY^j] \\
&= \int_0^1 \int_0^1 x^iy^j(x + y)\,dx\,dy \\
&= \frac{1}{(i+2)(j+1)} + \frac{1}{(i+1)(j+2)}, \qquad i = 0, 1, 2, \ldots, \qquad j = 0, 1, 2, \ldots.
\end{aligned}$$

Thus,

$$\begin{aligned}
E[X] &= \frac{1}{3 \cdot 1} + \frac{1}{2 \cdot 2} = \frac{7}{12} \\
E[XY] &= \frac{1}{3 \cdot 2} + \frac{1}{2 \cdot 3} = \frac{1}{3} \\
E[X^2] &= \frac{1}{4 \cdot 1} + \frac{1}{3 \cdot 2} = \frac{5}{12},
\end{aligned}$$

so that $\sigma_X{}^2 = E[X^2] - (E[X])^2 = \frac{11}{144}$.

*Example 5.2.5.* Suppose that $(X, Y)$ is discrete with probability function

$$p_{X,Y}(x, y) = \tfrac{1}{3}, \qquad \text{for } (x, y) = (0, 0), (-1, 0), (0, 1)$$
$$= 0, \qquad \text{otherwise.}$$

Then the $ij$-th moment of $(X, Y)$ is

$$m_{i0} = E(X^i) = \sum_{\substack{\text{range}\\ \text{of } X}} \sum_{\substack{\text{range}\\ \text{of } Y}} x^i p_{X,Y}(x, y)$$
$$= (-1)^i(\tfrac{1}{3}), \qquad \text{for } i = 1, 2, 3, \ldots$$
$$m_{0j} = E(Y^j) = \sum_{\substack{\text{range}\\ \text{of } X}} \sum_{\substack{\text{range}\\ \text{of } Y}} y^j p_{X,Y}(x, y)$$
$$= \tfrac{1}{3}, \qquad \text{for } j = 1, 2, 3, \ldots$$
$$m_{ij} = E(X^i Y^j) = \sum_{\substack{\text{range}\\ \text{of } X}} \sum_{\substack{\text{range}\\ \text{of } Y}} x^i y^j p_{X,Y}(x, y)$$
$$= 0, \qquad \text{for all } i = 1, 2, 3, \ldots, \quad j = 1, 2, 3, \ldots.$$

Thus $\mu_X = -\frac{1}{3}$, $\mu_Y = \frac{1}{3}$, $\sigma_X{}^2 = \sigma_Y{}^2 = \frac{2}{9}$.

The most frequently used mixed moment is $m_{11}$ which occurs in the definition of both the covariance and the correlation of $(X, Y)$. The covariance is defined as follows.

DEFINITION 5.2.4. The *covariance* of $X$ and $Y$ is defined to be

$$\sigma_{XY} = E[(X - \mu_X)(Y - \mu_Y)].$$

The covariance gives a measure of how $X$ and $Y$ tend to vary together. Since it is the expected value of the product of the deviation of $X$ from its mean times the deviation of $Y$ from its mean, we can see that $\sigma_{XY}$ will be positive if when $X$ is larger than its mean so is $Y$; if $X - \mu_X$ and $Y - \mu_Y$ have opposite signs with high probability, then $\sigma_{XY}$ will be negative. If $X - \mu_X$ is negative as frequently as it is positive when $Y - \mu_Y$ is positive or negative, we would expect $\sigma_{XY} = 0$. It is in this general sense that $\sigma_{XY}$ measures the covariability of $X$ and $Y$. The actual magnitude of $\sigma_{XY}$ does not have much meaning without knowledge of $\sigma_X{}^2$ and $\sigma_Y{}^2$ as well; for one probability law we might have $\sigma_{XY} = 1$ and for another $\sigma_{XY} = 100$. We cannot say that the tendency for $X$ and $Y$ to vary together is greater in the latter case than in the former because $X$ and $Y$ themselves might be extremely variable in the latter case, which tends to increase the numerical value of $\sigma_{XY}$. The correlation coefficient $\rho_{XY}$, defined below, gives a measure of covariability which also accounts for the variances of $X$ and of $Y$.

DEFINITION 5.2.5. The *correlation* between $X$ and $Y$ is defined to be

$$\rho_{XY} = \frac{\sigma_{XY}}{\sigma_X \sigma_Y}.$$

($\rho_{XY}$ is called the correlation coefficient of $X$ and $Y$.)

A simpler computational formula for $\sigma_{XY}$ (and thus also for $\rho_{XY}$) than is given by the definition is derived as follows.

$$\begin{aligned} \sigma_{XY} &= E[(X - \mu_X)(Y - \mu_Y)] \\ &= E[XY - \mu_X Y - X\mu_Y + \mu_X\mu_Y] \\ &= E[XY] - \mu_X E[Y] - \mu_Y E[X] + \mu_X\mu_Y \\ &= E[XY] - \mu_X\mu_Y. \end{aligned}$$

Note the similarity between this equation and the computational formula for a variance derived in Chapter 3.

*Example 5.2.6.* For the random variables $X$ and $Y$ discussed in Example 5.2.3, we found $E[XY] = E[X] = E[Y] = 1$ and thus, for that case, $\sigma_{XY} = 1 - 1 \cdot 1 = 0$ so that $\rho_{XY} = 0$. For the random variables discussed in Example 5.2.4, we find that $E[XY] = \frac{1}{3}$, $E[X] = E[Y] = \frac{7}{12}$, $E[X^2] = E[Y^2] = \frac{5}{12}$ so that

$$\begin{aligned} \sigma_{XY} &= \tfrac{1}{3} - \tfrac{49}{144} = -\tfrac{1}{144} \\ \sigma_X^{\,2} &= \tfrac{5}{12} - \tfrac{49}{144} = \tfrac{11}{144} = \sigma_Y^{\,2} \\ \rho_{XY} &= \frac{-\frac{1}{144}}{\sqrt{\frac{11}{144} \cdot \frac{11}{144}}} = -\tfrac{1}{11}. \end{aligned}$$

For the random variables $X$ and $Y$ discussed in Example 5.2.5, we have $\sigma_{XY} = 0 - (-\frac{1}{3})(\frac{1}{3}) = \frac{1}{9}$, $\sigma_X^{\,2} = \sigma_Y^{\,2} = \frac{2}{9}$, $\rho_{XY} = \frac{1}{2}$.

The reader will note that, of the three correlation coefficients derived in Example 5.2.6, one was negative, one was positive, and one was zero. That $|\rho_{XY}| \leq 1$ for all random variables is proved as Theorem 5.2.2.

***Theorem* 5.2.2.** For any random variables $X$ and $Y$, $-1 \leq \rho_{XY} \leq 1$.

*Proof:* The proof is a special case of the Schwarz inequality which finds application in many branches of mathematics. Suppose that $W$ and $Z$ are random variables. Then $(aW - Z)^2$ is a random variable which takes on only nonnegative values, no matter what the (real) value of $a$. Then its mean is nonnegative:

$$E[(aW - Z)^2] \geq 0;$$

that is,

$$E[a^2W^2 - 2aWZ + Z^2] \geq 0$$

so that

$$a^2E[W^2] - 2aE[WZ] + E[Z^2] \geq 0, \qquad \text{for all } a.$$

If in particular

$$a = \frac{E[WZ]}{E[W^2]},$$

this becomes

$$\frac{-(E[WZ])^2}{E[W^2]} + E[Z^2] \geq 0;$$

then

$$1 \geq \frac{(E[WZ])^2}{E[W^2]E[Z^2]}.$$

If we now let $W = X - \mu_X$, $Z = Y - \mu_Y$, we have

$$1 \geq \frac{(E[(X - \mu_X)(Y - \mu_Y)])^2}{E[(X - \mu_X)^2]E[(Y - \mu_Y)^2]};$$

that is

$$1 \geq \frac{{\sigma_{XY}}^2}{{\sigma_X}^2{\sigma_Y}^2} = {\rho_{XY}}^2,$$

which implies $|\rho_{XY}| \leq 1$ ◀

If in fact $|\rho_{XY}| = 1$, then there is a perfect linear relation between $X$ and $Y$; i.e., $Y$ is a linear function of $X$. If the equality holds in the proof of Theorem 5.2.2, then $E[(aW - Z)^2] = 0$. But the mean value of a non-negative random variable can only be zero if the random variable is equal to zero with probability 1. That is, we must have $P[(aW - Z)^2 = 0] = 1$ which implies that $P[aW - Z = 0] = 1$, or $P[Z = aW] = 1$, and thus

$$P[Y - \mu_Y = a(X - \mu_X)] = 1$$
$$P[Y = \mu_Y - a\mu_X + aX] = 1,$$

as was claimed. The fact that $Y = \mu_Y - a\mu_Y + aX$ with probability 1 means that in the $(x, y)$ plane of possible values for $(X, Y)$, only those points lying on $y = \mu_Y - a\mu_X + ax$ can have positive probability; all points off this line have probability zero. Since the sign of $\rho_{XY}$ is determined by the sign of

$$\sigma_{XY} = E[(X - \mu_X)(Y - \mu_Y)] = aE[(X - \mu_X)^2],$$

we find that $\rho_{XY} = 1$ if $a$ is positive and $\rho_{XY} = -1$ if $a$ is negative (see Figure 5.2). The "closer" that $Y$ is to a linear function of $X$, that is, the "closer" that all of the probability is to lying on a straight line in the $(x, y)$ plane, the closer that $|\rho_{XY}|$ is to 1. (See problems 7 and 8 below.) On the other hand, $\rho_{XY} = 0$ only if $\sigma_{XY} = 0$, which would imply that the covariability of $X$ and $Y$ is zero. (That is, $X$ and $Y$ do not tend to vary together.) Note, however, that $\sigma_{XY}$ is a measure of only linear covariability and it is in

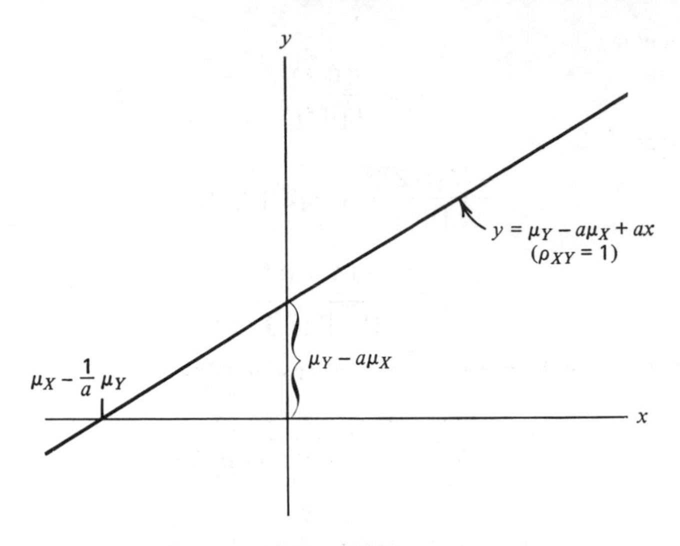


**Figure 5.2.**

fact possible that $Y$ can be completely determined by $X$ (nonlinearly) and still the correlation is zero. Example 5.2.7 presents a case of this sort.

*Example 5.2.7.* Suppose that $X$ is a discrete random variable with probability function

$$\begin{aligned} p_X(x) &= \tfrac{1}{4}, \qquad \text{at } x = -2, -1, 1, 2 \\ &= 0, \qquad \text{otherwise.} \end{aligned}$$

Define $Y = X^2$ and then $Y$ certainly is a function of $X$ and all the probability will lie on a line in the $(x, y)$ plane but not on a straight line; in fact, we have

$$\begin{aligned} p_{X,Y}(x, y) &= \tfrac{1}{4}, \qquad \text{at } (x, y) = (-2, 4), (-1, 1), (1, 1), (2, 4) \\ &= 0, \qquad \text{otherwise.} \end{aligned}$$

We find $\mu_X = 0$, $\mu_Y = 2\frac{1}{2}$, $E[XY] = 0$ and thus $\sigma_{XY} = 0$ and $\rho_{XY} = 0$, despite the fact that $Y$ is a quadratic function of $X$.

## EXERCISE 5.2.

**1.** Suppose that $+1$ is painted on one side of a fair coin and $-1$ is painted on the other. The coin is flipped 2 times; let $X$ be the number that occurs on the first flip and let $Y$ be the number that occurs on the second. Derive $p_{X,Y}(x, y)$, $\mu_X$, $\mu_Y$, and $E(X/Y)$.

**2.** Students at a large university are required to take an entrance exam when entering the school (scored on a basis of 100 points possible). Let $X$ be the score made by a particular incoming student (who does go on to graduate) and let $Y$ be his graduating quality point ratio (4 points $= A$). The joint density function of $X$ and $Y$ is observed to be

$$\begin{aligned} f_{X,Y}(x, y) &= \tfrac{1}{50}, \quad \text{for } 2 < y < 4, \quad 25(y-1) < x < 25y \\ &= 0, \quad \text{otherwise.} \end{aligned}$$

Find $\mu_X$, $\mu_Y$, and $m_{11} = E(XY)$.

**3.** In problem 2, compute $\sigma_X^2$, $\sigma_Y^2$, $\sigma_{XY}$, and $\rho_{XY}$.

**4.** Given

$$\begin{aligned} p_{W,Z}(w, z) &= \frac{1}{n}, \quad \text{for } (w, z) = (1, 1), (2, 2), \ldots, (n, n) \\ &= 0, \quad \text{otherwise,} \end{aligned}$$

evaluate $E[W/Z]$, $E[Z/W]$, $E[W^2 + Z^2]$, and $E[W^3/Z^2]$.

**5.** Given

$$\begin{aligned} f_{U,V}(u, v) &= 6(1 - u - v), \quad 0 < u < 1, \quad 0 < v < 1 - u \\ &= 0, \quad \text{otherwise,} \end{aligned}$$

compute $E[U]$, $E[V]$, $E[UV]$, and $\sigma_{UV}$.

**6.** If $X$ and $Y$ are jointly distributed random variables, show that $m_{X,Y}(t, 0) = m_X(t)$ and $m_{X,Y}(0, t) = m_Y(t)$.

**7.** $X$ and $Y$ are jointly discrete random variables with probability function

$$\begin{aligned} p_{X,Y}(x, y) &= \tfrac{1}{4}, \quad \text{at } (x, y) = (-3, -5), (-1, -1), (1, 1), (3, 5) \\ &= 0, \quad \text{otherwise.} \end{aligned}$$

Compute $\rho_{XY}$.

**8.** Suppose that $X$ and $Y$ are jointly discrete with probability function

$$\begin{aligned} p_{X,Y}(x, y) &= \tfrac{1}{4}, \quad \text{at } (x, y) = (-3, -4), (-1, -1), (1, 1), (3, 4) \\ &= 0, \quad \text{otherwise.} \end{aligned}$$

Compute $\rho_{XY}$. (Compare the plot of points with positive probability for this problem versus problem 7. Note that the correlation is higher in problem 8 because the points with positive probability are closer to lying on a straight line.)

**9.** Define a function $\psi_{X,Y}(t_1, t_2)$ which would generate factorial moments for $(X, Y)$. (Note that you should have $\psi_{X,Y}(t, 1) = \psi_X(t)$ and $\psi_{X,Y}(1, t) = \psi_Y(t)$.)

## 5.3. Independent Random Variables and Conditional Distributions

One of the most important concepts for statistical (as well as most other) applications of probability theory is that of independent random variables. As will be shown below, independent random variables are a generalization of independent events, discussed in Chapter 2. If two or more random variables can be assumed to be independent, then many problems are greatly simplified. The definition we shall adopt follows.

DEFINITION 5.3.1. (1) $X$ and $Y$ will be called *independent* discrete random variables if and only if

$$p_{X,Y}(x, y) = p_X(x)p_Y(y), \qquad \text{for all } (x, y).$$

(2) $X$ and $Y$ will be called *independent* continuous random variables if and only if

$$f_{X,Y}(x, y) = f_X(x)f_Y(y), \qquad \text{for all } (x, y).$$

Notice first of all that the idea of independence of $X$ and $Y$ requires that the joint density or probability function of $X$ and $Y$ will factor into the product of the marginals. Thus, if $X$ and $Y$ are independent, knowledge of their marginals is equivalent to knowledge of their joint distribution, a statement which is not generally true for all random variables as we have seen. In practice, then, if we are able to ascertain that the two random variables are independent, we need only investigate their individual marginal distributions to be able to construct their joint distribution.

Many texts take the following statement as a definition of independent random variables.

$$P[X \in A, Y \in B] = P[X \in A]P[Y \in B], \qquad \text{for all } A \text{ and } B \tag{5.1}$$

where $A$ is any set of numbers from the range of $X$ and $B$ is any set of numbers from the range of $Y$. This can be shown to be equivalent to Definition 5.3.1 and thus would serve to define independent random variables. It is this equivalent statement that possibly best illustrates the fact that independent random variables derive directly from independent events (or vice versa). The reader will recall that $A$ and $B$ are independent events if and only if

$$P(A \cap B) = P(A)P(B),$$

an equation which is very similar to 5.1. Since Equation 5.1 is equivalent to Definition 5.3.1, we can say that $X$ and $Y$ are independent random variables if and only if the value which $X$ has taken on has no influence on the value which $Y$ takes on (a very similar statement was made for independent events). Before looking at some specific examples, let us note another equivalent statement involving the distribution function for $X$ and $Y$ in order that $X$ and $Y$ be independent random variables. This statement is as follows. $X$ and $Y$ are independent if and only if

$$F_{X,Y}(t_1, t_2) = F_X(t_1)F_Y(t_2), \qquad \text{for all } (t_1, t_2).$$

That is, the distribution function for $X$ and $Y$ must factor into the product of the marginal distribution functions at all points if $X$ and $Y$ are to be independent. Since we know the density or probability function factors for independent random variables, this equivalent factorization of distribution functions follows very easily.

*Example 5.3.1.* If $X$ and $Y$ are jointly discrete with probability function

$$\begin{aligned} p_{X,Y}(x, y) &= \tfrac{1}{4}, \qquad \text{for } (x, y) = (-1, -1), (-1, 1), (1, -1), (1, 1) \\ &= 0, \qquad \text{otherwise,} \end{aligned}$$

then

$$\begin{aligned} p_X(x) &= \tfrac{1}{2}, \qquad \text{for } x = -1, 1 \\ &= 0, \qquad \text{otherwise} \end{aligned}$$

and

$$\begin{aligned} p_Y(y) &= \tfrac{1}{2}, \qquad \text{for } y = -1, 1 \\ &= 0, \qquad \text{otherwise.} \end{aligned}$$

Thus

$$p_{X,Y}(x, y) = p_X(x)p_Y(y), \qquad \text{for all } (x, y)$$

and $X$ and $Y$ are independent. However, if

$$\begin{aligned} p_{X,Y}(x, y) &= \tfrac{1}{4}, \qquad \text{for } (x, y) = (-1, -1), (-1, 1), (1, -1), (1, 2), \\ &= 0, \qquad \text{otherwise,} \end{aligned}$$

then

$$\begin{aligned} p_X(x) &= \tfrac{1}{2}, \qquad x = -1, 1 \\ &= 0, \qquad \text{otherwise} \end{aligned}$$

and

$$\begin{aligned} p_Y(y) &= \tfrac{1}{2}, \qquad y = -1 \\ &= \tfrac{1}{4}, \qquad y = 1, 2 \\ &= 0, \qquad \text{otherwise.} \end{aligned}$$

In this case $X$ and $Y$ are not independent since $p_{X,Y}(1, 1) \neq p_X(1)p_Y(1)$. (We have found one particular point in the plane for which the joint probability function does not equal the product of the marginals; this is sufficient to show that they are not independent.)

*Example 5.3.2.* If $U$ and $V$ are jointly continuous with density function

$$\begin{aligned} f_{U,V}(u, v) &= \tfrac{1}{4}, \qquad \text{for } -1 < u < 1, \qquad -1 < v < 1 \\ &= 0, \qquad \text{otherwise,} \end{aligned}$$

then we find

$$\begin{aligned} f_U(u) &= \tfrac{1}{2}, \qquad -1 < u < 1 \\ &= 0, \qquad \text{otherwise,} \end{aligned}$$

and

$$\begin{aligned} f_V(v) &= \tfrac{1}{2}, \qquad -1 < v < 1 \\ &= 0, \qquad \text{otherwise.} \end{aligned}$$

$U$ and $V$ are then independent since $f_{U,V}(u, v) = f_U(u)f_V(v)$ for all $(u, v)$. (We would say that $U$ and $V$ are uniformly distributed over the square with corners $(-1, -1)$, $(-1, 1)$, $(1, -1)$, and $(1, 1)$ because their joint density is nonzero and constant over that square; in such a case, $U$ and $V$ are independent.) If, however, $U$ and $V$ are uniformly distributed over a circle with radius 1 centered at the origin, then

$$\begin{aligned} f_{U,V}(u, v) &= \frac{1}{\pi}, \qquad \text{for } u^2 + v^2 < 1 \\ &= 0, \qquad \text{otherwise,} \end{aligned}$$

and we find

$$\begin{aligned} f_U(u) &= \frac{2\sqrt{1 - u^2}}{\pi}, \qquad \text{for } -1 < u < 1 \\ &= 0, \qquad \text{otherwise} \end{aligned}$$

$$\begin{aligned} f_V(v) &= \frac{2\sqrt{1 - v^2}}{\pi}, \qquad \text{for } -1 < v < 1 \\ &= 0 \qquad \text{otherwise.} \end{aligned}$$

Then $U$ and $V$ are not independent since, in particular, $f_{U,V}(0, 0) \neq f_U(0)f_V(0)$.

*Example 5.3.3.* One electronic device in a missile consists of 2 integrated circuits; in order that the missile perform successfully, it is necessary that this particular device operate correctly for 2 minutes. Assume that the 2 circuits in the device operate independently (that is, the failure of the first circuit has no effect on the second and vice versa). Let $S$ be the time to failure of circuit 1 and $T$ be the time to failure of circuit 2. It is known that $S$ is an exponential random variable with parameter $\lambda$ and $T$ is an exponential random variable with parameter $\beta$. Then, since the 2 integrated circuits operate independently in the electronic device, the joint distribution of the 2 failure times is:

$$\begin{aligned} f_{S,T}(s, t) &= f_S(s)f_T(t) \\ &= \lambda e^{-\lambda s}\beta e^{-\beta t}, \qquad s > 0, \quad t > 0 \\ &= 0, \qquad \text{otherwise.} \end{aligned}$$

If we are given that the device fails if either of the 2 circuits fail, then the probability that the device operates correctly for at least 2 minutes is

$$\begin{aligned} P(S > 2, T > 2) &= \int_2^\infty \int_2^\infty \lambda e^{-\lambda s}\beta e^{-\beta t}\, ds\, dt \\ &= e^{-2\lambda} \cdot e^{-2\beta} \end{aligned}$$

If $\lambda$ and $\beta$ were known, we could then get a numerical answer for this probability; the way in which we might realistically gather information on $\lambda$ and $\beta$ is a statistical question taken up in succeeding chapters.

Example 5.3.3 gives one of the most frequent uses of the concept of independent random variables. The known independence (derived from an analysis of the situation) then implies the joint distribution to be given by the product of the marginals.

Theorem 5.3.1 gives a very important consequence of the independence of two random variables.

***Theorem 5.3.1.*** If $X$ and $Y$ are independent random variables, then $E[H(X)G(Y)] = E[H(X)]E[G(Y)]$ where $H$ and $G$ are any two functions of a real variable.

*Proof:* Let us indicate the proof for jointly continuous random variables. If $X$ and $Y$ are jointly discrete, the reasoning is identical, integrations being replaced by summations. By definition,

$$\begin{aligned} E[H(X)G(Y)] &= \int_{-\infty}^{\infty}\int_{-\infty}^{\infty} H(x)G(y)f_{X,Y}(x, y)\,dx\,dy \\ &= \int_{-\infty}^{\infty}\int_{-\infty}^{\infty} H(x)G(y)f_X(x)f_Y(y)\,dx\,dy \\ &= \left\{\int_{-\infty}^{\infty} H(x)f_X(x)\,dx\right\}\left\{\int_{-\infty}^{\infty} G(y)f_Y(y)\,dy\right\} \\ &= E[H(X)]E[G(Y)]. \end{aligned}$$

◀

*Example 5.3.4.* Theorem 5.3.1 implies then that if $X$ and $Y$ are independent, all joint moments $m_{ij}$ are the product of the $i$-th moment of $X$ times the $j$-th moment of $Y$ since

$$m_{ij} = E[X^i Y^j] = E[X^i]E[Y^j] = m_{X_i} m_{Y_j}.$$

It also implies that the joint moment generating function for $(X, Y)$ factors into the product of the two moment generating functions because

$$\begin{aligned} m_{X,Y}(t_1, t_2) &= E[e^{t_1X+t_2Y}] \\ &= E[e^{t_1X}e^{t_2Y}] \\ &= E[e^{t_1X}]E[e^{t_2Y}] \\ &= m_X(t_1)m_Y(t_2). \end{aligned}$$

The following theorem gives one more particular statement of some importance implied by Theorem 5.3.1.

***Theorem 5.3.2.*** If $X$ and $Y$ are independent random variables, then $\sigma_{XY} = \rho_{XY} = 0$.

*Proof:* As was shown in Section 5.2,

$$\sigma_{XY} = E[XY] - E[X]E[Y];$$

but if $X$ and $Y$ are independent, then in particular

$$E[XY] = E[X]E[Y]$$

and thus $\sigma_{XY} = 0$. Since

$$\rho_{XY} = \frac{\sigma_{XY}}{\sigma_X \sigma_Y}$$

we also have $\rho_{XY} = 0$. ◀

This theorem says that independence implies a zero correlation; the converse is not true in general. In Example 5.2.7 we saw a case in which $Y$ was a function of $X$ and yet the correlation between them was zero, so the converse of Theorem 5.3.2 is not true.

The conditional distribution of a random variable $Y$, given that a second (jointly distributed with $Y$) random variable $X$ has taken on a particular value $x$, will occasionally be of interest in later applications. We recall from Chapter 2 that the conditional probability of an event $A$, given that $B$ had occurred, was defined to be

$$P(A \mid B) = \frac{P(A \cap B)}{P(B)}, \qquad \text{for } P(B) > 0.$$

Accordingly, if $X$ and $Y$ are jointly discrete random variables, we can think of $A$ as being the event that $Y = y$ and $B$ as being the event that $X = x$, and we would have

$$P(Y = y \mid X = x) = \frac{P(Y = y, X = x)}{P(X = x)} = \frac{p_{X,Y}(x, y)}{p_X(x)}, \qquad \text{for } p_X(x) > 0.$$

This ratio defines what is called the conditional probability function for $Y$, given that $X = x$, as we see in the following definition.

DEFINITION 5.3.2. (1) If $X$ and $Y$ are jointly discrete random variables, the *conditional probability function* for $Y$, given $X = x$, is defined to be

$$\begin{aligned} p_{Y \mid X}(y \mid x) &= \frac{p_{X,Y}(x, y)}{p_X(x)}, && \text{for } p_X(x) > 0 \\ &= 0, && \text{if } p_X(x) = 0. \end{aligned}$$

$p_{X \mid Y}(x \mid y)$ is defined analogously.

(2) If $X$ and $Y$ are jointly continuous random variables, the *conditional density* for $Y$, given $X = x$, is defined to be

$$\begin{aligned} f_{Y \mid X}(y \mid x) &= \frac{f_{X,Y}(x, y)}{f_X(x)}, && \text{for } f_X(x) > 0 \\ &= 0, && \text{if } f_X(x) = 0. \end{aligned}$$

$f_{X \mid Y}(x \mid y)$ is defined analogously.

The reader is asked in Exercise 5.3.7 to verify that these definitions are in fact consistent with those of one-dimensional probability functions and density functions.

The second part of this definition could use a little explanation. In many situations where we have two jointly distributed continuous random variables, we may be able to observe the value that $X$ takes on, say $x$, and then want to compute the probability that $Y$ lies in some interval. For example, in firing a two-stage rocket we might let $X$ be the length of time the first stage burns and let $Y$ be the time the second stage burns. Then, given that the first stage has burned for 80 seconds, we might be interested in the probability that $Y$ would lie between 70 and 75 seconds. The conditional density of $Y$, given $X = 80$, would be used to compute this probability. Or, suppose we had the joint density function describing the graduating quality point average in college, $X$, and the annual salary, $Y$, of the same individual 5 years after graduation. Then, given a person with a graduating quality point average of 2.75, we might ask the probability that this person will have an annual salary of 10,000 dollars or more 5 years after graduation. Such a probability would be computed from the conditional density function of $Y$ given $X = 2.75$.

Let us now briefly discuss why the definition of $f_{X \mid Y}(y \mid x)$ takes the particular form given above. The joint density function for $X$ and $Y$, $f_{X,Y}$, can be thought of as defining a surface over the $(x, y)$ plane. $f_{X,Y}$ is nonzero (above the plane) only over that set of points in the plane which has positive probability; the probability of occurrence of a region of points in the plane is given by the volume under $f_{X,Y}$ over the region. If we are given the information that $X = x$, then we are certain that some outcome was observed corresponding to points on the straight line $X = x$. It would seem reasonable to cut $f_{X,Y}$ with the plane $X = x$ and use the resulting cross-sectional curve to compute probabilities of $Y$ lying in various possible intervals. This is exactly what Definition 5.3.2(2) accomplishes; the cross-sectional curve resulting from this cut is $f_{X,Y}(x, y)$ where now $x$ is a fixed constant value and only $y$ is varying. Since

$$\int_{-\infty}^{\infty} f_{X,Y}(x, y)\, dy = f_X(x),$$

we must divide $f_{X,Y}(x, y)$ by $f_X(x)$ in order that the integral of $f_{Y \mid X}(y \mid x)$ over the range of $Y$ will equal 1 (so that we have a density function).

*Example 5.3.5.* Let us suppose that you and several of your friends have dealt frequently with the same used-car dealer. You pool your information and define $X$ to be the number of miles on the speedometer of a car this dealer will sell and define $Y$ to be the number of "trouble-free" miles remaining for the same car. Your pooled information indicates that $X$ and $Y$ have density function:

$$f_{X,Y}(x, y) = \tfrac{1}{85}, \qquad 3 < x < 21, \qquad \tfrac{1}{2} < y < 11\tfrac{1}{2} - \tfrac{1}{2}x$$
$$= 0, \qquad \text{otherwise}$$

where both $X$ and $Y$ are measured in thousands of miles. The marginal density of $X$ then is

$$f_X(x) = \int_{1/2}^{11\frac{1}{2}-x/2} \frac{1}{85}\,dy = \left(11 - \frac{x}{2}\right)\frac{1}{85}, \qquad \text{for } 3 < x < 21$$
$$= 0, \qquad \text{otherwise,}$$

and the conditional density of $Y$, given $X = x$, is

$$f_{Y\mid X}(y \mid x) = \frac{\dfrac{1}{85}}{\left(11 - \dfrac{x}{2}\right)\dfrac{1}{85}} = \frac{2}{22 - x}, \qquad \tfrac{1}{2} < y < 11\tfrac{1}{2} - \tfrac{1}{2}x$$
$$= 0, \qquad \text{otherwise.}$$

If you choose a car with 15 (thousand) miles on its speedometer, the conditional density of the number of miles of "trouble-free" driving remaining is

$$f_{Y\mid X}(y \mid 15) = \tfrac{2}{7}, \qquad \tfrac{1}{2} < y < 4$$
$$= 0, \qquad \text{otherwise,}$$

and you have probability $\frac{2}{7}$ of getting at least 3 (thousand) miles of "trouble-free" driving from the car. If, however, you choose a car with only 4 (thousand) miles on its speedometer, you have probability $\frac{13}{18}$ of getting at least 3 (thousand) "trouble-free" miles of driving.

*Example 5.3.6.* If $X$ and $Y$ are jointly distributed random variables, the mean value of $Y$, given that $X = x$, is called the regression of $Y$ on $X$. That is, assuming $X$ and $Y$ to be jointly continuous,

$$f_{Y\mid X}(y \mid x) = \frac{f_{X,Y}(x, y)}{f_X(x)}$$

and the regression of $Y$ on $X$ then is given by

$$E[Y \mid X = x] = \int_{-\infty}^{\infty} y\,\frac{f_{X,Y}(x, y)}{f_X(x)}\,dy$$
$$= \frac{1}{f_X(x)}\int_{-\infty}^{\infty} y f_{X,Y}(x, y)\,dy.$$

Notice that if we take the expected value of the regression of $Y$ on $X$ (acting like the given value of $X$ is a random variable), with respect to $X$ we get simply $\mu_Y$; i.e.,

$$E[E[Y \mid X = x]] = \int_{-\infty}^{\infty} \quad Y \mid X = x] f_X(x)\, dx$$

$$= \int_{-\infty}^{\infty} \int_{-\infty}^{\infty} y f_{X,Y}(x, y)\, dy\, dx = E[Y].$$

The same result holds for discrete random variables.

*Example 5.3.7.* This example illustrates how conditional probability functions and marginal probability functions can be used to construct a joint probability function as well as the computation of the regression of one variable on the other. Suppose the number of telephone calls that a busy executive receives per hour is a Poisson random variable $X$ with mean $\lambda = 10$. Also, this executive is busy with internal office affairs and, on the average, is away from his phone 15 minutes out of every hour. Thus he is willing to assume that if a call for him comes in the probability is $\frac{45}{60} = \frac{3}{4}$ that he will receive it. Furthermore, he is willing to assume that the fact he does or does not receive any particular call has no effect on whether or not he receives any other call (the receipts are independent). This last assumption would only be reasonable if he were popping in and out of the office for very short periods of time; if he were gone for relatively large blocks of time, he would certainly miss all arriving calls in that period and their receipts or nonreceipts would not be independent.

With this independence assumption, it is clear that $Y$, the number of calls he receives per hour, given that $X = x$ ($x$ calls were made to him), is a binomial random variable with parameters $x$ and $p = \frac{3}{4}$. Thus,

$$p_{Y \mid X}(y \mid x) = \binom{x}{y} \left(\frac{3}{4}\right)^y \left(\frac{1}{4}\right)^{x-y}, \qquad y = 0, 1, 2, \ldots, x.$$

The mean number of calls he would receive per hour then, given $x$ were made, is the regression of $Y$ on $X$ and we obviously have

$$E[Y \mid X = x] = \tfrac{3}{4}x.$$

The joint probability function for $X$ and $Y$ is

$$\begin{aligned} p_{X,Y}(x, y) &= p_{Y \mid X}(y \mid x) p_X(x) \\ &= \binom{x}{y} \left(\frac{3}{4}\right)^y \left(\frac{1}{4}\right)^{x-y} \frac{10^x}{x!} e^{-10}, \qquad x = 0, 1, 2, \ldots; \qquad y = 0, 1, \ldots, x \\ &= \frac{1}{y!} \left(\frac{30}{4}\right)^y e^{-10} \frac{1}{(x-y)!} \left(\frac{10}{4}\right)^{x-y}. \end{aligned}$$

Then, summing to get the marginal for $y$,

$$\begin{aligned}
p_Y(y) &= \sum_{x=y}^{\infty} \frac{1}{y!}\left(\frac{30}{4}\right)^y e^{-10} \frac{1}{(x-y)!}\left(\frac{10}{4}\right)^{x-y} \\
&= \frac{1}{y!}\left(\frac{30}{4}\right)^y e^{-10} \sum_{x-y=0}^{\infty} \frac{1}{(x-y)!}\left(\frac{10}{4}\right)^{x-y} \\
&= \frac{1}{y!}\left(\frac{30}{4}\right)^y e^{-10}\, e^{10/4} \\
&= \frac{1}{y!}\left(\frac{30}{4}\right)^y e^{-30/4}, \qquad y = 0, 1, 2, \ldots .
\end{aligned}$$

Thus, $Y$ itself is a Poisson random variable with parameter $\frac{30}{4} = 7\frac{1}{2}$ and we have

$$E[Y] = 7\tfrac{1}{2}.$$

As was shown to always be the case in Example 5.3.6,

$$\begin{aligned}
E[E[Y \mid X]] &= E[\tfrac{3}{4}X] \\
&= \tfrac{3}{4}E[X] \\
&= 7\tfrac{1}{2}.
\end{aligned}$$

We shall end this section with a theorem concerning independent random variables. The reader will recall that if two events $A$ and $B$ are independent, then the conditional probability of $A$ given $B$ is simply the probability of $A$; the analogous statement is true for independent random variables.

***Theorem 5.3.3.*** If $X$ and $Y$ are independent continuous random variables, then $f_{X\mid Y}(x \mid y) = f_X(x)$ for all $x$ and $y$ such that $f_Y(y) > 0$, and $f_{Y\mid X}(y \mid x) = f_Y(y)$ for all $y$ and $x$ such that $f_X(x) > 0$. (The analogous statements are true for probability functions for discrete random variables.)

*Proof:* By definition, if $X$ and $Y$ are independent, then

$$f_{X,Y}(x, y) = f_X(x) f_Y(y), \qquad \text{for all } (x, y).$$

Then

$$\begin{aligned}
f_{X\mid Y}(x \mid y) &= \frac{f_{X,Y}(x, y)}{f_Y(y)}, \qquad \text{if } f_Y(y) > 0 \\
&= \frac{f_X(x) f_Y(y)}{f_Y(y)} = f_X(x).
\end{aligned}$$

The second conclusion is justified in the same manner. ◀

## EXERCISE 5.3.

**1.** Given $X$ and $Y$ are jointly discrete random variables with

$$p_{X,Y}(x, y) = \frac{1}{n^2}, \qquad x = 1, 2, \ldots, n, \qquad y = 1, 2, \ldots, n$$
$$= 0, \qquad \text{otherwise,}$$

verify that $X$ and $Y$ are independent.

**2.** Given $X$ and $Y$ are jointly discrete random variables with

$$p_{X,Y}(x, y) = \frac{2}{n(n + 1)}, \qquad x = 1, 2, \ldots, n, \qquad y = 1, 2, \ldots, x$$
$$= 0, \qquad \text{otherwise,}$$

show that $X$ and $Y$ are not independent. Compute the regression of $Y$ on $X$ and the regression of $X$ on $Y$.

**3.** $X$ and $Y$ are jointly continuous with density function

$$f_{X,Y}(x, y) = 4, \qquad 0 < x < 1, \qquad 0 < y < \tfrac{1}{4}$$
$$= 0, \qquad \text{otherwise.}$$

Verify that $X$ and $Y$ are independent.

**4.** Suppose that $X$ and $Y$ have joint density

$$f_{X,Y}(x, y) = \tfrac{3}{2}, \qquad 0 < x < 1, \qquad -(x - 1)^2 < y < (x - 1)^2$$
$$= 0, \qquad \text{otherwise.}$$

Show that $X$ and $Y$ are not independent and compute the regression of $Y$ on $X$ and the regression of $X$ on $Y$.

**5.** Consider the population of people living in California. Suppose we select a person at random from California and let $X$ be the weight and $Y$ be the height of the selected person. Do you think $X$ and $Y$ are independent random variables?

**6.** Show that if $X$ and $Y$ are independent, then the regression of $Y$ on $X$ is a constant (not depending on $x$) and evaluate that constant value. If the regression of $Y$ on $X$ is a constant, are $X$ and $Y$ necessarily independent? If both the regression of $Y$ on $X$ and the regression of $X$ on $Y$ are constants, are $X$ and $Y$ necessarily independent?

**7.** Verify that the conditional densities and probability functions, as defined in Definition 5.3.2, satisfy the requirements for density and probability functions, respectively.

**8.** If the joint probability function for $(X, Y)$ is nonzero at exactly 3 points, what must be true for $X$ and $Y$ to be independent?

**9.** If $X$ and $Y$ are independent, jointly distributed, binomial random variables, each with parameters $n$ and $p$, write down their joint probability function $p_{X,Y}(x, y)$.

**10.** Assume that the number of deaths from automobile accidents in Montana is a Poisson random variable with parameter $\lambda = 2$ per day. If we let $X$ be the number of deaths that occur on Monday of a certain week and let $Y$ be the number of deaths that occur on Tuesday of the same week, what is the joint probability function for $X$ and $Y$?

**11.** Assuming that $X$ and $Y$ are independent normal random variables with parameters $\mu_X$, $\sigma_X$, $\mu_Y$, and $\sigma_Y$, respectively, what is the joint moment generating function for $(X, Y)$?

**12.** Let $X$ be the time you get out of bed in the morning (measured in fractions of an hour past 7 a.m.) and let $Y$ be the length of time it takes you to get to your place of business (in fractions of an hour) after getting up. Assume that the conditional density of $Y$ given $X = x$ is

$$
\begin{aligned}
f_{Y \mid X}(y \mid x) &= \frac{2y}{(1-x)^2}, \qquad 0 < x < \tfrac{2}{3}, \qquad 0 < y < 1 - x \\
&= 0, \qquad \text{otherwise}
\end{aligned}
$$

while the marginal of $X$ is

$$
\begin{aligned}
f_X(x) &= \tfrac{81}{26}(1-x)^2, \qquad 0 < x < \tfrac{2}{3} \\
&= 0, \qquad \text{otherwise.}
\end{aligned}
$$

Given that it took you 30 minutes to get to your place of business one morning (thus $Y = \frac{1}{2}$), compute the probability that you were out of bed by 7:15 a.m. that morning. Given that it took you 50 minutes to get there, what is the probability that you got up later than 7:20 a.m.?

**13.** Assume that $X_1$ and $X_2$ are independent random variables, each with mean zero and variance $\sigma^2$. Define

$$
\begin{aligned}
Y &= a_1X_1 + a_2X_2 \\
Z &= b_1X_1 + b_2X_2
\end{aligned}
$$

and compute $\mu_Y$, $\mu_Z$, $\sigma_Y^2$, $\sigma_Z^2$, $\sigma_{YZ}$, and the correlation between $Y$ and $Z$.

### 5.4. $n$-Dimensional Random Variables and the Multinomial Random Variable

As mentioned earlier, once the extension from one random variable to two has been made, the additional extension to any number, $n$, of jointly distributed random variables is easily made.

DEFINITION 5.4.1. Given an experiment, the $n$-tuple $(X_1, X_2, \ldots, X_n)$ is called an *$n$-dimensional random variable* if each $X_i$, $i = 1, 2, \ldots, n$, associates a real number with every experimental outcome $\omega \in S$.

Thus, an $n$-dimensional random variable is simply a rule associating an $n$-tuple of real numbers with every $\omega \in S$ (or a collection of $n$ rules each associating a single number with every $\omega \in S$). The study of $n$-dimensional random variables is in itself a large body of material. We shall concentrate mainly on the idea of independent random variables because these are very useful in our study of statistics. As we might imagine, we can speak of $n$-dimensional discrete random variables or $n$-dimensional continuous random variables (mixtures of the two can easily be defined but we shall not be concerned with them). The $n$-dimensional discrete random variable will be described by an $n$-dimensional probability function, denoted by

$$p_{X_1, X_2, \ldots, X_n}(x_1, x_2, \ldots, x_n)$$

(defined quite analogously to Definition 5.1.2), while the $n$-dimensional continuous random variable is described by a density function

$$f_{X_1, X_2, \ldots, X_n}(x_1, x_2, \ldots, x_n)$$

(analogous to Definition 5.1.3).

We shall continue to subscript the $f$ and $p$ with symbols representing the appropriate random variables, but to conserve space this notation will be abbreviated as follows: $\underline{X}$ will be written instead of $X_1, X_2, \ldots, X_n$ as a subscript for the $f$ or $p$. The number of variables $\underline{X}$ represents will be clear from the context. Similarly, instead of writing $(x_1, x_2, \ldots, x_n)$ as the $n$ arguments of $f_{\underline{X}}$ or $p_{\underline{X}}$, we shall use $(\underline{x})$. Thus in the remainder of this book

$$f_{\underline{X}}(\underline{x}) = f_{X_1, X_2, \ldots, X_n}(x_1, x_2, \ldots, x_n)$$

$$p_{\underline{X}}(\underline{x}) = p_{X_1, X_2, \ldots, X_n}(x_1, x_2, \ldots, x_n)$$

where $n$ will be clear from the context. As before, probability statements about discrete random variables are evaluated by summing the probability function appropriately, while probability statements about continuous random variables are evaluated by integrating the density function. The operation of computing the expected value of a function of an $n$-dimensional random variable extends readily from the two-dimensional case as well. Let us illustrate some of the foregoing discussion by considering the multinomial random variable.

DEFINITION 5.4.2. A *multinomial* trial, with parameters $p_1, p_2, \ldots, p_k$, is a trial which results in one of $k$ possible outcomes (these outcomes are called classes). The probability of the $i$-th class occurring on a single trial is $p_i$, $i = 1, 2, \ldots, k$; thus $0 \le p_i \le 1$, $i = 1, 2, \ldots, k$, and $\sum_{i=1}^{k} p_i = 1$.

A single roll of a single die is a multinomial trial (with $k = 6$) since every roll results in 1 of the 6 faces being uppermost. The grade a student gets in a statistics course can be thought of as a multinomial trial with $k = 5$ (if the only grades he may receive are A, B, C, D, and F). Clearly, a multinomial trial is simply a generalization of a binomial trial, having an arbitrary $k$ rather than just 2 possible outcomes. The multinomial random variable is defined as follows:

DEFINITION 5.4.3. Given an experiment which consists of $n$ repeated, independent, multinomial trials with parameters $p_i$, $i = 1, 2, \ldots, k$, let $X_i$ be the number of trials which result in outcomes in the $i$-th class, $i = 1, 2, \ldots, k$. $(X_1, X_2, \ldots, X_k)$ is called the *multinomial* random variable with parameters $n, p_1, p_2, \ldots, p_k$.

Notice that we have allowed some redundancy in the definition of the multinomial random variable. Since every trial must result in exactly 1 of the $k$ possible classes occurring, $\sum_{i=1}^{k} X_i = n$, so that inclusion of all $k$ is redundant. If we know the number of trials, $n$, and the values of $X_1, X_2, \ldots, X_{k-1}$, the value of $X_k$ is specified. (The listing of both the number of successes and the number of failures for repeated Bernoulli trials would be redundant in the same manner.)

***Theorem 5.4.1.*** If $(X_1, X_2, \ldots, X_k)$ is the multinomial random variable with parameters $n, p_1, p_2, \ldots, p_k$, then

$$p_{\underline{X}}(\underline{x}) = \frac{n!}{x_1!\, x_2! \cdots x_k!}\, p_1^{x_1} p_2^{x_2} \cdots p_k^{x_k},$$

for $x_i = 0, 1, 2, \ldots, n$, $i = 1, 2, \ldots, k$, such that $\sum_{i=1}^{k} x_i = n$ and the probability function is zero, otherwise. (Note that here $\underline{X}$ stands for $X_1, X_2, \ldots, X_k$ and $\underline{x}$ stands for $x_1, x_2, \ldots, x_k$.)

*Proof:* We are given an experiment consisting of $n$ repeated, independent, multinomial trials. Let $s_1, s_2, \ldots, s_k$ represent the outcome of a single trial falling into classes 1 through $k$, respectively. Then, if we define $S_i = \{s_1, s_2, \ldots, s_k\}$, $i = 1, 2, \ldots, n$, $S_i$ is clearly a sample space for a single trial and the sample space for the $n$ trials is $S = S_1 \times S_2 \times \cdots \times S_n$; the probability assigned to a single $n$-tuple in $S$ is the product of the probabilities of the appropriate outcomes on the individual trials. Let $X_i$ be the number of outcomes in the $i$-th class, $i = 1, 2, \ldots, k$. Then

$$P(X_1 = x_1, X_2 = x_2, \ldots, X_k = x_k) = p_{\underline{X}}(\underline{x})$$

is gotten by counting the number of $n$-tuples in $S$ containing exactly $x_1$ $s_1$'s,

$x_2$ $s_2$'s, . . . and exactly $x_k$ $s_k$'s and multiplying by $p_1^{x_1} p_2^{x_2} \cdots p_k^{x_k}$, the common probability of any such $n$-tuple occurring (given by our product rule for assigning probabilities to single-element events in $S$). The total number of such $n$-tuples is given by counting the number of ways we could lay out $n$ things in a row of which $x_1$ are of one kind, $x_2$ are of a second kind, . . . $x_k$ are of a $k$-th kind. The number of ways we could choose $x_1$ positions for the $s_1$'s is $\binom{n}{x_1}$; after having put the $s_1$'s in their positions, the number of ways we could choose positions for the $s_2$'s is $\binom{n - s_1}{s_2}$, etc. The total number of $n$-tuples with $x_1$ $s_1$'s, $x_2$ $s_2$'s, . . . , $x_k$ $s_k$'s then is

$$\binom{n}{s_1}\binom{n - s_1}{s_2}\binom{n - s_1 - s_2}{s_3} \cdots \binom{n - s_1 - s_2 \cdots - s_{k-1}}{x_k} = \frac{n!}{x_1!\, x_2! \cdots x_k!}$$

and we have

$$p_{\underline{X}}(\underline{x}) = \frac{n!}{x_1!\, x_2! \cdots x_k!} p_1^{x_1} p_2^{x_2} \cdots p_k^{x_k},$$

as was to be proved. ◀

*Example 5.4.1.* Suppose 6 fair dice are rolled 1 time. Let $X_i$ be the number of $i$'s that occur, $i = 1, 2, \ldots, 6$. Then $(X_1, X_2, \ldots, X_6)$ is the multinomial random variable with parameters

$$n = 6, \qquad p_1 = p_2 = \cdots = p_6 = \tfrac{1}{6}.$$

The probability that each number occurs exactly once then is

$$p_{\underline{X}}(1, 1, \ldots, 1) = \frac{6!}{(1!)^6}\left(\frac{1}{6}\right)^6 = \frac{6!}{6^6} = \frac{5}{54}.$$

The probability that exactly four 1's and two 2's occur is

$$p_{\underline{X}}(4, 2, 0, 0, 0, 0) = \frac{6!}{4!\,2!\,(0!)^4}\left(\frac{1}{6}\right)^4\left(\frac{1}{6}\right)^2 = \frac{5}{2592};$$

the probability that exactly two 3's, two 4's, and two 5's occur is

$$p_{\underline{X}}(0, 0, 2, 2, 2, 0) = \frac{6!}{(2!)^3(0!)^3}\left(\frac{1}{6}\right)^2\left(\frac{1}{6}\right)^2\left(\frac{1}{6}\right)^2 = \frac{15}{432}.$$

*Example 5.4.2.* A college sophomore is taking 5 courses in a semester. He is willing to assume that he will get an A with probability .1, a B with probability .8, and a C with probability .1 in each of the courses. He defines $X_1$, $X_2$, $X_3$ to be the number of A's, B's, and C's, respectively, that he gets in this semester. Then $X_1$,

$X_2, X_3$ is a multinomial random variable with parameters $n = 5, p_1 = .1, p_2 = .8, p_3 = .1$, and

$$p_{\underline{X}}(\underline{x}) = \frac{5!}{x_1!\,x_2!\,x_3!}(.1)^{x_1}(.8)^{x_2}(.1)^{x_3}.$$

The probability that he gets all B's then is

$$p_{\underline{X}}(0, 5, 0) = (.8)^5;$$

the probability that he gets 2 A's and 3 B's is

$$p_{\underline{X}}(2, 3, 0) = \frac{5!}{2!\,3!}(.1)^2(.8)^3.$$

The probability he gets no C's (all A's and B's) is $(.9)^5$.

***Theorem 5.4.2.*** If $X_1, X_2, \ldots, X_k$ is a multinomial random variable with parameters $n, p_1, p_2, \ldots, p_k$, then $X_1$ is a binomial random variable with parameters $n$ and $p_1$; similarly $X_i$ is binomial with parameters $n$ and $p_i$, $i = 2, 3, \ldots, k$.

*Proof:* Our experiment consists of $n$ repeated, independent trials, each with probability $p_1$ of class 1 occurring (class 1 is a success). Thus $X_1$, the number of trials resulting in a success, is a binomial random variable with parameters $n$ and $p_1$. The marginal distributions of $X_2, X_3, \ldots, X_k$ are derived in the same manner. ◀

As mentioned earlier, the concept of independent, $n$-dimensional random variables is quite important. These are defined as follows.

DEFINITION 5.4.4. (1) $X_1, X_2, \ldots, X_n$ are *independent* discrete *random variables* if and only if

$$p_{\underline{X}}(\underline{x}) = \prod_{i=1}^{n} p_{X_i}(x_i), \qquad \text{for all } (\underline{x}).$$

(2) $X_1, X_2, \ldots, X_n$ are *independent* continuous *random variables* if and only if

$$f_{\underline{X}}(\underline{x}) = \prod_{i=1}^{n} f_{X_i}(x_i), \qquad \text{for all } (\underline{x}).$$

This is a direct extension of the definition of independence of two random variables; their joint probability or density function must be equal to the product of their marginals. Thus, if $X_1, X_2, \ldots, X_n$ are independent, we need only know their marginal probability laws to be able to construct their joint probability law.

*Example 5.4.3.* The number of telephone calls a particular individual receives per day is a Poisson random variable with parameter $\lambda s = 8$. Let $X_1$ be the number of calls he receives on Monday, $X_2$ the number of calls he receives on Tuesday, and $X_3$ the number of calls he receives on Wednesday of a certain week. Since the numbers of occurrences in a Poisson process for nonoverlapping intervals are independent (see Definition 4.3.1), $X_1$, $X_2$, and $X_3$ are independent random variables; their joint probability function then is

$$\begin{aligned} p_{\underline{X}}(\underline{x}) &= p_{X_1}(x_1)p_{X_2}(x_2)p_{X_3}(x_3) \\ &= \frac{8^{x_1}}{x_1!}e^{-8}\frac{8^{x_2}}{x_2!}e^{-8}\frac{8^{x_3}}{x_3!}e^{-8} \\ &= \frac{8^{x_1+x_2+x_3}}{x_1!\,x_2!\,x_3!}e^{-24}, \qquad \text{for } x_1 = 0, 1, 2, \ldots; \\ x_2 &= 0, 1, 2, \ldots; \qquad x_3 = 0, 1, 2, \ldots. \end{aligned}$$

The probability that he receives exactly 1 call on each of these 3 days then is

$$p_{\underline{X}}(1, 1, 1) = \frac{8^3}{(1!)^3}e^{-24} = .193 \times 10^{-7};$$

the probability that he gets exactly 7 calls on Monday, 8 calls on Tuesday, and 9 calls on Wednesday is

$$p_{\underline{X}}(7, 8, 9) = \frac{8^{24}}{7!\,8!\,9!}e^{-24} = .00242.$$

*Example 5.4.4.* The "true" weight of a package of butter produced by a certain dairy is a normal random variable with parameters $\mu = 1.01$ pounds, $\sigma = .02$ pound. When doing your weekly grocery shopping, you buy 4 pounds of this dairy's butter. Let $X_1, X_2, X_3, X_4$ be the "true" weights of the 4 packages you buy. If we assume that each of these 4 has the normal probability law mentioned above and that the weights are independent (weights of successive packages produced vary independently), the joint probability density function of the weights of the 4 packages you buy is

$$\begin{aligned} f_{\underline{X}}(\underline{x}) &= f_{X_1}(x_1)f_{X_2}(x_2)f_{X_3}(x_3)f_{X_4}(x_4) \\ &= \frac{1}{\sigma\sqrt{2\pi}}e^{-(x_1-\mu)^2/2\sigma^2} \cdot \frac{1}{\sigma\sqrt{2\pi}}e^{-(x_2-\mu)^2/2\sigma^2} \\ &\quad \cdot \frac{1}{\sigma\sqrt{2\pi}}e^{-(x_3-\mu)^2/2\sigma^2} \cdot \frac{1}{\sigma\sqrt{2\pi}}e^{-(x_4-\mu)/2\sigma^2} \\ &= \left(\frac{1}{\sigma\sqrt{2\pi}}\right)^4 e^{-\sum_{i=1}^{4}(x_i-\mu)^2/2\sigma^2} \end{aligned}$$

where $\mu = 1.01$, $\sigma = .02$. The probability that all 4 packages weigh at least a pound is

$$\begin{aligned} P(X_1 \geq 1, X_2 \geq 1, X_3 \geq 1, X_4 \geq 1) &= P(X_1 \geq 1)P(X_2 \geq 1)P(X_3 \geq 1)P(X_4 \geq 1) \\ &= (.6915)^4. \end{aligned}$$

*Example 5.4.5.* We saw in Theorem 5.4.2 that if $X_1, X_2, \ldots, X_k$ is the multinomial random variable with parameters $n, p_1, p_2, \ldots, p_k$, then $X_i$ is a binomial random variable with parameters $n$ and $p_i$, $i = 1, 2, \ldots, k$. Then we know immediately that $X_1, X_2, \ldots, X_k$ are not independent random variables since

$$\begin{aligned} p_{\underline{X}}(\underline{x}) &= \frac{n!}{x_1! \cdots x_k!} p_1^{x_1} \cdots p_k^{x_k} \\ &\neq \binom{n}{x_1} p_1^{x_1}(1 - p_1)^{n-x_1} \cdots \binom{n}{x_k} p_k^{x_k}(1 - p_k)^{n-x_k} \\ &= \prod_{i=1}^{k} p_{X_i}(x_i). \end{aligned}$$

In Examples 5.4.3 and 5.4.4, the random variables (within the same example) had the same marginal probability laws. These are examples of identically distributed random variables; the definition of this concept follows.

DEFINITION 5.4.5. (1) $X_1, X_2, \ldots, X_n$ are *identically distributed* discrete *random variables* if and only if

$$p_{X_i}(x) = p_X(x), \qquad \text{for all } x, \qquad i = 1, 2, \ldots, n.$$

(2) $X_1, X_2, \ldots, X_n$ are *identically distributed* continuous *random variables* if and only if

$$f_{X_i}(x) = f_X(x), \qquad \text{for all } x, \qquad i = 1, 2, \ldots, n.$$

Thus, in brief, a group of random variables are identically distributed if and only if each of their marginals is the same (then all of their moment generating functions are also equal). When we consider the concept of sampling, identically distributed random variables will play a central role.

The following theorem gives the form of the moment generating function of the sum of independent, identically distributed random variables.

***Theorem 5.4.3.*** Suppose that $X_1, X_2, \ldots, X_n$ are independent, identically distributed (discrete or continuous) random variables and define $Y = \sum_{i=1}^{n} X_i$. The moment generating function for $Y$ then is

$$m_Y(t) = [m_X(t)]^n,$$

where $m_X(t)$ is the common moment generating function of each of $X_1, X_2, \ldots, X_n$.

*Proof:* We have, by definition,

$$\begin{aligned} m_Y(t) &= E[e^{tY}] \\ &= E[e^{t\sum_{i=1}^{n} X_i}] \\ &= E[e^{t\sum_{i=1}^{n} X_i}] = E[e^{tX_1}e^{tX_2}\cdots e^{tX_n}] \\ &= E[e^{tX_1}]E[e^{tX_2}]\cdots E[e^{tX_n}] \\ &= m_{X_1}(t)m_{X_2}(t)\cdots m_{X_n}(t) = [m_X(t)]^n, \end{aligned}$$

since $m_{X_i}(t) = m_X(t)$, $i = 1, 2, \ldots, n$. ◀

Using this result, coupled with Theorem 3.4.1 (which says that the moment generating function uniquely characterizes the distribution function for the random variable), we can easily establish several distributional results for sums of independent, identically distributed random variables.

***Theorem 5.4.4.*** $X_1, X_2, \ldots, X_n$ are independent Bernoulli random variables, each with parameter $p$. Then $Y = \sum_{i=1}^{n} X_i$ is a binomial random variable with parameters $n$ and $p$.

*Proof:* The Bernoulli moment generating function is $m_X(t) = q + pe^t$ (see Section 4.1). Then, from Theorem 5.4.2,

$$m_Y(t) = [m_X(t)]^n = (q + pe^t)^n.$$

But this is the moment generating function of the binomial random variable with parameters $n$ and $p$ (see Section 4.1). Thus $Y$ is a binomial random variable with parameters $n$ and $p$. ◀

***Theorem 5.4.5.*** $X_1, X_2, \ldots, X_n$ are independent, identically distributed, normal random variables, each with parameters $\mu$ and $\sigma^2$. Then $Y = \sum_{i=1}^{n} X_i$ is a normal random variable with parameters $n\mu$ and $n\sigma^2$.

*Proof:* The normal moment generating function is

$$m_X(t) = e^{t\mu + t^2\sigma^2/2}$$

(see Section 4.5). Then, by Theorem 5.4.2,

$$m_Y(t) = [e^{t\mu + t^2\sigma^2/2}]^n = e^{t(n\mu) + (t^2/2)(n\sigma^2)}.$$

The moment generating function of $Y$ is therefore of the form of a normal moment generating function with parameters $n\mu$ and $n\sigma^2$, from which the conclusion follows. ◀

The result given in Theorem 5.4.5 is called the reproductive property of the normal probability law; sums of independently distributed normal random variables reproduce the same form of probability law. This property is also shared by the binomial and the Poisson random variables, as you are asked to show in the problems below.

*Example 5.4.6.* Suppose that, as in Example 5.4.4, the weight of a package of butter produced by a certain dairy is a normal random variable with mean $\mu = 1.01$ pounds and $\sigma = .02$ pound. The dairy ships the butter to retailers in boxes of 48 packages. Let $X_1, X_2, \ldots, X_{48}$ be the weights of the individual packages which will be put in the same box. Then the total weight of the box of 48 packages is $Y = \sum_{i=1}^{48} X_i$. If we assume that the weights of the individual boxes are independent, the total weight of the box, $Y$, is also a normal random variable with parameters $\mu_Y = n\mu = 48(1.01) = 48.48$ and $\sigma_Y{}^2 = 48(.02)^2 = .0192$. Thus, $\sigma_Y = \sqrt{.0192} = .139$. Therefore, the probability that a box of 48 packages weighs at least 48 pounds is $P(Y > 48) = .9997$. The probability that such a box weighs less than $48\frac{1}{2}$ pounds is $P(Y < 48.5) = .5557$.

Let us now make note of a theorem regarding the mean and the variance of a linear function of independent random variables. (The random variable $Y = \sum_{i=1}^{n} a_i X_i$ is called a linear function of $X_1, X_2, \ldots, X_n$ if $a_1, a_2, \ldots, a_n$ are any arbitrary set of constants.)

***Theorem 5.4.6.*** $X_1, X_2, \ldots, X_n$ are independent random variables with means $\mu_1, \mu_2, \ldots, \mu_n$ and variances $\sigma_1{}^2, \sigma_2{}^2, \ldots, \sigma_n{}^2$, respectively. If $Y = \sum_{i=1}^{n} a_i X_i$, where the $a_i$'s are arbitrary constants, then $\mu_Y = \sum_{i=1}^{n} a_i\mu_i$ and $\sigma_Y{}^2 = \sum_{i=1}^{n} a_i{}^2\sigma_i{}^2$.

*Proof:*

$$\begin{aligned}\mu_Y = E[Y] &= E\left[\sum_{i=1}^{n} a_i X_i\right]\\ &= \sum_{i=1}^{n} E[a_i X_i]\\ &= \sum_{i=1}^{n} a_i E[X_i]\\ &= \sum_{i=1}^{n} a_i \mu_i.\end{aligned}$$

(Note that this part of the theorem does not depend on the assumed independence.)

$$\begin{aligned}
\sigma_Y{}^2 = E[(Y-\mu_Y)^2] &= E\left[\left(\sum_{i=1}^{n} a_i X_i - \sum_{i=1}^{n} a_i\mu_i\right)^2\right] \\
&= E\left[\left(\sum_{i=1}^{n} a_i(X_i-\mu_i)\right)^2\right] \\
&= E\left[\sum_{i=1}^{n} a_i{}^2(X_i-\mu_i)^2 + 2\sum_{\substack{i \\ i<j}}\sum_{j} a_i a_j (X_i-\mu_i)(X_j-\mu_j)\right] \\
&= \sum_{i=1}^{n} a_i{}^2 E[(X_i-\mu_i)^2] + 2\sum_{\substack{i \\ i<j}}\sum_{j} a_i a_j E[(X_i-\mu_i)(X_j-\mu_j)] \\
&= \sum_{i=1}^{n} a_i{}^2\sigma_i{}^2.
\end{aligned}$$

◀

Every one of the terms of the double sum of cross products is zero since each is a constant times the covariance of 2 of the variables in $X_1, X_2, \ldots, X_n$. Each of these covariances is zero because of the independence of $X_1, X_2, \ldots, X_n$.

*Example 5.4.7.* Suppose that $X_1, X_2, \ldots, X_n$ are independent, identically distributed random variables, each with mean $\mu$ and variance $\sigma^2$. Let $\bar{X}$ be the arithmetic average of these $n$ random variables; that is,

$$\bar{X} = \frac{1}{n}\sum_{i=1}^{n} X_i \left( = \sum_{i=1}^{n} \frac{1}{n} X_i \right).$$

Thus $\bar{X}$ is a linear function of $X_1, X_2, \ldots, X_n$ with $a_i = 1/n$ for all $i$. Then, from Theorem 5.4.5,

$$\mu_{\bar{X}} = \sum_{i=1}^{n} \frac{1}{n}\mu = \mu$$

$$\sigma_{\bar{X}}{}^2 = \sum_{i=1}^{n} \left(\frac{1}{n}\right)^2 \sigma^2 = \frac{\sigma^2}{n}.$$

Thus the expected value of $\bar{X}$ is the same as the common expected value of the $X_i$'s and the variance of $\bar{X}$ is $1/n$ times the common variance of the $X$'s, which means that $\bar{X}$ is considerably less variable than the original $X_i$'s. This fact has considerable impact on the theory of statistics.

**EXERCISE 5.4.**

**1.** Suppose that you and a friend gamble for coffee on each of 20 successive days. Define:

$$\begin{aligned} X_i &= 1, \quad \text{if you win on the } i\text{-th day} \\ &= 0, \quad \text{if you lose on the } i\text{-th day,} \end{aligned}$$

for $i = 1, 2, \ldots, 20$. Then $Y = \sum_{i=1}^{20} X_i$ is the total number of times you win in the 20 days. Assuming the game you are playing is fair (that is the probability of your winning is $\frac{1}{2}$ for each day) and that the results are independent from day to day, compute the probability that $Y \geq 10$ and that $Y \geq 15$.

**2.** (a) If $X_1, X_2, X_3$ are independent Poisson random variables, each with parameter $\lambda s$, write down their joint probability function $p_{X_1,X_2,X_3}(x_1, x_2, x_3)$.

(b) If $X_1, X_2, X_3$ are independent Poisson random variables with parameters $\lambda_1 s, \lambda_2 s, \lambda_3 s$, respectively, write down their joint probability function

$$p_{X_1,X_2,X_3}(x_1, x_2, x_3).$$

**3.** The distance a football referee covers in 1 step is a normal random variable with mean $\mu = 1.05$ yards and standard deviation $\sigma = .1$ yard. Assume that this referee steps off a 5-yard penalty against your team (the distances covered by successive steps are independent). What is the probability that the total distance your team is penalized exceeds 5 yards?

**4.** Assume that $X_1, X_2, \ldots, X_n$ are independent random variables with moment generating functions $m_{X_1}(t), m_{X_2}(t), \ldots, m_{X_n}(t)$, respectively. If $Y = \sum_{i=1}^{n} X_i$, show that $m_Y(t) = \prod_{i=1}^{n} m_{X_i}(t)$.

**5.** Using the result of problem 4, show that if $X_1, X_2, \ldots, X_n$ are independent normal random variables with parameters $\mu_1, \sigma_1^2, \mu_2, \sigma_2^2, \ldots, \mu_n, \sigma_n^2$, respectively, then $Y = \sum_{i=1}^{n} X_i$ is a normal random variable with parameters $\mu_Y = \sum_{i=1}^{n} \mu_i$, $\sigma_Y^2 = \sum_{i=1}^{n} \sigma_i^2$.

**6.** If $X_1, X_2, \ldots, X_k$ are independent binomial random variables with parameters $n_1, p, n_2, p, \ldots, n_k, p$, respectively, show that $Y = \sum_{i=1}^{k} X_i$ is a binomial random variable with parameters $\sum_{i=1}^{k} n_i$ and $p$. (Use the result of problem 4.)

**7.** Show that if $X_1, X_2, \ldots, X_n$ are independent Poisson random variables with parameters $\lambda_1 s_1, \lambda_2 s_2, \ldots, \lambda_n s_n$, respectively, then $Y = \sum_{i=1}^{n} X$ is a Poisson random variable with parameter $\sum_{i=1}^{n} \lambda_i s_i$. (Use problem 4.)

**8.** Suppose that $X_1, X_2, \ldots, X_n$ are independent, identically distributed, exponential random variables with parameter $\lambda$. Show that $Y = \sum_{i=1}^{n} X_i$ has moment generating function $m_Y(t) = \lambda^n/(\lambda - t)^n$. (Use Theorem 5.4.2; this is the moment-generating function of a gamma random variable with parameters $n$ and $\lambda$. See Appendix 4.)

**9.** $X_1$ and $X_2$ are independent, identically distributed random variables, each with mean $\mu$ and variance $\sigma^2$. Show that $V = X_1 - X_2$ has $\mu_V = 0$, $\sigma_V^2 = 2\sigma^2$. (Use Theorem 5.4.5.)

**10.** $X_1$ and $X_2$ are independent, identically distributed random variables, each with mean $\mu$ and variance $\sigma^2$. Define $Y = a_1X_1 + a_2X_2$. Show that if we restrict $a_1$ and $a_2$ to values such that $\mu_Y = \mu$, then the variance of $Y$ is minimized with $a_1 = a_2 = \frac{1}{2}$. (The condition that $\mu_Y = \mu$ implies $a_1 + a_2 = 1$; i.e., $a_2 = 1 - a_1$. Use this value for $a_2$ in ${\sigma_Y}^2$ and differentiate with respect to $a_1$.)

**11.** Generalize problem 10. If $X_1, X_2, \ldots, X_n$ are independent random variables, each with mean $\mu$ and variance $\sigma^2$, show that the variance of $Y = \sum_{i=1}^{n} a_iX_i$ is minimized with $a_i = 1/n$, $i = 1, 2, \ldots, n$, if we require that $\mu_Y = \mu$.

**12.** Given:

$$\begin{aligned} E[X_i] &= \mu_i, & i &= 1, 2, \ldots, n \\ E[(X_i - \mu_i)^2] &= \sigma_i^2, & i &= 1, 2, \ldots, n \\ \operatorname{cov}(X_i, X_j) &= \sigma_{ij}, & i &\neq j, \end{aligned}$$

show that

$$\mu_Y = \sum_{i=1}^{n} a_i\mu_i$$

$${\sigma_Y}^2 = \sum_{i=1}^{n} {a_i}^2{\sigma_i}^2 + 2\sum_{i<j}\sum a_ia_j\sigma_{ij}$$

where

$$Y = \sum_{i=1}^{n} a_iX_i.$$

**13.** Suppose that $Y = X_1 + X_2$; is it possible to have ${\sigma_Y}^2 < {\sigma_{X_1}}^2$ and ${\sigma_Y}^2 < {\sigma_{X_2}}^2$?

## 5.5. The Chebychev Inequality and the Law of Large Numbers

The Chebychev inequality gives some insight into the fact that the standard deviation of a random variable is a rather natural unit for the probability law of a random variable. It was first derived by the Russian mathematician Chebychev in 1867. As we shall see, the inequality gives a bound on the probability that a random variable will be within $k$ standard deviations of its mean. The result is presented as Theorem 5.5.1.

***Theorem 5.5.1.*** (Chebychev) If $X$ is a random variable with mean $\mu_X$ and standard deviation $\sigma_X$, then

$$P(|X - \mu_X| < k\sigma_X) \geq 1 - \frac{1}{k^2}$$

and

$$P(|X - \mu_X| \geq k\sigma_X) \leq \frac{1}{k^2}.$$

(The second of these two statements is simply the complement of the first.)

*Proof:* We shall give the proof for the continuous case only. For a discrete random variable the same line of approach is appropriate. By definition,

$$\begin{aligned}\sigma_X^2 &= \int_{-\infty}^{\infty} (x - \mu_x)^2 f_X(x)\,dx \\ &= \left\{\int_{-\infty}^{\mu_X - k\sigma_X} + \int_{\mu_X - k\sigma_X}^{\mu_X + k\sigma_X} + \int_{\mu_X + k\sigma_X}^{\infty}\right\}(x - \mu_X)^2 f_X(x)\,dx \\ &\geq \int_{-\infty}^{\mu_X - k\sigma_X} (x - \mu_X)^2 f_X(x)\,dx + \int_{\mu_X + k\sigma_X}^{\infty} (x - \mu_X)^2 f_X(x)\,dx,\end{aligned}$$

where the inequality is appropriate because, by deleting the middle of these three integrals, we have decreased the right-hand side ($k$ is any positive constant). In the second of the two remaining integrals the range of integration is $x > \mu_X + k\sigma_X$, which is equivalent to $x - \mu_X > k\sigma_X$ or $(x - \mu_X)^2 > k^2\sigma_X^2$. In the first of the two remaining integrals the range of integration is $x < \mu_X - k\sigma_X$, which is equivalent to $x - \mu_X < -k\sigma_X$ or again $(x - \mu_X)^2 > k^2\sigma_X^2$ (the inequality is reversed with squaring since the quantities are both negative). Thus, in each of the two remaining integrals we have $(x - \mu_X)^2 > k^2\sigma_X^2$. Therefore, the inequality is preserved if we replace $(x - \mu_X)^2$ by $k^2\sigma_X^2$, which yields

$$\begin{aligned}\sigma_X^2 &\geq \int_{-\infty}^{\mu_X - k\sigma_X} k^2\sigma_X^2 f_X(x)\,dx + \int_{\mu_X + k\sigma_X}^{\infty} k^2\sigma_X^2 f_X(x)\,dx \\ &= k^2\sigma_X^2\left\{\int_{-\infty}^{\mu_X - k\sigma_X} f_X(x)\,dx + \int_{\mu_X + k\sigma_X}^{\infty} f_X(x)\,dx\right\} \\ &= k^2\sigma_X^2 P[X \leq \mu_X - k\sigma_X \text{ or } X \geq \mu_X + k\sigma_X] \\ &= k^2\sigma_X^2 P[|X - \mu_X| \geq k\sigma_X].\end{aligned}$$

Thus we have established

$$\sigma_X^2 \geq k^2\sigma_X^2 P[|X - \mu_X| \geq k\sigma_X]$$

$$\frac{1}{k^2} \geq P[|X - \mu_X| \geq k\sigma_X].$$

Subtracting both sides from 1 yields

$$P[|X - \mu_X| < k\sigma_X] \geq 1 - \frac{1}{k^2},$$

and the result is established. ◀

Notice in particular that we make no distributional assumption about $X$ when deriving the Chebychev inequality (other than the assumption that $\mu_X$ and $\sigma_X^2$ exist), and we are still able to place a bound on the probability that

$X$ is within $k$ standard deviations of its mean. The following example compares some exact probabilities for the normal and exponential laws with the bound given by the Chebychev inequality.

*Example 5.5.1.* Suppose that $X$ is a normal random variable with mean $\mu_X$ and variance $\sigma_X{}^2$. Then

$$\begin{aligned} P[|X - \mu_X| < k\sigma_X] &= P\left[\frac{|X - \mu_X|}{\sigma_X} < k\right] \\ &= N_Z(k) - N_Z(-k) \\ &= 2N_Z(k) - 1. \end{aligned}$$

If $X$ is an exponential random variable with parameter $\lambda$, then $\mu_X = \sigma_X = 1/\lambda$ and

$$\begin{aligned} P(|X - \mu_X| < k\sigma_X) &= \int_{(1-k)/\lambda}^{(1+k)/\lambda} \lambda e^{-\lambda x}\, dx \\ &= e^{-(1-k)} - e^{-(1+k)}, \qquad \text{for } 0 \le k \le 1 \end{aligned}$$

and

$$\begin{aligned} P(|X - \mu_X| < k\sigma_X) &= \int_0^{(1+k)/\lambda} \lambda e^{-\lambda x}\, dx \\ &= 1 - e^{-(1+k)}, \qquad \text{for } k > 1. \end{aligned}$$

The following table compares these two exact probabilities with the bound provided by the Chebychev inequality for selected values of $k$. For these two

$P(|\, X - \mu_X| \ < k_X\sigma)$

| $k$ | Normal | Exponential | Chebychev Bound |
|---|---|---|---|
| 1 | .6826 | .8647 | 0 |
| 1.5 | .8664 | .9179 | .5556 |
| 2 | .9546 | .9502 | .75 |
| 3 | .9974 | .9817 | .8889 |
| 4 | 1.0000 | .9933 | .9334 |

probability laws the exact probability is considerably higher than the bound, especially for relatively small values of $k$. But then the exact probabilities are resting on the stronger assumption of knowing the exact probability law for $X$. Problem 2 in Exercise 5.5 shows a case in which the Chebychev bound is actually the exact probability; thus the bound is attained in some cases.

We should note that the Chebychev inequality says that

$$P(|X - \mu_X| < k\sigma_X) \ge 0$$

for any $k \le 1$, a statement we already knew to be true. Thus, for any $k \le 1$

the bound does not add anything useful for describing random variables. For $k \geq 1$, however, it is at times of use.

Let us now discuss the law of large numbers. If we were to toss a fair coin a large number of times, $n$, and let $X_n$ be the number of heads that we observe, it seems that $X_n/n$ should be close to $\frac{1}{2}$. In fact, it seems that the larger that $n$ becomes, the closer this ratio should be to $\frac{1}{2}$. We cannot say with certainty that $X_n/n$ will ever be $\frac{1}{2}$; we can, however, show that the probability it differs from $\frac{1}{2}$ by any small amount tends to zero as $n$ increases. This is a special case of what is called the law of large numbers, presented as Theorem 5.5.2.

***Theorem 5.5.2.*** (Law of Large Numbers) Suppose that $X_1, X_2, \ldots, X_k, \ldots$ is a sequence of independent, identically distributed random variables, each with mean $\mu$ and variance $\sigma^2$. Define the new sequence of $\bar{X}_i$ values by

$$\bar{X}_n = \frac{1}{n}\sum_{i=1}^{n} X_i, \qquad n = 1, 2, 3, \ldots$$

(Thus, $\bar{X}_1 = X_1$, $\bar{X}_2 = \frac{1}{2}(X_1 + X_2)$, $\bar{X}_3 = \frac{1}{3}(X_1 + X_2 + X_3)$, etc.) Then,

$$\lim_{n\to\infty} P(|\bar{X}_n - \mu| > \varepsilon) = 0, \qquad \text{for any } \varepsilon > 0.$$

*Proof:* We saw in Section 5.4 that for any finite $n$

$$\mu_{\bar{X}_n} = E[\bar{X}_n] = \mu, \qquad \sigma_{\bar{X}_n}{}^2 = E[(\bar{X}_n - \mu)^2] = \frac{\sigma^2}{n}.$$

By Chebychev's inequality,

$$P[|\bar{X}_n - \mu_{\bar{X}_n}| > k\sigma_{\bar{X}_n}] \leq \frac{1}{k^2}$$

for any $k > 0$. Then

$$P\left[|\bar{X}_n - \mu| > \frac{k\sigma}{\sqrt{n}}\right] \leq \frac{1}{k^2}.$$

Choose $k = \varepsilon\sqrt{n}/\sigma$ for any given $\varepsilon > 0$. Then

$$P[|\bar{X}_n - \mu| > \varepsilon] \leq \frac{\sigma^2}{\varepsilon^2 n}$$

and

$$\lim_{n\to\infty} P[|\bar{X}_n - \mu| > \varepsilon] \leq \lim_{n\to\infty} \frac{\sigma^2}{\varepsilon^2 n} = 0.$$

Since all of the terms in this sequence of probabilities must be nonnegative, we clearly have

$$\lim_{n\to\infty} P[|\bar{X}_n - \mu| > \varepsilon] = 0,$$

no matter how small $\varepsilon$ may be. ◀

It is important to realize that the law of large numbers gives the limit of a sequence of probability statements, not the limit of a sequence of random variables. It does not, for example, say that $\bar{X}_n$ is necessarily getting closer and closer to $\mu$; it does say that the probability tends to zero that $\bar{X}_n$ differs from $\mu$ by more than $\varepsilon$, no matter how small $\varepsilon$ may be. Let us consider the application of this theorem to the tossing of a fair coin, as mentioned earlier.

*Example 5.5.2.* Suppose that we are given a fair coin. Conceivably we could flip this coin any number of times that we like. In particular, we could define $X_i = 1$ if we get a head on the $i$-th flip and $X_i = 0$ if we get a tail on the $i$-th flip for $i = 1, 2, 3, \ldots$. Then we have defined a sequence of independent Bernoulli random variables, $X_1, X_2, \ldots$, each with parameter $p = \frac{1}{2}$. We can then define the sequence of proportions of heads that we observe:

$$\bar{X}_n = \frac{1}{n}\sum_{i=1}^{n} X_i, \qquad \text{for } n = 1, 2, 3, \ldots.$$

In this case $\mu_{\bar{X}_n} = p = \frac{1}{2}$ and $\sigma_{\bar{X}}^2 = pq/n = 1/4n$. Then, by the law of large numbers, $\lim_{n\to\infty} P[|\bar{X}_n - \frac{1}{2}| > \varepsilon] = 0$; that is, the probability that the proportion of flips on which we observe a head differs from $\frac{1}{2}$ by more than $\varepsilon$ tends to 0 as the number of flips increases indefinitely. We do not claim that the proportion itself will necessarily get closer and closer to $\frac{1}{2}$ because, in particular, we might get a head on every flip. The probability of this latter event happening is shrinking to zero, however.

*Example 5.5.3.* Suppose that we have a number $n$ of vacuum tubes, all made by the same manufacturer and quite identical. If the time to failure, $X_i$, for the $i$-th tube is assumed to be an exponential random variable with parameter $\lambda$ for each $i$, then the following is true. If we put all $n$ of these tubes on test until they failed (independently) and we defined

$$\bar{X} = \frac{1}{n}\sum_{i=1}^{n} X_i,$$

then

$$\lim_{n\to\infty} P\left[\left|\bar{X} - \frac{1}{\lambda}\right| > \varepsilon\right] = 0.$$

Thus, if we tested a large number $n$ of tubes and computed the arithmetic average $\bar{X}$ of their failure times (actually we could only compute $\bar{x}$, the observed value of $\bar{X}$), the probability is high that $|\bar{X} - 1/\lambda|$ is small. If we didn't know $\lambda$, it would seem plausible to use $1/\bar{X}$ as a guess of its value.

**EXERCISE 5.5.**

**1.** If $X$ is uniformly distributed on the interval (0, 1), compare $P(|X - \mu_X| < k\sigma_X)$ with the values given by the Chebychev inequality for $k = 1\frac{1}{4}, 1\frac{1}{2}, 1\frac{3}{4}$, and 2.

**2.** For any value of $k \geq 1$, we can define a discrete random variable $X$ to have probability function

$$p_X(x) = \frac{k^2 - 1}{k^2}, \qquad x = 0$$
$$= \frac{1}{2k^2}, \qquad x = -k, k$$
$$= 0, \qquad \text{otherwise.}$$

Compute $\mu_X$ and $\sigma_X$, and compare the exact probability $P(|X - \mu_X| < k\sigma_X)$ with the bound given by Chebychev's inequality.

**3.** Suppose that $X_1, X_2, X_3, \ldots$ are independent, identically distributed random variables each with probability function

$$p_X(x) = \tfrac{1}{2}, \qquad \text{for } x = -1, 1$$
$$= 0, \qquad \text{otherwise.}$$

Define the sequence of averages

$$\bar{X}_n = \frac{1}{n}\sum_{i=1}^{n} X_i$$

and show that the probability function for $\bar{X}_n$ is

$$p_{\bar{X}_n}(x) = \frac{\binom{n}{\frac{n}{2}(x+1)}}{2^n}, \qquad \text{for } x = \pm\frac{1}{n}, \pm\frac{3}{n}, \ldots, \pm\frac{n-2}{n}, \pm 1,$$
$$= 0, \qquad \text{otherwise.}$$

(Assume that $n$ is an odd number.)

**4.** Derive the mean and standard deviation for the probability function for $\bar{X}_n$ given in problem 3. (**Hint:** Try $n = 1, 3, 5$ to get an idea of the general form.)

**5.** If you assume that $X_1, X_2, \ldots, X_n$ are independent, identically distributed, Poisson random variables, each with parameter $\lambda s$ (where $n$ is quite large), what function of $X_1, X_2, \ldots, X_n$ might you guess would be close in value to $\lambda s$? (See Example 5.5.3.)

**6.** If you assume that $X_1, X_2, \ldots, X_n$ ($n$ large) are independent, identically distributed, geometric random variables with parameter $p$, what function of $X_1, X_2, \ldots, X_n$ would you guess should be close in value to $p$?

## 5.6. The Central Limit Theorem; Approximations

We have already seen the normal probability density function and several examples of its occurrence. Even if the normal random variable never occurred in practice, the normal distribution function would still be one of the most important of all probability laws because of its role in the central limit theorem. It can be shown that, under a wide range of conditions, the distribution function for the sum of random variables approaches the normal distribution function as the number of variables in the sum increases. One version of the central limit theorem is given as Theorem 5.6.1. (The assumptions regarding the random variables $X_1$, $X_2$, $X_3$, ... can be weakened in several ways and still the conclusion follows.)

***Theorem 5.6.1.*** (Central Limit Theorem) $X_1$, $X_2$, $X_3$, ... is a sequence of independent, identically distributed random variables, each with mean $\mu$ and variance $\sigma^2$. Define the sequence of random variables $Z_1, Z_2, Z_3, \ldots$ by

$$Z_n = \frac{\bar{X}_n - \mu}{\sigma/\sqrt{n}}, \qquad n = 1, 2, 3, \ldots,$$

where

$$\bar{X}_n = \frac{1}{n} \sum_{i=1}^{n} X_i.$$

Then, for all real $t$,

$$\lim_{n \to \infty} F_{Z_n}(t) = N_Z(t)$$

where $N_Z$ is the standard normal distribution function.

*Proof:* The following heuristic argument is presented for the case in which the moment generating function, $m_X(t)$, exists for the random variables $X_i$ in the original sequence. We shall show that $\lim_{n\to\infty} m_{Z_n}(t) = e^{t^2/2}$, the moment generating function of the standard normal distribution function. It can be shown that if the moment generating functions converge to a certain form (such as the above), the distribution functions of the $Z_n$ converge to the corresponding distribution function. Now, for any $n$, we have seen that $E[\bar{X}_n] = \mu$, $E[(\bar{X}_n - \mu)^2] = \sigma^2/n$. Thus $E[Z_n] = 0$, $E[Z_n{}^2] = 1$, for any $n$ where $Z_n$ is the standard form for $\bar{X}_n$ (see Definition 3.5.1). Now,

$$\begin{aligned}
m_{Z_n}(t) &= E[e^{tZ_n}] \\
&= E[e^{t(\bar{X}_n - \mu)/(\sigma/\sqrt{n})}] \\
&= E\left[\prod_{i=1}^{n} e^{t(X_i - \mu)/\sigma\sqrt{n}}\right] \\
&= \prod_{i=1}^{n} E[e^{t(X_i - \mu)/\sigma\sqrt{n}}] \\
&= \left[m_{(X-\mu)/\sigma}\left(\frac{t}{\sqrt{n}}\right)\right]^n
\end{aligned}$$

where $m_{(X-\mu)/\sigma}$ is the moment generating function for the standard form for $X$. Then

$$\log_e m_{Z_n}(t) = n \log_e \left[ m_{(X-\mu)/\sigma}\left(\frac{t}{\sqrt{n}}\right)\right].$$

The standard form for $X$ has mean 0 and variance 1, so

$$m_{(X-\mu)/\sigma}(t) = 1 + \frac{t^2}{2} + m_3\frac{t^3}{3!} + m_4\frac{t^4}{4!} + \cdots,$$

where $m_3$, $m_4$, etc., are the third, fourth, etc., moments of the standard form for $X$. Then

$$\begin{aligned} m_{(X-\mu)/\sigma}\left(\frac{t}{\sqrt{n}}\right) &= 1 + \frac{t^2}{2n} + \frac{m_3}{3!}\left(\frac{t}{\sqrt{n}}\right)^3 + \frac{m_4}{4!}\left(\frac{t}{\sqrt{n}}\right)^4 + \cdots \\ &= 1 + a(t) \end{aligned}$$

and $\log_e m_{Z_n}(t) = n \log_e [1 + a(t)]$. It is shown in calculus courses that

$$\log(1+b) = b - \frac{b^2}{2} + \frac{b^3}{3} - \frac{b^4}{4} + \cdots, \qquad \text{for } |b| < 1.$$

If we choose $t$ close to zero, so $|a(t)| < 1$, then

$$\log_e m_{Z_n}(t) = [a(t) - \tfrac{1}{2}a^2(t) + \tfrac{1}{3}a^3(t) - \cdots].$$

By inspection

$$\lim_{n\to\infty} na(t) = \frac{t^2}{2}$$

$$\lim_{n\to\infty} na^k(t) = 0, \qquad k = 2, 3, 4, \ldots.$$

Thus

$$\lim_{n\to\infty} \log_e m_{Z_n}(t) = \frac{t^2}{2}$$

which implies that

$$\lim_{n\to\infty} m_{Z_n}(t) = e^{t^2/2},$$

the moment generating function for the standard normal random variable. Then

$$\lim_{n\to\infty} F_{Z_n}(t) = N_Z(t), \qquad \text{for all } t.$$ ◄

Since $\lim_{n\to\infty} F_{Z_n}(t) = N_Z(t)$, we would expect that, for large $n$, $F_{Z_n}(t) \doteq N_Z(t)$ ($\doteq$ is meant to stand for approximate equality). But for any $n$,

$$F_{\bar{X}_n}(t) = F_{Z_n}\left(\frac{t-\mu}{\sigma/\sqrt{n}}\right),$$

and thus for large $n$

$$F_{\bar{X}_n}(t) \doteq N_Z\left(\frac{t-\mu}{\sigma/\sqrt{n}}\right).$$

That is, the distribution function of the arithmetic average of a large number of independent, identically distributed random variables is approximately equal to the standard normal distribution function (appropriately adjusted), regardless of the common distribution of the individual random variables averaged together. This fact is the basis of the normal approximation to a great number of probability laws. The accuracy of the approximation varies for different probability laws for the same value of $n$. It is not possible in general to say how large is large, that is, how large $n$ should be for a given accuracy for a given probability law. The following examples illustrate applications of the normal approximation.

*Example 5.6.1.* (Normal Approximation to the Binomial) We saw in Section 5.4 that if $X_1, X_2, \ldots, X_n$ are independent Bernoulli random variables, each with parameter $p$, then $Y = \sum_{j=1}^{n} X_i$ is a binomial random variable with parameters $n$ and $p$. By the central limit theorem, the distribution function for $Y/n$ tends to the normal form, and for large $n$ we have

$$F_{\frac{Y}{n}}(t) \doteq N_Z\left(\frac{t-p}{\sqrt{pq/n}}\right)$$

or

$$F_Y(t) = F_{\frac{Y}{n}}\left(\frac{t}{n}\right) \doteq N_Z\left(\frac{t/n-p}{\sqrt{pq/n}}\right) = N_Z\left(\frac{t-np}{\sqrt{npq}}\right).$$

As an example of the use of this to approximate the binomial distribution function, suppose that we drop 49 coins onto a table and let $Y$ be the number of coins that land heads up. Then $Y$ is a binomial random variable with parameters 49 and $\frac{1}{2}$. The probability that at most 28 land heads up is

$$F_Y(28) \doteq N_Z\left(\frac{28 - 24\frac{1}{2}}{3\frac{1}{2}}\right) = .8413$$

(the exact value of this probability is .8736). The probability that between 20 and 25, inclusive, will show heads is

$$F_Y(25) - F_Y(19) \doteq N_Z\left(\frac{25 - 24\frac{1}{2}}{3\frac{1}{2}}\right) - N_Z\left(\frac{19 - 24\frac{1}{2}}{3\frac{1}{2}}\right)$$

$$= N_Z(.14) - N_Z(-1.57) = .4975$$

(the exact value is .5361).

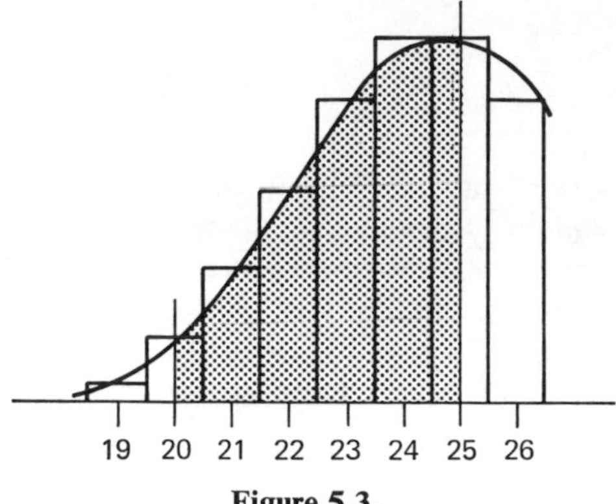


**Figure 5.3.**

In the above example the approximation was off by about .04 in both instances. We can materially improve the approximation of discrete probabilities by the continuous normal curve by realizing that if $X$ is discrete, then

$$P(X = a) = P(a - \tfrac{1}{2} < X \leq a + \tfrac{1}{2}).$$

If we use the normal approximation on these adjusted statements, we shall get a better approximation. The reason for this is sketched in Figure 5.3. Briefly, suppose that we want the exact probability that $20 \leq X \leq 25$. This is given by the sum of the areas of the bars numbered 20, 21, 22, 23, 24, and 25. If we take the area under the smooth normal curve only between 20 and 25 to approximate this value, we have in effect ignored half the area for both bars 20 and 25; the area under the continuous curve between $19\frac{1}{2}$ and $25\frac{1}{2}$ will correct this omission. (Extending the limits on both ends by $\frac{1}{2}$ is called the continuity correction.)

Another point is illustrated in Figure 5.4. Suppose that we want the

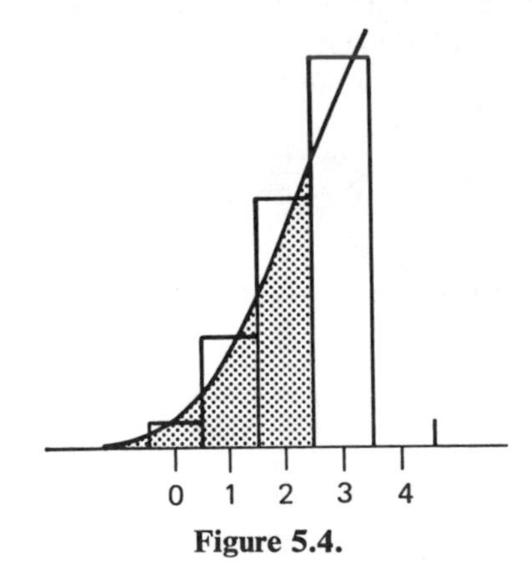


**Figure 5.4.**

probability that $X$ is no larger than 2. The exact probability is the sum of the areas of the bars labelled 0, 1, and 2. Rather than taking the area under the smooth curve between $-\frac{1}{2}$ and $2\frac{1}{2}$ to approximate this, we will in general get a closer value by taking the area under the smooth curve from $2\frac{1}{2}$ all the way back to $-\infty$. Thus, when approximating discrete tail probabilities, the normal approximation should not be truncated at the end of the last bar.

*Example 5.6.2.* (Binomial Approximation with Continuity Correction) Let us apply the continuity correction to the two probabilities computed in Example 5.6.1. As before, $Y$ is binomial with parameters 49 and $\frac{1}{2}$.

$$\begin{aligned} F_Y(28) &= F_Y(28\tfrac{1}{2}) \\ &\doteq N_Z\left(\frac{28\frac{1}{2} - 24\frac{1}{2}}{3\frac{1}{2}}\right) = .8729 \\ P(20 \leq Y \leq 25) &= F_Y(25\tfrac{1}{2}) - F_Y(19\tfrac{1}{2}) \\ &\doteq N_Z\left(\frac{25\frac{1}{2} - 24\frac{1}{2}}{3\frac{1}{2}}\right) - N_Z\left(\frac{19\frac{1}{2} - 24\frac{1}{2}}{3\frac{1}{2}}\right) \\ &= .5377. \end{aligned}$$

Comparison of these values with the exact probabilities given in Example 5.6.1 shows a considerable improvement in the accuracy of the approximation.

It can be shown that the normal approximation to the binomial is best for $p$ close to $\frac{1}{2}$ for any $n$, and that as $p$ gets further from $\frac{1}{2}$, $n$ must be larger to get the same accuracy. Many different writers recommend various rules-of-thumb for deciding when the normal gives a good approximation to the binomial. Generally, if $n \geq 30$ and $np \geq 5$, the approximation should be quite good.

*Example 5.6.3.* (Normal Approximation to the Poisson) We saw in Section 5.4 that if $X_1, X_2, \ldots, X_n$ are independent Poisson random variables, each with parameter $\lambda$, then $Y = \sum_{i=1}^{n} X_i$ is a Poisson random variable with parameter $n\lambda$. Then, for large $n$,

$$F_Y(t) \doteq N_Z\left(\frac{t - n\lambda}{\sqrt{n\lambda}}\right),$$

giving the basis for approximating Poisson probabilities by the normal. (We shall find that again the continuity correction discussed for the binomial gives an improved approximation since the Poisson is also a discrete random variable.) Let us assume that a college hospital receives an average of 5 cases of tetanus per month and that it is reasonable to assume that incidences of tetanus among the students are independently distributed in time. Then, for a particular school year, let $X_1, X_2, \ldots, X_9$ be the number of cases that occur in the 9 individual months and assume each $X_i$ is a Poisson random variable with parameter 5. The total number of cases for the year then is

$$Y = \sum_{i=1}^{9} X_i$$

and $Y$ is also Poisson with parameter $9 \cdot 5 = 45$. Then, using the continuity correction, the probability that between 40 and 50 cases, inclusive, occur in this year is

$$\begin{aligned} F_Y(50\tfrac{1}{2}) - F_Y(39\tfrac{1}{2}) &\doteq N_Z\left(\frac{50\frac{1}{2} - 45}{\sqrt{45}}\right) - N_Z\left(\frac{39\frac{1}{2} - 45}{\sqrt{45}}\right) \\ &= N_Z(.82) - N_Z(-.82) \\ &= .5878. \end{aligned}$$

(The exact probability is .5879.) The probability that exactly 45 cases occur is

$$\begin{aligned} F_Y(45\tfrac{1}{2}) - F_Y(44\tfrac{1}{2}) &\doteq N_Z\left(\frac{45\frac{1}{2} - 45}{\sqrt{45}}\right) - N_Z\left(\frac{44\frac{1}{2} - 45}{\sqrt{45}}\right) \\ &= .0558. \end{aligned}$$

(The exact probability is .0594.)

The normal approximation to the Poisson gets better the larger the value of $n\lambda$, the parameter of the sum. Thus, if $\lambda$, the common parameter of the individual variables making up the sum, is relatively small the number of variables in the sum must be relatively large to achieve a given degree of accuracy. If $\lambda$ is quite large, then $n$ can be correspondingly reduced.

*Example 5.6.4.* A father and his 12-year old son are going to build a dog house. The plans call (in part) for 10 $\frac{1}{2}$ inch by 4 inch boards, each 18 inches long. The father has purchased a single 18-foot length of $\frac{1}{2}$ inch by 4 inch board and asks his son to saw off the 10 pieces. Assume that the final length of each of the first 9 boards cut by the son is a random variable equally likely to lie between 17.7 inches and 18.2 inches, and that the lengths are independent. (The length of the tenth board is of course determined by the first 9.) Thus we can think of the 9 lengths he cuts as being 9 independent, identically distributed random variables, each uniformly distributed on the interval 17.7 to 18.2.

Let us use the central limit theorem to approximate the probability that the average length of the 9 boards he cuts exceeds 18 inches. We let $X_1, X_2, \ldots, X_9$ represent the 9 lengths and assume each of the $X_i$'s is uniformly distributed on the interval 17.7 to 18.2. Thus, each individual $X_i$ has mean 17.95 and standard deviation .144 (see Section 4.4). The arithmetic average of the $X_i$'s is

$$\bar{X} = \tfrac{1}{9}\sum_{i=1}^{9} X_i,$$

which has mean 17.95 and standard deviation .048(= $.144/\sqrt{9}$). Then

$$P(\bar{X} > 18) = 1 - F_{\bar{X}}(18) \doteq 1 - N_Z\left(\frac{18 - 17.95}{.048}\right)$$
$$= .1492.$$

We would find that this approximation to the probability that $\bar{X} > 18$ is quite good. It is shown in more advanced courses that the sum of uniformly distributed random variables converges quite rapidly to the normal (as $n$ increases) and that the exact density function of the sum of as few as 6 uniform random variables is very close to the normal.

## EXERCISE 5.6.

**1.** A pair of dice are rolled 180 times an hour (approximately) at a craps table in Las Vegas. What is the (approximate) probability that 25 or more rolls have a sum of 7 during the first hour? What is the (approximate) probability that between 700 and 750 rolls have a sum of 7 during 24 hours?

**2.** A lot of 1000 articles is to be accepted if a sample of 100 contains 3 or fewer defectives. If we assume that the process producing the article has a probability of .95 of making a nondefective article, what is the probability that the lot will be accepted?

**3.** A man leaving for work each morning is equally likely to step out of his door at any time between 7:15 a.m. and 7:20 a.m. If he works 250 days a year, compute the (approximate) probability that the average time he leaves for work (averaged over the year) lies between 7:18 a.m. and 7:20 a.m.

**4.** It is known that a particular method of producing vacuum tubes will yield tubes whose operating life is an exponential random variable with mean life of 1000 hours. Approximate the probability that the arithmetic average of the lengths of lives of 100 tubes will exceed 950 hours.

**5.** The number of incoming telephone calls at a particular industrial switchboard between 11 a.m. and 12 a.m. is known to be a Poisson random variable with parameter $\lambda = 120$ calls per hour. Approximate the probability that the total number of incoming telephone calls over 100 days, between 11 a.m. and 12 a.m., will exceed 12,500.

**6.** Due to the idiosyncrasies of the machine cutting them, the "true" length of a yardstick made by a particular firm is equally likely to take on any value in the interval from 35.95 inches to 36.10 inches. Approximate the probability that the average length of 100 of these yardsticks will lie between 36.00 inches and 36.02 inches.

**7.** A nursery man plants 115 cuttings of ivy in every flat he prepares. Assume that the probability that an individual cutting will develop roots is .9 and approximate the probability that the average number of rooted cuttings (per flat) in 50 flats is less than 100.

**8.** Experience has shown that the number of accidents to occur along a particular 10-mile stretch of 2-lane highway is a Poisson random variable with a mean of 2 per week. What is the (approximate) probability of there being less than 100 accidents on this stretch of road in a year?

**9.** The length of a continuous nylon filament that can be drawn without a break occurring is an exponential random variable with mean 5000 feet. What is the (approximate) probability that the average length of 100 filaments lies between 4750 feet and 5550 feet?

**10.** The probability that a person survives an attack of cholera (with good medical help) is assumed to be .4. What is the probability that at least half of 100 patients with cholera will survive?

# 6

# Sampling and Statistics

In this chapter we shall begin our study of statistics and statistical methods. We have to this point been concerned only with the mathematical theory of probability; this theory is the backbone of statistics and provides the rationale for deriving the statistical methods used in answering many types of questions.

As has already been mentioned, statistics is generally concerned with problems of inference. For example, prior to the actual election, a candidate for the presidency of the United States is vitally interested in the proportion $p$ of eligible voters who will vote for him. It would be extremely expensive for him to query all the voters on this point, so such a candidate relies on an opinion poll for this sort of information. The opinion poll consists of a selected sample or portion of all the voters who give an indication of how they will vote. Then, based on the results of this sample, the candidate tries to infer the value of $p$ in the total population of voters (actually the people running the poll frequently make this inference to be passed on to the candidate). This, then, is one example of a problem of statistical inference.

Or, to take another example, a company that sells rifle shells to the United States Army will generally claim that the proportion of supplied shells that will misfire is no bigger than .0001, for instance. The only way in which such a claim could be definitively checked would be by firing all the shells and computing the proportion that did misfire; the difficulty with this approach is apparent. What is done to check the claim is to select a sample or portion of the shells, fire them, and on the basis of the results decide to accept or reject the manufacturer's claim. Again a question of statistical inference is involved: based on the sample used, should the claim be granted or not?

In either of the examples just quoted, probability theory can be used to give the probabilities of occurrence of various possible sample outcomes, assuming that the proportion $p$ of voters favoring the candidate is known, in the first example, or assuming the true probability, $p$, of a misfire is known in the second. But if these quantities were known, there would be no reason to take the sample. The field of statistics is concerned with the use of observed sample results (which are observed values of random variables) in making statements about the unknown value of $p$ (or of some other unknown feature of the phenomenon sampled). Such an unknown quantity is called a parameter; probability theory is used in measuring the "correctness" of our inferences about parameters and in comparing different ways of making the same inference from the same observed sample results.

## 6.1. Random Sampling

A very basic concept to most statistical methods is that of a random sample. By sample we mean portion, only a part of the whole, and by random we mean essentially that the portion taken (the sample) is determined haphazardly or nonsystematically. For example, in taking a sample of voters to determine how many might vote for a particular candidate, we want to pick the voters for the sample at random and independently; we do not want to systematically include only those that will or only those that won't vote for the given candidate. Similarly, in selecting a sample of rifle shells to investigate the proportion of misfires among a given manufacturer's output, we select the shells for the sample in a haphazard way and do not systematically include only those that we are sure will or will not fire. Heuristically, it is expected that a haphazardly selected sample will give results representative of the population sampled from; mathematically we shall see that a haphazardly selected sample permits considerable simplification of the theory.

The samples that we shall be concerned with will in general be quantified; that is, if we have a sample of size $n$, then we have ended up with $n$ numbers and these constitute our sample results. Our concern then will be with what we might do with the $n$ numbers to answer the questions that motivated the sample in the first place. To return to the example of the presidential candidate, suppose it had been decided to sample, randomly, 2000 voters and from their responses to try to infer how many in the total electorate would vote for a given candidate. To keep the example simple, let us suppose that the only response a person in the sample can give is a simple yes or no (that he will vote the specified way). Then, to quantify these responses, we can simply record a 1 for each yes response and a zero for each no response (or any two distinct numbers could be used). Thus, after taking the sample we would have available 2000 numbers, each either a zero or a 1. If we let $p$ be the proportion of voters in the electorate that will vote for the given candidate, then we would want to know what we might do with these 2000 numbers to get a good guess of the unknown value of $p$.

Similarly, in the example of the rifle shell manufacturer's claim, suppose we decided to fire 5000 of his shells (chosen randomly from the whole group of shells he produced). We could then record a zero for all shells that fired correctly and a 1 for all that misfired, ending up with a sample of 5000 zeros and 1's. The question then becomes one of accepting or rejecting his claim, based on these particular numbers; what is it we should look for in terms of acceptance versus rejection of his claim?

Since we shall always be able to quantify the results of our sampling, we can think of the numbers which occur in the sample as being the observed values of random variables. Specifically, if we take a sample of size $n$ (we shall end up with $n$ numbers when the sampling is finished), we could let $X_1$ represent the first number we observe (measurement we make), $X_2$ the second, $X_3$ the third, etc., up to letting $X_n$ represent the $n$-th number we observe. Then $X_1, X_2, \ldots, X_n$ are $n$ jointly distributed random variables defined on the sample space consisting of all possible sample results we might observe. Let us consider some specific examples.

*Example 6.1.1.* A manufacturer makes rifle shells; assume there is a constant probability $p$ that any particular shell he makes will be defective (result in a misfire). Suppose we take a sample of $n = 3$ of his shells (just to keep the discussion simple) and define

$$X_1 = \begin{cases} 1 & \text{if the first shell misfires} \\ 0 & \text{if the first shell does not misfire} \end{cases}$$

$$X_2 = \begin{cases} 1 & \text{if the second shell misfires} \\ 0 & \text{if the second shell does not misfire} \end{cases}$$

$$X_3 = \begin{cases} 1 & \text{if the third shell misfires} \\ 0 & \text{if the third shell does not misfire.} \end{cases}$$

Then each of the $X_i$'s is a Bernoulli random variable with the same parameter $p$. That is,

$$\begin{aligned} p_{X_i}(x_i) &= p^{x_i}(1 - p)^{1-x_i}, & x_i &= 0, 1 \\ &= 0, & &\text{otherwise} \end{aligned}$$

for $i = 1, 2, 3$. Assuming that the 3 shells were selected at random, this would imply that $X_1, X_2, X_3$ are independent random variables and we have

$$\begin{aligned} p_{\underline{X}}(\underline{x}) &= \prod_{i=1}^{3} p_{X_i}(x_i) \\ &= p^{\Sigma x_i}(1 - p)^{3-\Sigma x_i} \end{aligned}$$

for $x_i = 0, 1$, $i = 1, 2, 3$. This joint probability function $p_{\underline{X}}$ then describes the probabilities of occurrence of the possible observed samples $(x_1, x_2, x_3)$, $x_i = 0, 1$, $i = 1, 2, 3$.

*Example 6.1.2.* Suppose that the total electorate for a given election consists of $N$ voters and that of these, $M$ favor (will vote for) candidate A. We randomly select 2 people without replacement from the electorate and ask them if they will vote for candidate A. Then we can quantify our sample results by defining

$$X_i = \begin{cases} 1 & \text{if } i\text{-th person says he will vote for A} \\ 0 & \text{if } i\text{-th person says he will not vote for A} \end{cases}$$

for $i = 1, 2$. Then our possible sample results are $(x_1, x_2) = (0, 0), (1, 0), (0, 1), (1, 1)$. The joint probability function for $X_1$ and $X_2$ is

$$\begin{aligned} p_{\underline{X}}(\underline{x}) &= \frac{\binom{N-M}{2}}{\binom{N}{2}}, \quad \text{for } (x_1, x_2) = (0, 0) \\ &= \frac{\binom{N-M}{1}\binom{M}{1}}{2\binom{N}{2}}, \quad \text{for } (x_1, x_2) = (1, 0) \text{ or } (0, 1) \\ &= \frac{\binom{M}{2}}{\binom{N}{2}}, \quad \text{for } (x_1, x_2) = (1, 1). \end{aligned}$$

Note that the marginal probability function for $X_1$ is

$$\begin{aligned} p_{X_1}(0) &= \frac{2\binom{N-M}{2} + \binom{N-M}{1}\binom{M}{1}}{2\binom{N}{2}} \\ &= \frac{(N-M)(N-1)}{N(N-1)} = \frac{N-M}{N} = 1 - \frac{M}{N} \\ p_{X_1}(1) &= \frac{\binom{N-M}{1}\binom{M}{1} + 2\binom{M}{2}}{2\binom{N}{2}} \\ &= \frac{M(N-1)}{N(N-1)} = \frac{M}{N}. \end{aligned}$$

Similarly the marginal probability function for $X_2$ is

$$p_{X_2}(0) = \frac{2\binom{N-M}{2} + \binom{N-M}{1}\binom{M}{1}}{2\binom{N}{2}}$$

$$= 1 - \frac{M}{N}$$

$$p_{X_2}(1) = \frac{\binom{N-M}{1}\binom{M}{1} + 2\binom{M}{2}}{2\binom{N}{2}}$$

$$= \frac{M}{N}$$

so that, as we would expect, the marginal probability functions for the two responses are identical. Note, however, that

$$p_{X_1,X_2}(0, 0) \neq p_{X_1}(0)p_{X_2}(0)$$

so that $X_1$ and $X_2$ are not independent random variables.

Note that in Example 6.1.1 we could define the single random variable $X$ to have probability function

$$p_X(x) = p^x(1-p)^{1-x}, \qquad x = 0, 1$$
$$= 0, \qquad \text{otherwise}$$

and that the marginals of the random variables in the sample, $X_1$, $X_2$, $X_3$, are each of exactly this form. Furthermore, $X_1$, $X_2$, $X_3$ are independent random variables and their joint probability function is given by the product of the marginals. In such a case as this, we shall say that $X_1$, $X_2$, $X_3$ is a random sample of size 3 of the random variable $X$, since they are independent, identically distributed random variables with marginals equal to the probability function for $X$. Throughout the remainder of this book we shall be concerned with random samples of random variables, so the following formal definition is quite important.

DEFINITION 6.1.1. (1) $X_1, X_2, \ldots, X_n$ is a *random sample* of a discrete random variable $X$ if and only if the joint probability function for the sample is

$$p_{\underline{X}}(\underline{x}) = \prod_{i=1}^{n} p_X(x_i), \qquad \text{for all } (x_1, x_2, \ldots, x_n).$$

(2) $X_1, X_2, \ldots, X_n$ is a *random sample* of a continuous random variable $X$ if and only if the joint density function for the sample is

$$f_{\underline{X}}(\underline{x}) = \prod_{i=1}^{n} f_X(x_i), \qquad \text{for all } (x_1, x_2, \ldots, x_n).$$

In either case (1) or (2), $X_1, X_2, \ldots, X_n$ are called the *elements* of the sample.

We note immediately, then, that in Example 6.1.2 the random variables $X_1$ and $X_2$ would not be called a random sample of a random variable $X$ since they are not independent. We do have what would be called a random sample from the electorate, but since the sampling was done without replacement the two responses are not independent. If the sampling had been done with replacement, $X_1$ and $X_2$ would be a random sample of a random variable $X$ since they would then be independent. For all practical purposes, of course, $N$ is generally quite large; we could just as well define the joint probability function of the sample to be the product of the marginals, because it does not matter much whether the 2 voters in the sample are chosen with or without replacement when $N$ is large.

Whenever we assume that we have a random sample of a random variable $X$, the probability function or density function for the sample values is derived quite easily as the product of the marginals of the elements of the sample. Thus, if the probability law of $X$ has an unknown parameter $\theta$, so will the probability law of the random sample of $X$. Many problems of statistics are centered around rational development of manipulations of the observed sample values to answer various queries about $\theta$. Thus, in Example 6.1.1, if $p$ were unknown we might want to use $x_1, x_2, x_3$ in some manner to guess at its unknown value. Or, in Example 6.1.2, $M$ would generally be unknown and again we might like to use $x_1$ and $x_2$ in some manner to guess at its value or, equivalently, to guess at the value of the ratio $M/N$.

To get a clear picture of statistical methods and their properties, it is important to remember the distinction between a random variable and the observed value of a random variable. Prior to taking a sample the elements $X_1, X_2, \ldots, X_n$ are random variables; after we observe the sample results, what we have available is $x_1, x_2, \ldots, x_n$, their observed values. We shall continue the notation used in the first half of this volume; capital letters will represent random variables (rules defined on the elements of a sample space) while lower case letters will represent their observed values (particular sample outcomes).

*Example 6.1.3.* Suppose that the length of life of a light bulb made a certain way is a normal random variable $X$ with mean $\mu$ and variance $\sigma^2$. If we randomly select 10 of these bulbs and let them burn until they fail, we can let $X_i$ be the length of life of the $i$-th bulb, $i = 1, 2, \ldots, 10$; then $X_1, X_2, \ldots, X_{10}$ is a random sample of size 10 of $X$ and the density function for the sample lives is

$$f_{\underline{X}}(\underline{x}) = \prod_{i=1}^{10} f_X(x_i)$$

$$= \prod_{i=1}^{10} \frac{1}{\sigma\sqrt{2\pi}} e^{-(x_i-\mu)^2/2\sigma^2}$$

$$= \frac{1}{\sigma^{10}(2\pi)^5} e^{-\sum_{i=1}^{10}(x_i-\mu)^2/2\sigma^2}$$

The probability that all 10 bulbs would burn at least 100 hours then is

$$\left[1 - N_Z\left(\frac{100 - \mu}{\sigma}\right)\right]^{10};$$

the probability that none of the bulbs would burn more than 1000 hours is

$$\left[N_Z\left(\frac{1000 - \mu}{\sigma}\right)\right]^{10}.$$

Note that the density function for the sample results depends on $\mu$ and $\sigma^2$, the mean and variance of $X$.

*Example 6.1.4.* Assume that deaths from traffic accidents on a nonholiday weekend (Saturday and Sunday only) occur at a constant rate $\lambda$ and that their occurrences satisfy the assumptions for a Poisson process. If we take a sample of the number of deaths occurring on 5 nonholiday weekends and let $X_i$ be the number of deaths on the $i$-th weekend, $i = 1, 2, 3, 4, 5$, then $X_1, X_2, \ldots, X_5$ is a random sample of a Poisson random variable $X$ with parameter $\lambda$. The probability function for the sample results is

$$p_{\underline{X}}(\underline{x}) = \prod_{i=1}^{5} p_X(x_i)$$

$$= \prod_{i=1}^{5} \frac{\lambda^{x_i}}{x_i!} e^{-\lambda}$$

$$= \frac{\lambda^{\Sigma x_i}}{\Pi x_i!} e^{-5\lambda}.$$

The probability that there are exactly 50 deaths on each of the 5 weekends is

$$\frac{\lambda^{250}}{(50!)^5} e^{-5\lambda};$$

the probability that we observe a total of 100 deaths over the 5 weekends is

$$\frac{(5\lambda)^{100}}{100!} e^{-5\lambda}.$$

(This latter problem of determining probabilities for the total number of deaths to be observed is referring to the random variable

$$Y = X_1 + X_2 + X_3 + X_4 + X_5.$$

Since each of the $X_i$'s is a Poisson random variable with parameter $\lambda$, and they are independent, we know from Chapter 5 that $Y$ is also Poisson with parameter $5\lambda$.)

## EXERCISE 6.1.

**1.** It is known that a certain process makes light bulbs whose lifetimes are normally distributed with average lifetime of 200 hours and standard deviation of 20 hours. If we randomly select 3 bulbs made by this process, what is the probability that all 3 lifetimes exceed 195 hours?

**2.** Assume that vacuum tubes of a certain type have lifetimes that are exponentially distributed with parameter $\lambda$. If we take a random sample of $n$ of these tubes and let $X_i$ = length of life of $i$-th tube, $i = 1, 2, \ldots, n$, what is the joint density function of the sample?

**3.** Suppose that $X$ is uniformly distributed on the interval from zero to 1. If we take a random sample of 5 observations of $X$, what is the joint density function for the sample?

**4.** Assume that weights of college students are normally distributed. If we take a random sample of 20 students and record the weight of each, what is the joint density function for the sample weights? As a function of $\mu$ and $\sigma$, what is the probability that the total of the weights exceeds 4000 pounds?

**5.** It is known that scores on a certain aptitude test are normally distributed with mean $\mu$ and variance $\sigma^2$. If we select 10 people at random to take this test, what is the joint density function for their scores? What is the probability that the average of the 10 scores is less than $\mu$?

**6.** In problem 5, assume that $\mu = 50$, $\sigma^2 = 25$, and that anyone who scores more than 60 on the test is certain to make a good pilot. What is the probability that there are no good pilots in the group of 10 people taking the test? What is the probability that there is one or more good pilots in the group of 10?

**7.** An urn contains 4 white and 1 red balls. We are going to draw balls, with replacement, from the urn until we get the white ball. Let $X$ be the number of draws required. If we take a random sample of 5 observations of $X$, what is the joint probability function of the sample?

**8.** Assume that a certain college football team has a constant probability $p$ of winning every game it plays and that all games it plays are independent. In any year this team plays exactly 10 games. Let $X_1, X_2, X_3, X_4, X_5$ be a random sample of 5 successive years of this team's performance (i.e., $X_i$ is the number of games won in the $i$-th year, $i = 1, 2, \ldots, 5$). What is the joint probability function for the sample?

**9.** Assume that all burglars have a constant probability $p$ of being caught while doing a job and let $X$ be the number of jobs done until the burglar is first caught. If we take a random sample of 10 burglars, what is the joint probability function of the number of jobs each accomplished before first being caught?

### 6.2. Statistics

We shall in this section introduce the technical meaning of the word statistic and look at some commonly used statistics. The definition of statistic follows.

DEFINITION 6.2.1. Any function of the elements of a random sample, which does not depend on unknown parameters, is called a *statistic*.

Thus, if $X_1, X_2, \ldots, X_n$ are a random sample of some random variable $X$, the quantities

$$X_1 + X_2, \qquad \frac{X_3}{X_4}, \qquad \frac{1}{n}\sum_{i=1}^{n} X_i, \qquad \sum_{i=1}^{n} X_i^2, \qquad \prod_{i=1}^{n} X_i$$

are each statistics. If $X$ is a random variable with mean $\mu$ and variance $\sigma^2$, then such quantities as

$$X_1 - \mu, \qquad \frac{X_2^2}{\sigma^2}, \qquad \sum_{i=1}^{n} (X_i - \mu)^2, \qquad \frac{X_1 + X_2 - 2\mu}{\sigma}$$

are not statistics if $\mu$ and $\sigma^2$ are unknown but they are statistics if the values of $\mu$ and $\sigma^2$ are known.

Since a statistic is a function of the elements of a random sample of a random variable, a statistic is itself a random variable. From one sample to another the value a statistic takes on will vary, because the particular values of the random variables will vary. The probability function or density function of a statistic may be derived from the probability function or density function of the sample; then the statistic may be used to make inferences regarding unknown parameters in the probability law of $X$. This type of problem will be examined in subsequent chapters.

Let us now define some common statistics that will be of use in these inference problems. The reader will recall that the $k$-th moment of a random variable $X$ is $m_k = E(X^k)$ (see Definition 3.4.1). By interpreting the operation of expectation as being averaging (with respect to the probability law of $X$), we might say that the $k$-th moment of the random variable $X$ is the average value of $X^k$. Analogous to this, if $X_1, X_2, \ldots, X_n$ are a random sample of $X$, then the average of the $k$-th powers of the elements of the sample is called the $k$-th sample moment. This is formally stated as follows.

DEFINITION 6.2.2. If $X_1, X_2, \ldots, X_n$ is a random sample of $X$, the $k$-th *sample moment* is:

$$M_k = \frac{1}{n}\sum_{i=1}^{n} X_i^k, \qquad \text{for } k = 1, 2, 3, \ldots.$$

Notice first of all that each sample moment (no matter what the value of $k$) then is a statistic and is itself a random variable. Once we actually take the sample and observe the values of $X_1, X_2, \ldots, X_n$, we can then compute the observed value of $M_k$ for any $k$. This observed value of $M_k$ will be denoted by $\hat{m}_k$; we use the lower case $m$ to be consistent with our notation of capital letters for random variables and lower case letters for their observed values. The carat ($\hat{}$) is placed over the lower case $m$ to distinguish this observed sample value from the like numbered moment of $X$ (the notation used has been $m_k = E[X^k]$). Clearly, the observed value, $\hat{m}_2$ say, is not necessarily the same number as $m_2 = E[X^2]$. Just as the moments of the random variable measure certain aspects of the probability law of the random variable, we shall see that the sample moments measure certain aspects of the sample.

*Example 6.2.1.* Assume that heights of college students are normally distributed with mean $\mu$ and variance $\sigma^2$. We take a random sample of 5 students, measure their heights, and find them to be 70, 74, 66, 69, and 72 (measured in inches). Once we have actually done the sampling and observed the sample values that occurred, we are able to compute the observed values of any sample moments in which we might be interested. Thus

$$\hat{m}_1 = \tfrac{1}{5}(70 + 74 + 66 + 69 + 72) = 70.2$$

$$\hat{m}_2 = \tfrac{1}{5}[(70)^2 + (74)^2 + (66)^2 + (69)^2 + (72)^2] = 4935.4$$

and

$$\hat{m}_3 = \tfrac{1}{5}[(70)^3 + (74)^3 + (66)^3 + (69)^3 + (72)^3] = 347{,}495.4.$$

The first sample moment, $M_1$, is simply the arithmetic average of the elements of the sample,

$$M_1 = \frac{1}{n}\sum_{i=1}^{n} X_i,$$

and is called the sample mean; we shall henceforth use $\bar{X}$ to represent the sample mean (the reader will recall this usage from Example 5.4.6). Since $\bar{X}$ (or $M_1$) is the arithmetic average of the elements of the sample, it is a measure of the "middle" of the sample values in a center of gravity sense; it is rather analogous to $\mu$, the mean of $X$, but is not to be confused with $\mu$. The mean of the probability law ($\mu$) from which we are sampling is a fixed number and is generally unknown to us, while $\bar{X}$ is simply the average of the elements of the sample and is a random variable whose value can be computed once the sample has been taken. Its computed value will vary from one sample to another.

Other aspects of the sample (besides locating its middle) are reflected in the values of the other moments, $M_2, M_3, \ldots$. Just as we used $m_2 = E[X^2]$ to build a measure of the variability of a probability law (by defining the variance), so can we use $M_2$ to build a measure of the variability of the sample. In fact, let us now define the sample variance and show how it is related to $M_2$.

DEFINITION 6.2.3. Given $X_1, X_2, \ldots, X_n$ is a random sample of $X$, the *sample variance* is defined to be

$$S^2 = \frac{1}{n-1} \sum_{i=1}^{n} (X_i - \bar{X})^2$$

where $\bar{X}$ is the sample mean.

Note that $S^2$ is essentially the average of the squares of the deviations of the elements of the sample about their mean value. The reason for using $n - 1$ rather than $n$ as a divisor will be made apparent as we proceed.

***Theorem 6.2.1.*** If $S^2$ is the sample variance, then

$$S^2 = \frac{n}{n-1} [M_2 - \bar{X}^2]$$

where $n$ is the sample size.

*Proof:* By definition,

$$\begin{aligned} S^2 &= \frac{1}{n-1} \sum_{i=1}^{n} (X_i - \bar{X})^2 \\ &= \frac{1}{n-1} \sum_{i=1}^{n} [X_i^2 - 2X_i\bar{X} + \bar{X}^2] \\ &= \frac{1}{n-1} [\sum X_i^2 - 2\bar{X} \sum X_i + n\bar{X}^2] \\ &= \frac{1}{n-1} [\sum X_i^2 - 2\bar{X}(n\bar{X}) + n\bar{X}^2] \\ &= \frac{1}{n-1} [\sum X_i^2 - n\bar{X}^2] = \frac{n}{n-1} [M_2 - \bar{X}^2], \end{aligned}$$

as was to be proved. ◄

The quantity $S = \sqrt{S^2}$ is called the *sample standard deviation* and will also prove quite useful.

*Example 6.2.2.* For the sample of heights given in Example 6.2.1, we find the observed value of the sample mean to be $\bar{x} = 70.2$; the observed value of the sample variance is

$$s^2 = \tfrac{5}{4}[4935.4 - (70.2)^2] = 9.2$$

and the observed value of the sample standard deviation is

$$s = \sqrt{9.2} = 3.033.$$

Notice that we are following the convention of using capital letters to denote random variables and lower case letters to denote their observed values (as computed from the sample results).

*Example 6.2.3.* A group of 20 students took an examination graded from zero to 100. The 20 scores which they achieved were as follows: 99, 87, 94, 86, 89, 79, 84, 90, 86, 86, 81, 98, 85, 91, 87, 89, 95, 88, 85, and 94. Looking at these numbers as though they are the observed values of a random sample of size 20 of a random variable $X$, we find the observed value of the sample mean to be

$$\bar{x} = \frac{1}{20}\sum_{i=1}^{20} x_i = 88.65$$

and the observed value of the sample variance to be

$$s^2 = \frac{1}{19}\sum_{i=1}^{20} (x_i - \bar{x})^2 = \frac{1}{19}[\sum x_i^2 - 20(\bar{x})^2] = 27.71;$$

the observed value of the sample standard deviation is $s = \sqrt{27.71} = 5.26$.

Let us now define some additional statistics which will prove of use in particular problems.

DEFINITION 6.2.4. Given a random sample $X_1, X_2, \ldots, X_n$ of a random variable $X$, rank the elements in order of increasing numerical magnitude, yielding $X_{(1)}, X_{(2)}, \ldots, X_{(n)}$, where $X_{(1)}$ is the smallest and $X_{(n)}$ the largest of $X_1, X_2, \ldots, X_n$. $X_{(i)}$ is called the $i$-th *order statistic*, $i = 1, 2, \ldots, n$. ($X_{(1)}$ is also referred to as the *minimum* sample value and $X_{(n)}$ as the *maximum*.)

Let us immediately look at an example of computations of order statistics

*Example 6.2.4.* The ranked sample of heights given in Example 6.2.1 is 66, 69' 70, 72, and 74. Thus, for this sample, $x_{(1)} = 66$, $x_{(5)} = 74$. For the sample of 20 scores given in Example 6.2.3, we find that $x_{(1)} = 79$, $x_{(20)} = 99$, $x_{(19)} = 98$, $x_{(4)} = 85$, etc.

Thus, to compute the observed values of the order statistics, we need merely rank the values that occur in the sample. We shall see several uses of these statistics for making inferences as we proceed. Two additional statistics are easily defined in terms of the order statistics; these are the sample median and the sample range. The sample median is defined to be the "middle" value of the order statistics, which is fine as a definition if the size of the sample, $n$, is an odd number. If the sample size is even, we simply define the sample median to be midway between the two middle order statistics. The sample range is defined to be the difference between the largest and the smallest order statistics. These are formally stated in the following definition.

DEFINITION 6.2.5. Given a random sample $X_1, X_2, \ldots, X_n$ of $X$, the *sample median* is defined to be

$$M_0 = X_{(\frac{n+1}{2})}, \qquad \text{if } n \text{ is odd}$$
$$= \tfrac{1}{2}[X_{(\frac{n-1}{2})} + X_{(\frac{n+1}{2})}] \qquad \text{if } n \text{ is even;}$$

the sample range is defined to be $R_a = X_{(n)} - X_{(1)}$.

The sample median and the sample range are alternative ways of measuring the middle and the variability, respectively, of a sample (alternative to the sample mean and the sample standard deviation). The following example illustrates their computation.

*Example 6.2.5.* For the ranked heights (first given in Example 6.2.1) we find $m_0 = 70$, $r_a = 8$. For the exam scores (first given in Example 6.2.3) we find $m_0 = 87.5$, $r_a = 20$.

Let us close this section by discussing the sample distribution function. As we saw in Chapter 3, the distribution function of a random variable evaluated at $t$ gives the probability that the random variable is less than or equal to $t$; thus $F_X(t)$ gives the proportion of the time that $X \leq t$. Analogously, the sample distribution function evaluated at $t$ gives the proportion of the sample values which are less than or equal to $t$; the following definition defines $F^*(t)$, the sample distribution function.

DEFINITION 6.2.6. Given that $X_1, X_2, \ldots, X_n$ is a random sample of $X$ (and $X_{(1)}, X_{(2)}, \ldots, X_{(n)}$ are the order statistics), the *sample distribution function* is

$$F^*(t) = 0, \qquad \text{for } t < x_{(1)}$$
$$= \frac{i}{n}, \qquad \text{if } i \text{ sample values are less than or equal to } t \text{ for } i = 1, 2, \ldots, n-1$$
$$= 1, \qquad \text{for } t \geq x_{(n)}.$$

The following example and accompanying graphs illustrate the computation of the sample distribution function.

*Example 6.2.6.* For the sample of heights given in Example 6.2.1, we have

$$
\begin{aligned}
F^*(t) &= 0, & t &< 66 \\
&= \tfrac{1}{5}, & 66 \le t &< 69 \\
&= \tfrac{2}{5}, & 69 \le t &< 70 \\
&= \tfrac{3}{5}, & 70 \le t &< 72 \\
&= \tfrac{4}{5}, & 72 \le t &< 74 \\
&= 1, & t &\ge 74.
\end{aligned}
$$

The graph of this function is given in Figure 6.1. Notice that the jumps occur at the sample values. The test scores given in Example 6.2.3 include some repeated scores and thus $x_{(1)}, x_{(2)}, \ldots, x_{(20)}$ are not all distinct. By checking the numbers that occurred, we see that $m = 14$ distinct values occur. Then the sample distribution function is

$$
\begin{aligned}
F^*(t) &= 0, & t &< 79 \\
&= \tfrac{1}{20}, & 79 \le t &< 81 \\
&= \tfrac{2}{20}, & 81 \le t &< 84 \\
&= \tfrac{3}{20}, & 84 \le t &< 85 \\
&= \tfrac{5}{20}, & 85 \le t &< 86 \\
&\;\;\vdots & &\;\;\vdots \\
&= 1, & t &\ge 99.
\end{aligned}
$$

The graph of this sample distribution function is given in Figure 6.2.

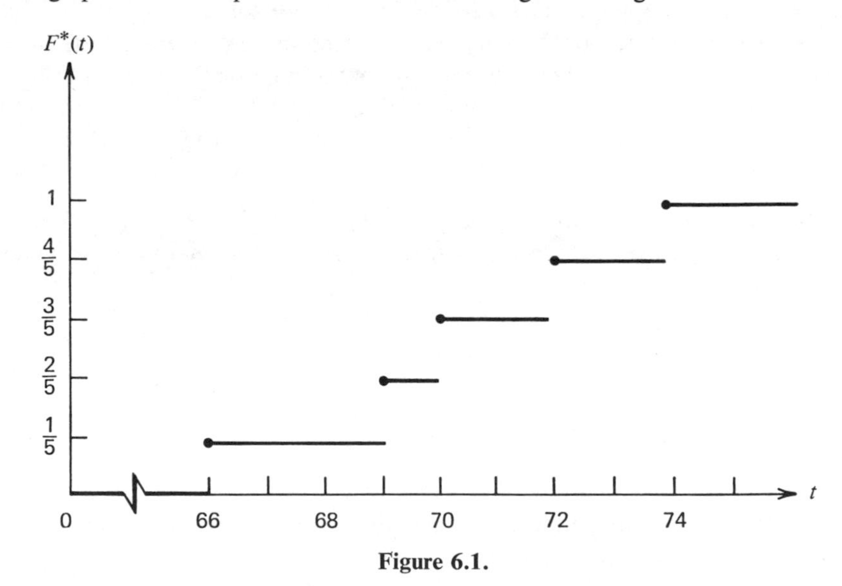


**Figure 6.1.**

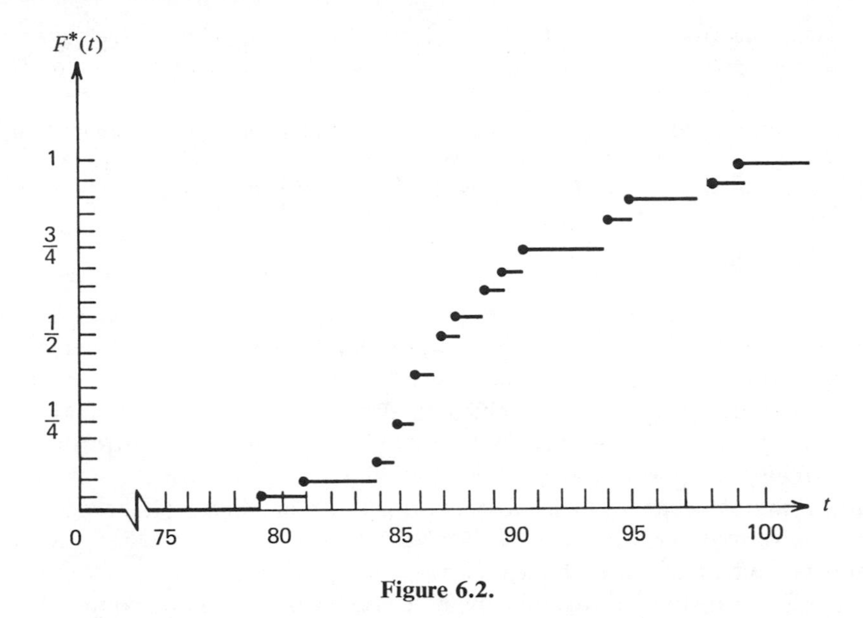


**Figure 6.2.**

This section has been devoted to introducing the reader to a number of statistics, each of whose values can be computed from the observation of the elements of a random sample. The reasons why these particular functions are of importance will be taken up in Chapter 7.

## EXERCISE 6.2.

**1.** A builder of sports cars took 6 identical cars of 1 of his models out to a track to see how the top speed attainable varied from car to car. The maximum speeds observed for the 6 were 118, 121, 120, 110, 125, and 116 (measured in miles per hour). Assume that these numbers are the observed values of a random sample of a random variable and compute the observed values of the first 3 sample moments, of $\overline{X}$, of $S^2$, of $M_0$, of $X_{(4)}$, and of $R_a$.

**2.** A college sprinter competed in 8 100-yard dash races during spring; the times recorded for his races were 10.2, 9.6, 9.8, 9.8, 10.0, 9.6, 9.7, and 9.6 (in seconds). Assuming that these numbers are the observed values of a random sample of a random variable, compute $\bar{x}$, $s^2$, the observed values of the order statistics, and the sample distribution function.

**3.** A child-development study included, among other things, observing the age at which babies are first able to sit up alone. The ages at which 5 of the babies first sat up alone were 155, 162, 150, 180, and 171 (days). Compute $\bar{x}$, $s^2$, $s$, $m_0$, and $r_a$.

**4.** Four sons from the same family all attended the same college. Their grade-point averages, at graduation, were 3.63, 3.10, 2.75, and 3.22 (4.0 is an A). Compute $\bar{x}$, $s$, $m_0$, and $r_a$.

**5.** Five students enrolled in the same 6-month typing course. Upon completion, the numbers of words per minute which they could type were 55, 70, 72, 38, and 46. Compute $\bar{x}$, $s^2$, $m_0$, and the observed values of the order statistics.

### 6.3. Sampling Distributions

It was stressed in the last chapter that the elements of a random sample are random variables; thus, from one sample to another the actual numbers that we observe will vary. We have seen how to use the probability law for the random variable being sampled to derive the probability law of the sample results. This probability law of the sample could be used to make probability statements about things that might happen when the sample is taken (that is, what we might observe prior to the actual sampling; once we have observed the sample results, of course, we know whether various possible outcomes did or did not occur).

Since a statistic is a function of the elements of a random sample, it is also a random variable; therefore there must be some probability law that describes its behavior. In this section we shall consider the probability laws for some specific statistics to see how these statistics may vary from one sample to another.

Let us first make a remark about the distributions of the sample moments.

***Theorem 6.3.1.*** Let $X_1, X_2, \ldots, X_n$ be a random sample of $X$. Then

$$E[M_k] = m_k, \qquad k = 1, 2, 3, \ldots$$

where $M_k$ is the $k$-th moment of the sample and $m_k$ is the $k$-th moment of $X$.

*Proof:* It will be recalled that the marginal (density or probability) function for $X_i$, $i = 1, 2, \ldots, n$, is identical with that of $X$. Then

$$\begin{aligned} E[M_k] &= E\left[\frac{1}{n}\sum_{i=1}^{n} X_i^k\right] \\ &= \frac{1}{n}\sum_{i=1}^{n} E[X_i^k] \\ &= \frac{1}{n}\sum_{i=1}^{n} m_k = m_k, \qquad \text{for } k = 1, 2, \ldots, \end{aligned}$$

as was to be proved. ◀

Thus the expected value of the $k$-th sample moment is the $k$-th moment of $X$; this says that if we were repeatedly to select random samples of size $n$, and for each compute the observed value of $M_k$, then the average of these observed values would be $m_k$. In particular, for any single sample we take, the observed value of $M_k$ may or may not itself be "close" to $m_k$. But, over an infinite number of repetitions, the average equals $m_k$. When we study estimation in Chapter 7, this property will be called unbiasedness.

As a special case then of Theorem 6.3.1, we know that $E[\bar{X}] = \mu$ where $\mu$ is the mean of the random variable sampled. The reader will recall from Example 5.4.6 that the variance of $\bar{X}$ is $\sigma^2/n$ where $\sigma^2$ is the variance of the random variable sampled. That is,

$$E[(\bar{X} - \mu)^2] = E[\bar{X}^2] - \mu^2 = \frac{\sigma^2}{n}$$

and thus

$$E[\bar{X}^2] = \frac{\sigma^2}{n} + \mu^2.$$

Using this result, we find the expected value of $S^2$ as follows:

$$\begin{aligned} E[S^2] &= \frac{n}{n-1} E[M_2 - \bar{X}^2] \\ &= \frac{n}{n-1}\left[m_2 - \left(\frac{\sigma^2}{n} + \mu^2\right)\right] \\ &= \frac{n}{n-1}\left[m_2 - \mu^2 - \frac{\sigma^2}{n}\right] \\ &= \frac{n}{n-1}\left[\sigma^2 - \frac{\sigma^2}{n}\right] = \sigma^2 \end{aligned}$$

since $m_2 - \mu^2 = \sigma^2$. Thus, the average value of the sample variance $S^2$ is $\sigma^2$, the variance of the random variable sampled. It is for this reason that we used the divisor $n - 1$ rather than $n$, in defining $S^2$. Had we used the divisor $n$, the average value of the sample variance would have been $[(n - 1)/n]\sigma^2$ rather than $\sigma^2$.

A slight adaptation of Theorem 5.4.4 enables us to prove the following theorem.

***Theorem 6.3.2.*** Suppose that $X_1, X_2, \ldots, X_n$ is a random sample of a random variable $X$ which is normally distributed with mean $\mu$ and variance $\sigma^2$. Then $\bar{X}$ is normally distributed with mean $\mu$ and variance $\sigma^2/n$.

*Proof:*

$$\begin{aligned} m_{\bar{X}}(t) &= E[e^{t\bar{X}}] \\ &= E[e^{(1/n)t\Sigma X_i}] \\ &= \{E[e^{(t/n)X}]\}^n \\ &= \left[m_X\left(\frac{t}{n}\right)\right]^n \\ &= [e^{(t/n)\mu+(t^2/2n^2)\sigma^2}]^n \\ &= e^{t\mu+(t^2/2)(\sigma^2/n)} \end{aligned}$$

which is the moment generating function of a normal random variable with mean $\mu$ and variance $\sigma^2/n$. ◀

Thus, if we are sampling a normal random variable with known mean and known variance, the density function for $\bar{X}$ is completely known. If we are sampling a random variable $X$ that is not normal, then $\bar{X}$ will still be approximately normally distributed for a large sample size $n$ because of the central limit theorem (5.6.1).

*Example 6.3.1.* Suppose that it is known that lives of light bulbs made a certain way are normally distributed (that is, if $X$ is the length of life of 1 bulb, $X$ is a normal random variable) with unknown mean $\mu$ and known variance $\sigma^2 = 100$. If we randomly select 100 bulbs made in this way, then the sample mean $\bar{X}$ is a normal random variable with mean $\mu$ and variance $\frac{100}{100} = 1$. Then by referring to the normal table, we can say that

$$P(|\bar{X} - \mu| < 2) = .9546$$

$$P(|\bar{X} - \mu| < 3) = .9974;$$

that is, the probability that the sample mean is within 2 hours of $\mu$ is .9546 and that it is within 3 hours of $\mu$ is .9974.

*Example 6.3.2.* If $X_1, X_2, \ldots, X_n$ is a random sample of a Poisson random variable $X$, with parameter $\lambda$, then $n\bar{X}$ is a Poisson random variable with parameter $n\lambda$ (see problem 7, Exercise 5.4). This result can be used to make exact probability statements about $\bar{X}$.

***Theorem 6.3.3.*** If $X_1, X_2, \ldots, X_n$ are a random sample of a normal random variable $X$ with mean $\mu$ and variance $\sigma^2$, then

$$Y = \sum_{i=1}^{n} \frac{(X_i - \mu)^2}{\sigma^2}$$

is a $\chi^2$ random variable with $n$ degrees of freedom.

*Proof:* Clearly, $(X_i - \mu)^2/\sigma^2$ is a $\chi^2$ random variable with 1 degree of freedom, $i = 1, 2, \ldots, n$ (since it is the square of a standard normal random variable; see Section 4.5). As such, its moment generating function is

$$m_{(X_i-\mu)^2/\sigma^2}(t) = \frac{1}{(1-2t)^{1/2}}, \qquad i = 1, 2, \ldots, n.$$

(See problem 11, Exercise 4.5). Since $X_1, X_2, \ldots, X_n$ are independent,

$$m_Y(t) = \prod_{i=1}^{n} m_{(X_i-\mu)^2/\sigma^2}(t) = \prod_{i=1}^{n} \frac{1}{(1-2t)^{1/2}} = \frac{1}{(1-2t)^{n/2}};$$

but this is the moment generating function of the density

$$\begin{aligned} f_Y(y) &= \frac{1}{\Gamma(n/2)} \frac{1}{2^{n/2}} y^{(n/2)-1} e^{-y/2}, && \text{for } y > 0 \\ &= 0, && \text{otherwise,} \end{aligned}$$

which is the density of what is called the $\chi^2$ random variable with $n$ degrees of freedom (the only parameter). Hence

$$Y = \frac{\sum_{i=1}^{n} (X_i - \mu)^2}{\sigma^2}$$

is a $\chi^2$ random variable with $n$ degrees of freedom. ◀

This $\chi^2$ density function will occur in several inference problems which we shall study. Figure 6.3 gives some graphs of the $\chi^2$ density function for varying degrees of freedom. Note that if the number of degrees of freedom $n$ exceeds 2 the density function has a maximum value at $n - 2$. By differentiating the moment generating function, it is easy to see that the mean of a $\chi^2$ random variable is $n$, its degrees of freedom, and that its variance is $2n$. It can be shown that as the number of degrees of freedom, $n$, gets larger and larger, the $\chi^2$ density function gets more and more symmetric. In fact, its limit, as $n \to \infty$, is the normal density function. The following theorem gives a result which we shall later find of some use.

***Theorem 6.3.4.*** If $Y$ and $Z$ are independent $\chi^2$ random variables with $m$ and $n$ degrees of freedom, respectively, then $Y + Z$ is a $\chi^2$ random variable with $m + n$ degrees of freedom.

*Proof:* We saw in Chapter 5 that

$$m_{Y+Z}(t) = m_Y(t) m_Z(t)$$

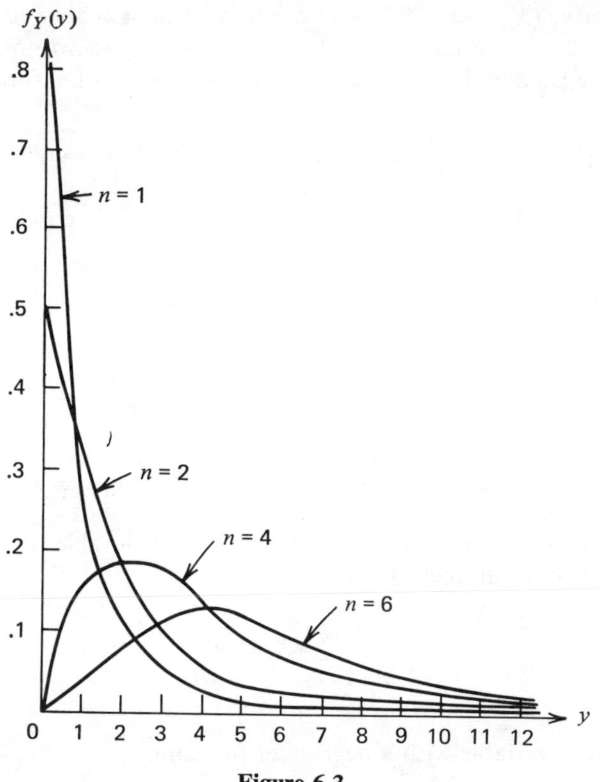


**Figure 6.3.**

if $Y$ and $Z$ are independent. Since each is a $\chi^2$ random variable, then

$$m_Y(t) = \frac{1}{(1 - 2t)^{m/2}}$$

$$m_Z(t) = \frac{1}{(1 - 2t)^{n/2}}$$

and

$$m_{Y+Z}(t) = \frac{1}{(1 - 2t)^{m/2}} \frac{1}{(1 - 2t)^{n/2}}$$

$$= \frac{1}{(1 - 2t)^{(m+n)/2}}$$

and the theorem is proved. ◄

It is easy to see that

$$\sum_{i=1}^{n}(X_i-\mu)^2=\sum_{i=1}^{n}[(X_i-\bar{X})+(\bar{X}-\mu)]^2$$
$$=\sum(X_i-\bar{X})^2+\sum(\bar{X}-\mu)^2+2(\bar{X}-\mu)\sum(X_i-\bar{X})$$
$$=\sum(X_i-\bar{X})^2+n(\bar{X}-\mu)^2$$

and thus

$$\frac{\sum(X_i-\mu)^2}{\sigma^2}=\frac{\sum(X_i-\bar{X})^2}{\sigma^2}+\frac{(\bar{X}-\mu)^2}{\sigma^2/n},$$

no matter what the values of $\mu$ and $\sigma^2$.

Now, we have just seen that the term on the left of the equality is a $\chi^2$ random variable with $n$ degrees of freedom if $X_1, X_2, \ldots, X_n$ are a random sample of a normal random variable $X$; the last term on the right is the square of a standard normal random variable and thus is a $\chi^2$ random variable with 1 degree of freedom. Perhaps it then is plausible that the first term on the right of the equality should be a $\chi^2$ random variable with $n-1$ degrees of freedom. This and another important result is stated without proof in the following theorem.

***Theorem 6.3.5.*** If $X_1, X_2, \ldots, X_n$ is a random sample of a normal random variable $X$ with mean $\mu$ and variance $\sigma^2$, then

$$Z=\frac{\sum_{1}^{n}(X_i-\bar{X})^2}{\sigma^2}$$

is a $\chi^2$ random variable with $n-1$ degrees of freedom and $Z$ and $\bar{X}$ are independent random variables.

This theorem has some interesting implications in problems of inference, as we shall see in Chapter 8. Some selected percentiles of the distribution function of a $\chi^2$ random variable, for varying degrees of freedom, are given in Table D, Appendix 5.

*Example 6.3.3.* Suppose that we have a random sample of 10 observations of a normal random variable with unknown mean $\mu$ and variance $\sigma^2$. Then $Y = 9S^2/\sigma^2$ is a $\chi^2$ random variable with 9 degrees of freedom. We see from Table D that $P(3.33 < Y < 16.9) = .9$. Thus $P(.37\sigma^2 < S^2 < 1.88\sigma^2) = .9$, which gives an indication of how likely the sample variance is to be close to $\sigma^2$ for a sample of this size.

Let us look now at the distributions of some of the order statistics, specifically the maximum and the minimum sample values.

***Theorem 6.3.6.*** Let $X_{(1)}, X_{(2)}, \ldots, X_{(n)}$ be the order statistics for a sample of size $n$ of $X$ with distribution function $F_X$. Then

$$F_{X_{(n)}}(t) = [F_X(t)]^n$$
$$F_{X_{(1)}}(t) = 1 - [1 - F_X(t)]^n.$$

*Proof:* $X_{(n)}$ is less than some number $t$ if and only if all $n$ $X_i$'s are less than $t$; thus

$$\begin{aligned} F_{X_{(n)}}(t) &= P(X_{(n)} \leq t) = P(X_1 \leq t, X_2 \leq t, \ldots, X_n \leq t) \\ &= P(X_1 \leq t)P(X_2 \leq t) \cdots P(X_n \leq t) \\ &= [F_X(t)]^n. \end{aligned}$$

Then, if $X$ is continuous so is $X_{(n)}$ and

$$f_{X_{(n)}}(t) = \frac{d}{dt} F_{X_{(n)}}(t) = n[F_X(t)]^{n-1} f_X(t).$$

Very similarly, we know that $X_{(1)}$ exceeds some number $t$ if and only if every one of the sample values exceeds $t$; thus

$$\begin{aligned} P(X_{(1)} > t) &= P(X_1 > t, X_2 > t, \ldots, X_n > t) \\ &= P(X_1 > t)P(X_2 > t) \cdots P(X_n > t) \\ &= [1 - F_X(t)]^n \end{aligned}$$

Then

$$\begin{aligned} F_{X_{(1)}}(t) &= 1 - P(X_{(1)} > t) \\ &= 1 - [1 - F_X(t)]^n; \end{aligned}$$

again if $X$ is continuous so is $X_{(1)}$ and we have

$$f_{X_{(1)}}(t) = n[1 - F_X(t)]^{n-1} f_X(t).$$ ◀

*Example 6.3.4.* Suppose that we have a random sample of 4 observations of an exponential random variable $X$ with parameter $\lambda$. Then

$$F_X(t) = 1 - e^{-\lambda t}, \qquad t > 0$$

and

$$F_{X_{(4)}}(t) = [1 - e^{-\lambda t}]^4, \qquad t > 0$$

and

$$f_{X_{(4)}}(t) = 4[1 - e^{-\lambda t}]^3 \lambda e^{-\lambda t}, \qquad t > 0.$$

Similarly the distribution and density functions of the smallest element in the sample are

$$F_{X_{(1)}}(t) = 1 - e^{-4\lambda t}, \qquad t > 0$$
$$f_{X_{(1)}}(t) = 4\lambda e^{-4\lambda t}, \qquad t > 0.$$

**EXERCISE 6.3.**

**1.** Suicides occur in a certain city at a rate of 1 per month. Suppose that we take a sample of the number of suicides per month for all of the months in a given year (thus our sample consists of $X_1, X_2, \ldots, X_{12}$). Compute the probability that $\bar{X}(= \frac{1}{12}\sum_{1}^{12} X_i)$ exceeds 1.1 and that $\bar{X} < .85$. (Assume that $X_1, X_2, \ldots, X_{12}$ are independent Poisson random variables with $\lambda = 1$; thus $12\bar{X}$ is Poisson with parameter 12.)

**2.** Manhattan short-sleeve dress shirts of neck size $a$ have neck sizes that are normally distributed with mean $a$ and standard deviation of .1 inch ($a$ equals 14 inches or $14\frac{1}{2}$ inches or 15 inches, etc.). Suppose that you select 5 of these shirts (of size $a$) for your wardrobe. What is the probability that the average neck size (of these 5) lies between $a - .05$ and $a + .05$ inch?

**3.** A fair coin is flipped 5 times. What is the exact probability that the sample proportion of heads is within .05 of $\frac{1}{2}$? That it is within .15 of $\frac{1}{2}$?

**4.** For the sample described in problem 2, find $a$ and $b$ such that

$$P(a\sigma^2 < S^2 < b\sigma^2) = .95.$$

**5.** What is the probability that the largest value of a sample of 10 of a uniformly distributed random variable on (0, 1) exceeds .9? What is the probability that it is less than $\frac{1}{2}$?

**6.** For the sample mentioned in problem 5, what is the probability that the smallest value is less than $\frac{1}{2}$? Greater than .9?

**7.** Assume that a baseball player has 4 times at bat in every game in which he participates. Also assume that his probability of getting a hit, each time he is at bat, is .3. If we have a sample of 8 days in which he plays, how might we evaluate the probability that the average number of hits he gets is less than $1\frac{1}{2}$? More than 2?

**8.** Given that $X_1, X_2, \ldots, X_n$ is a random sample of a Bernoulli random variable with parameter $p$, show that

$$P(X_{(n)} = 1) = 1 - (1 - p)^n$$
$$P(X_{(1)} = 1) = 1 - p^n.$$

**9.** Let $X_1, X_2, \ldots, X_n$ be a random sample of a normal random variable with parameters $\mu$ and $\sigma^2$. Derive the density functions for $X_{(1)}$ and $X_{(n)}$. Are either of these normally distributed?

**10.** Given that $X_1, X_2, \ldots, X_n$ is a random sample of a continuous random varible $X$, evaluate $E[F_X(X_{(n)})]$ and $E[F_X(X_{(1)})]$. $\left(\textbf{Hint:}\ \int n[F_X(t)]^n f_X(t)\,dt = \frac{n}{n+1}[F_X(t)]^{n+1}.\right)$

# 7

# Statistical Inference: Estimation

We now begin our study of problems of statistical inference. The first of these inference problems which we shall study is that of estimation of parameters. Suppose that we are quite sure a random variable $X$ has a normal probability law but we do not know the parameters of the probability law ($\mu$ and $\sigma^2$). Thus our assumption of a normal probability law has picked out a shape for the density of $X$ but we do not know which of the infinite number of possible values for $\mu$ and $\sigma^2$ seem appropriate. In this situation we might decide to take a sample of $n$ observations of $X$ and ask what function (statistic) of the sample values would give a good guess of the unknown values of $\mu$ or of $\sigma^2$.

Or, again, an analysis of the situation might make it reasonable to assume that a random variable $X$ has a Poisson probability function, but we might not know the value of $\lambda$ (which would uniquely specify the probability law). Given a random sample of $n$ observations of $X$, what function of the sample values would be a good guess for the unknown value of $\lambda$?

Many similar problems could be posed; each is called an estimation problem because we are asking what manipulation we might make of the observed values in a sample to get a good guess, or estimate, of the value of an unknown parameter or parameters. In this chapter we shall study two general methods of generating estimators of parameters, criteria frequently employed for comparing estimators, and then methods of interval estimation.

A word may be in order at this point regarding the notation that we shall use. As has been mentioned, in estimation problems we are trying to find a statistic (function of the elements of a random sample) which would be a good guess to make for the value of an unknown parameter. This statistic will be called the estimator of the unknown parameter; since the estimator is a statistic it is a random variable and will be denoted by a capital letter (frequently with a tilde ($\sim$) or carat ($\wedge$) over it to indicate the rationale used in deriving it). Thus, if our unknown parameter is denoted by $\gamma$, the estimator of $\gamma$ will be denoted by $\tilde{\Gamma}$ (if the method of moments was used) or by $\hat{\Gamma}$ (if maximum likelihood was used). The observed value of the estimator, once we have the observed values of the random variables, is called the estimate of the parameter and will be denoted by $\tilde{\gamma}$ or $\hat{\gamma}$.

### 7.1. The Method of Moments

The oldest general method for generating estimates of unknown parameters, given a sample, is called the method of moments. It was first proposed by Karl Pearson in 1894 and was used extensively for many years. It is a simple method to use, in most cases, and generates "reasonable" estimators. However, it does not share the universal good points of the maximum likelihood method which we shall study in the next section.

Briefly, the rationale behind the method of moments (and the method itself) proceeds as follows. We are given a random variable $X$ which has a distribution function $F_X$; this distribution function $F_X$ is indexed or specified by an unknown parameter $\gamma$. Generally, the first moment of $X$ (its mean) will depend in some relatively simple way upon $\gamma$, say $\mu_X = g(\gamma)$ ($g$ may be the identity function). Given a sample of $n$ values of $X$, we can define the first sample moment $\bar{X}$; the method of moments then specifies that we equate $\bar{X}$ to $g(\tilde{\Gamma})$ and solve for $\tilde{\Gamma}$. The resulting value of $\tilde{\Gamma}$ is called the method of moments estimator of $\gamma$.

*Example 7.1.1.* Suppose that $X$ is uniformly distributed over the interval $(0, \gamma)$. Thus the distribution function (and density function) of $X$ is specified, once we know $\gamma$. The mean of $X$ is

$$\int_0^\gamma x \frac{1}{\gamma}\, dx = \frac{\gamma}{2};$$

thus $g(\gamma) = \gamma/2$. Given a random sample of $n$ observations of $X$, we then set up the equation

$$\bar{X} = g(\tilde{\Gamma}) = \frac{\tilde{\Gamma}}{2}$$

whose solution immediately is $\tilde{\Gamma} = 2\bar{X}$.

Our method of moments estimator of $\gamma$ then is twice the sample mean; if our sample values consist of .5, .6, .1, 1.3, .9, 1.6, .7, .9, 1.0 we find that $\bar{x} = 7.6/9 = .84$ and $\tilde{\gamma} = 1.68$ (the estimate based on this sample). An obvious defect of this estimator of $\gamma$ is that it does not necessarily give rise to an estimate that generates an interval containing all of the sample values. Obviously, each sample value must lie between zero and $\gamma$, if our initial assumption is correct; then it would seem foolish to use as a guess of the value of $\gamma$ a number which did not exceed each of the sample values. For example, if we had taken a sample of 3 values of $X$ (for this same example) and found them to be 4, 6, and 50, our estimate would be $2(20) = 40$, yet we know that $\gamma$ cannot be smaller than 50. Thus, we see it is possible for this rather intuitively appealing procedure to give unsatisfactory results. Nonetheless, it is still generally simple to use and most frequently gives "good" estimators.

*Example 7.1.2.* Suppose we assume the number of traffic fatalities in San Francisco per day to be a Poisson random variable $X$ with parameter $\lambda$ (which is unknown to us). Then the mean of $X$ is $\lambda$ and thus $g(\lambda) = \lambda$; the method of moments estimator of $\lambda$, given a random sample, then is simply $g(\tilde{\Lambda}) = \tilde{\Lambda} = \bar{X}$. Suppose that we took a sample of 15 days and found the number of fatalities to be 0, 0, 2, 0, 0, 1, 0, 0, 0, 0, 0, 0, 0, 0, 0. Then $\bar{x} = \frac{3}{15} = .2$ and $\tilde{\lambda} = .2$.

As we have seen, many probability laws depend on two or more unknown parameters. Suppose then that $F_X$ is indexed by $\gamma$ and $\lambda$ (that is, values must be specified for both $\gamma$ and $\lambda$ to specify $F_X$ completely). Then, generally, both of the first two moments of $X$ will be functions of $\gamma$ and $\lambda$; i.e.,

$$\mu_X = g(\gamma, \lambda)$$
$$E(X^2) = h(\gamma, \lambda).$$

The method of moments then specifies that we should set the first two sample moments equal to $g$ and $h$, respectively, and solve the resulting equations simultaneously for $\tilde{\Gamma}$ and $\tilde{\Lambda}$, the method of moments estimators. (If $F_X$ contains more than two unknown parameters, then a correspondingly larger number of sample moments is used and an equal number of equations must be solved simultaneously; the procedure is identical to the two-parameter case which we now illustrate.)

*Example 7.1.3.* Assume that $X$, the distance that a discus will be tossed by a particular person, is normally distributed with unknown mean $\mu$ and unknown variance $\sigma^2$. Given a random sample of $n$ tosses this person has made, the method of moments estimators of $\mu$ and $\sigma^2$ are determined as follows. From our study of the normal probability law, we know that $\mu$ is the mean of $X$ and that $E[X^2] = \sigma^2 + \mu^2$. Then $g(\mu, \sigma^2) = \mu$, $h(\mu, \sigma^2) = \sigma^2 + \mu^2$, and the equations whose solutions determine the method of moments estimators are

$$\bar{X} = \tilde{\mathrm{M}}$$
$$M_2 = \tilde{\Sigma}^2 + \tilde{\mathrm{M}}^2.$$

(M is capital $\mu$, $\Sigma$ is capital $\sigma$.)

The solutions obviously are

$$\tilde{\mathrm{M}} = \bar{X}$$

$$\tilde{\Sigma}^2 = M_2 - \bar{X}^2 = \frac{n-1}{n} S^2.$$

*Example 7.1.4.* Let us now assume that $X$ is uniformly distributed on $(a, b)$ where neither $a$ nor $b$ is known. We recall that

$$\mu = g(a, b) = \frac{b + a}{2}$$

$$E[X^2] = h(a, b) = \frac{b^2 + ab + a^2}{3};$$

then given a random sample of $n$ observations of $X$ the equations which determine the method of moments estimators of $a$ and $b$ are

$$\bar{X} = \frac{\tilde{B} + \tilde{A}}{2}$$

$$M_2 = \frac{\tilde{B}^2 + \tilde{A}\tilde{B} + \tilde{A}^2}{3}.$$

Solving the first equation for $\tilde{A}$ and substituting this in the second results in the solutions:

$$\tilde{B} = \bar{X} + \sqrt{3(M_2 - \bar{X}^2)}$$

$$\tilde{A} = \bar{X} - \sqrt{(3M_2 - \bar{X}^2)}.$$

If we assume that the 9 sample values given in Example 7.1.1 are our observed values for the sample of this random variable, we find that $\bar{x} = .84$, $m_2 = .89$, $\sqrt{3(m_2 - \bar{x}^2)} = .744$, and thus $\tilde{b} = 1.584$ and $\tilde{a} = .096$. Note that the interval $(\tilde{a}, \tilde{b})$ does not include all of the sample values. If we had a sample of 3 whose values were 4, 6, and 50, then $\bar{x} = 20$, $m_2 = 850.67$, $\sqrt{3(m_2 - \bar{x}^2)} = 36.76$, and $\tilde{b} = 56.76$, $\tilde{a} = -16.76$. Notice that this interval does include all of the sample values.

Not a great deal can be definitely said about the method of moments for generating estimators of parameters. It generally yields estimators that are good but does occasionally give estimators that are obviously deficient, as illustrated in Examples 7.1.1 and 7.1.4. Since the sample moments are averages of identically distributed random variables, they are approximately normally distributed for large sample sizes. If the method of moments estimators always yielded estimators that were linear functions of the moments, we would be able to say that the method of moments estimators are approximately normal for large sample sizes. But such is not the case. For some probability laws the method of moments estimators will be approximately normally distributed, for large samples, and for others they may not.

## EXERCISE 7.1.

**1.** Assume that the amount of rainfall recorded at a certain station on a given date (for example, September 30) is uniformly distributed on the interval $(0, b)$. If a sample of 10 years' records shows that the following amounts were recorded on that date, compute the estimate $\tilde{b}$ of $b$: 0, 0, .7, 1, .1, 0, .2, .5, 0, .6. (Measurements are recorded in inches.)

**2.** Assume that the amount of growth in height of 1-foot tall Monterey pine trees in a year is a normal random variable with unknown mean and unknown variance. The growths of 5 trees were recorded as: 3 feet, 5 feet, 2 feet, 1.5 feet, 3.5 feet. Compute the method of moments estimates for $\mu$ and $\sigma^2$.

**3.** The number of cars arriving at a supermarket parking lot per hour is assumed to be a Poisson random variable with parameter $\lambda$. The following numbers of cars were observed arriving during the hour from 9 to 10 a.m. on 6 successive days: 50, 47, 82, 91, 46, 64. Compute $\tilde{\lambda}$.

**4.** $X$ is a geometric random variable with parameter $p$. Given a random sample of $n$ observations of $X$, what is the method of moments estimator of $p$?

**5.** Richard is allowed to shoot a basketball from the free-throw line until he makes a basket. He does this 5 times, with the number of shots necessary being 5, 1, 7, 4, and 9, respectively. Compute the method of moments estimate of $p$, the probability he will make a basket, shooting from the free-throw line. (Assume that the number of shots necessary is a geometric random variable.)

**6.** The number of hours an electron tube will work is assumed to be an exponential random variable with parameter $\lambda$. Given a sample of $n$ lifetimes for tubes of this sort, compute the method of moments estimator for $\lambda$.

**7.** $X$ is a Bernoulli random variable with parameter $p$. Given a random sample of $n$ observations of $X$, compute the method of moments estimator for $p$.

**8.** $X$ is a binomial random variable with parameters $n$ (known) and $p$. Given a random sample of $N$ observations of $X$, compute the method of moments estimator for $p$.

**9.** Suppose that $X$ is a normal random variable with mean $\mu = 5$ and unknown variance $\sigma^2$. What is the method of moments estimator for $\sigma^2$, based on a random sample of $n$ observations of $X$?

**10.** If $X$ is assumed to be uniformly distributed on the interval $(b - \frac{1}{4}, b + 5)$, what is the method of moments estimator for $b$ based on a random sample of $n$ observations?

**11.** $X$ is a binomial random variable with parameters $n$ and $p$, both unknown.

Given a random sample of $N$ observations of $X$, compute the method of moments estimators of $n$ and $p$.

### 7.2. Maximum Likelihood

The maximum likelihood method of generating estimators of unknown parameters, based on a sample of a random variable, was first introduced by Sir R. A. Fisher. It has proved itself a very powerful technique and one that generally gives "good" estimators, in a sense we shall be investigating. It frequently leads to the same estimator as does the method of moments but, in cases where the two methods do not agree, the maximum likelihood estimator is generally to be preferred.

The maximum likelihood method may be described as follows. Suppose that $X$ is a random variable with distribution function $F_X$ indexed by, or dependent on, an unknown parameter $\lambda$. Fisher proposed that the likelihood function $L(\lambda)$ be defined as the joint density function of the sample if $X$ is continuous or as the joint probability function of the sample if $X$ is discrete; in either case it should be evaluated at the observed sample values $x_1, x_2, \ldots, x_n$. $L(\lambda)$ then is a function of the single unknown parameter $\lambda$. Choosing that value of $\lambda$, say $\hat{\lambda}$, that maximizes $L(\lambda)$ yields the maximum likelihood estimate of $\lambda$; since $L(\lambda)$ has been evaluated at the observed sample values the maximizing value $\hat{\lambda}$ will generally be a function of the observed sample values. Let us formally define the likelihood function of the sample and the maximum likelihood estimator.

DEFINITION 7.2.1. (1) Let $X$ be a discrete random variable whose probability function $p_X$ depends on an unknown parameter $\lambda$. Let $X_1, X_2, \ldots, X_n$ be a random sample of $X$ and let $x_1, x_2, \ldots, x_n$ be the observed sample values. The *likelihood function* of the sample is defined to be

$$L(\lambda) = p_X(x_1)p_X(x_2) \cdots p_X(x_n).$$

(2) Let $X$ be a continuous random variable whose density function $f_X$ depends upon an unknown parameter $\lambda$. Let $X_1, X_2, \ldots, X_n$ be a random sample of $X$ and let $x_1, x_2, \ldots, x_n$ be the observed sample values. The *likelihood function* of the sample is defined to be

$$L(\lambda) = f_X(x_1)f_X(x_2) \cdots f_X(x_n).$$

DEFINITION 7.2.2. Given the likelihood function of the sample $L(\lambda)$, let $\hat{\lambda} = g(x_1, x_2, \ldots, x_n)$ be the maximizing value of $L(\lambda)$; i.e., $L(\hat{\lambda}) = \max_{\lambda} L(\lambda)$. The *maximum likelihood estimator* of $\lambda$ then is

$$\hat{\Lambda} = g(X_1, X_2, \ldots, X_n).$$

By basing our estimator on the value of $\lambda$ that maximizes the likelihood function, we have in a sense taken that value of $\lambda$ that maximizes the probability of occurrence of the sample results, an intuitively appealing thing to do. If we use the estimate $\hat{\lambda}$ as our guess of the unknown value of $\lambda$ (our estimate of $\lambda$), we have chosen that value which maximizes the probability of what we observed in our sample.

Before looking at some examples, let us take note of a calculus theorem which is almost always of use in finding $\hat{\lambda}$, the maximizing value of $\lambda$. Generally, since $L(\lambda)$ is either a product of probability functions or density functions, it will always be positive (for the range of possible values of $\lambda$). Thus, $K(\lambda) = \log_e L(\lambda)$ can always be defined; the theorem referred to merely states that the value of $\lambda$ that maximizes $K$ also maximizes $L$. It is almost always easier to maximize $K$ than $L$.

*Example 7.2.1.* Let $X$ be a Bernoulli random variable with unknown parameter $p$ and let $X_1, X_2, \ldots, X_n$ be a random sample of $X$. Then

$$p_X(x) = p^x(1-p)^{1-x}, \qquad x = 0, 1$$

and the likelihood function of the sample is

$$L(p) = \prod_1^n p_X(x_i) = \prod_1^n p^{x_i}(1-p)^{1-x_i}$$
$$= p^{\Sigma x_i}(1-p)^{n-\Sigma x_i}.$$

Then we define

$$K(p) = \log L(p) = \sum x_i \log p + (n - \sum x_i) \log (1-p).$$

$K$ is a continuous function of $p$ and, thus, if there exists a value of $p$ such that

$$\frac{dK}{dp} = 0, \qquad \frac{d^2K}{dp^2} < 0,$$

then this value makes $K$ a maximum. Now

$$\frac{dK}{dp} = \frac{\sum x_i}{p} - \frac{n - \sum x_i}{1-p};$$

setting this equal to zero and solving for $\hat{p}$ (the extreme point) yields

$$\hat{p} = \frac{\sum x_i}{n} = g(x_1, x_2, \ldots, x_n).$$

We note that

$$\frac{d^2K}{dp^2} = -\frac{\sum x_i}{p^2} - \frac{n - \sum x_i}{(1-p)^2}$$

which is negative for any value of $p$, so $\hat{p}$ does correspond to a maximum value of

$K$ (and $L$). The maximum likelihood estimator of $p$ then is

$$\hat{P} = g(X_1, X_2, \ldots, X_n) = \frac{\sum X_i}{n} = \bar{X}.$$

(Note that this is the same estimator as the method of moments yields in problem 7, Exercise 7.1.)

*Example 7.2.2.* Suppose that $X$ is a normal random variable whose mean $\mu$ is unknown and whose variance is equal to 1. If $X_1, X_2, \ldots, X_n$ is a random sample of $X$, then the likelihood function of the sample is

$$L(\mu) = \prod_{i=1}^{n} f_X(x_i) = \prod_{1}^{n} \frac{1}{\sqrt{2\pi}} e^{-(x_i-\mu)^2/2}$$

$$= \left(\frac{1}{2\pi}\right)^{n/2} e^{-\Sigma(x_i-\mu)^2/2};$$

its logarithm is

$$K(\mu) = \log L(\mu) = -\frac{n}{2} \log 2\pi - \sum \frac{(x_i - \mu)^2}{2}$$

and

$$\frac{dK}{d\mu} = \sum (x_i - \mu) = \sum x_i - n\mu$$

$$\frac{d^2K}{d\mu^2} = -n < 0.$$

Thus

$$\hat{\mu} = \frac{1}{n} \sum x_i = \bar{x}$$

maximizes $K$ (and $L$) and the maximum likelihood estimator is

$$\hat{\text{M}} = \frac{1}{n} \sum X_i = \bar{X}.$$

(This is also the value that the method of moments yields in this case.)

In each of the two examples discussed so far for maximum likelihood estimation we were able to take merely the derivative of the log of the likelihood function and equate it to zero to find the estimator. If we were to plot the likelihood function as a function of the unknown parameter in these two cases, we would find it had a graph like Figure 7.1. The likelihood function in these cases has a unique maximum value and the derivative is zero at that point. It is possible, however, for the likelihood function to have one of the shapes pictured in Figure 7.2; as illustrated in 7.2*a*, there may be two points at which the derivative is zero. The maximum likelihood estimator is the value corresponding to the absolute maximum value, as marked. For the shapes given in 7.2*b* and *c*, the derivative is not zero at the maximum value.

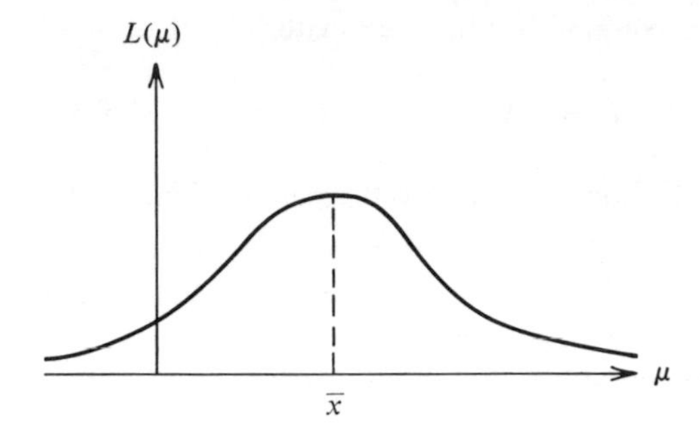


**Figure 7.1.**

In these cases other techniques must be relied upon to find the estimator. Example 7.2.3 gives an example in which the likelihood function has a shape like that illustrated in Figure 7.2*c*.

*Example 7.2.3.* Suppose that $X$ is a uniform random variable on the interval $(0, \gamma)$ where $\gamma$ is an unknown positive parameter. Then

$$f_X(x) = \frac{1}{\gamma}, \qquad \text{for } 0 < x < \gamma$$

and the likelihood function for a random sample of $n$ is

$$L(\gamma) = \frac{1}{\gamma^n}.$$

The graph of this function is given in Figure 7.2*c*. We note that the slope is not

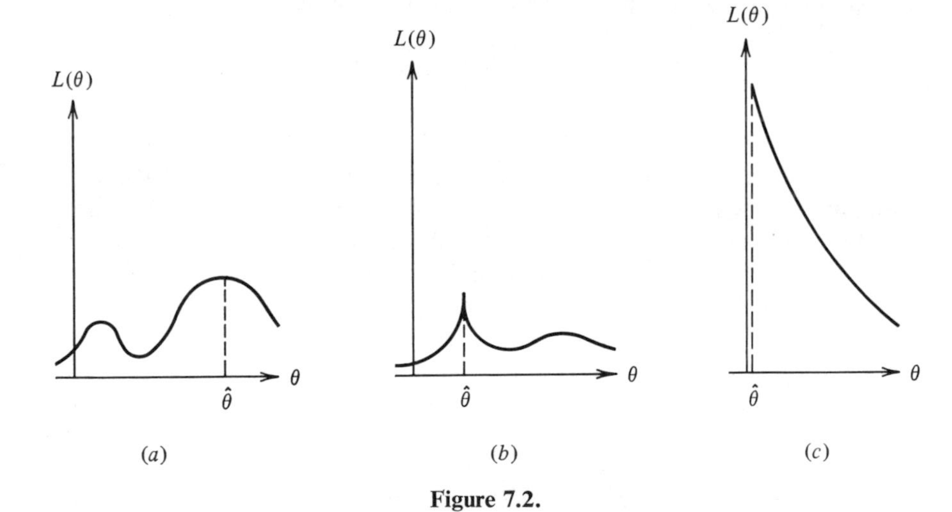


**Figure 7.2.**

zero anywhere, so there would be no point in taking the derivative of $L$ (or of $K$) and equating it to zero. However, we note that the likelihood function increases in value the closer that $\gamma$ is to zero; thus $L(\gamma)$ is maximized by setting $\gamma$ equal to the smallest value it could possibly equal. Clearly, $\gamma$ can be no smaller than the largest value that occurs in our sample (since each element of the sample is assumed to have the same density as $X$). Thus $\hat{\gamma} = x_{(n)}$, and the maximum likelihood estimator is $\hat{\Gamma} = X_{(n)}$, the maximum value in the sample. Note that this estimator would not be subject to the criticism levelled against the method of moments estimator for this parameter in Example 7.1.1.

If the random variable $X$ has a distribution function indexed by two parameters $\gamma$ and $\lambda$, for example, then the likelihood function is defined in exactly the same way but it is now a function of two arguments, $\lambda$ and $\gamma$, rather than one. The maximum likelihood estimators of the two unknown parameters are specified by the two coordinates $\hat{\gamma}$ and $\hat{\lambda}$ (each of which will be functions of the observed sample values) corresponding to the maximum value of $L(\gamma, \lambda)$.

If $F_X$ is indexed by more than two unknown parameters, the likelihood function $L$ is a function of more than two parameters. The maximum likelihood estimators are again specified simply by the coordinates, no matter how many there may be, of the maximum value of $L$. Let us consider a specific example.

*Example 7.2.4.* Suppose we assume that the weights of Valencia oranges are normally distributed with mean $\mu$ and variance $\sigma^2$, both of which are unknown. If we were to take a random sample of $n$ oranges of this type, we could use the method of maximum likelihood to estimate $\mu$ and $\sigma^2$ as follows. The likelihood function for the sample is a function of the two arguments, $\mu$ and $\sigma^2$, and is

$$\begin{aligned} L(\mu, \sigma^2) &= \prod_{i=1}^{n} \frac{1}{\sigma\sqrt{2\pi}} e^{-(x_i-\mu)^2/2\sigma^2} \\ &= \frac{1}{(2\pi\sigma^2)^{n/2}} e^{-\frac{1}{2}\Sigma(x_i-\mu)/\sigma^2}. \end{aligned}$$

Then, since $L$ is a continuous function of $\mu$ and $\sigma^2$, we might expect to find the maximizing values, $\hat{\mu}$ and $\hat{\sigma}^2$, by setting the two first partial derivatives of $L$ (or of $K = \log L$) equal to zero and solving. Thus,

$$K(\mu, \sigma^2) = -\frac{n}{2}\log 2\pi - \frac{n}{2}\log \sigma^2 - \frac{1}{2\sigma^2}\sum (x_i - \mu)^2$$

and

$$\begin{aligned} \frac{\partial K}{\partial \mu} &= \frac{1}{\sigma^2}\sum (x_i - \mu) \\ \frac{\partial K}{\partial \sigma^2} &= -\frac{n}{2\sigma^2} + \frac{1}{2(\sigma^2)^2}\sum (x_i - \mu)^2. \end{aligned}$$

The equations to be solved for the estimates then are

$$\frac{1}{\hat{\sigma}^2}\sum (x_i - \hat{\mu}) = 0$$

$$-\frac{n}{2\hat{\sigma}^2} + \frac{1}{2(\hat{\sigma}^2)^2}\sum (x_i - \hat{\mu})^2 = 0.$$

The first equation clearly has as solution

$$\hat{\mu} = \frac{1}{n}\sum x_i = \bar{x}.$$

Substituting this in the second equation and solving for $\hat{\sigma}^2$ yields

$$\hat{\sigma}^2 = \frac{1}{n}\sum (x_i - \bar{x})^2.$$

It can be shown that these two quantities do correspond to the maximum value of $L$. The maximum likelihood estimators of $\mu$ and $\sigma^2$ then are

$$\hat{\mathrm{M}} = \bar{X}$$

$$\hat{\Sigma}^2 = \frac{1}{n}\sum (X_i - \bar{X})^2 = \frac{n-1}{n}S^2.$$

These are the same estimators that the method of moments yields for this problem.

## EXERCISE 7.2.

**1.** Suppose that $X$ is a normal random variable with mean $\mu = 10$ and variance $\sigma^2$ unknown. What is the maximum likelihood estimator for $\sigma^2$, based on a random sample of $n$ observations of $X$?

**2.** Suppose that you buy a dozen Valencia oranges, weigh each of them, and find

$$\sum_{i=1}^{12} x_i = 66, \qquad \sum_{i=1}^{12} x_i^2 = 385$$

(in ounces). Compute the maximum likelihood estimates of $\mu$ and $\sigma^2$, assuming that the 12 oranges you have purchased are a random sample of all Valencia oranges (and their weights are normally distributed).

**3.** Suppose that $X$ is a Poisson random variable with parameter $\lambda$. Given a random sample of $n$ observations of $X$, what is the maximum likelihood estimator for $\lambda$?

**4.** Assume that the number of new car sales, per day, that a particular dealer makes is a Poisson random variable with parameter $\lambda$. Given that in 20 days the total number of sales he made was 30 cars, what is the maximum likelihood estimate of $\lambda$?

**5.** If $X$ is a geometric random variable with parameter $p$, what is the maximum likelihood estimator for $p$ based on a random sample of $n$ observations of $X$?

**6.** Suppose that $X$ were uniformly distributed on the interval $b - \frac{1}{2}$ to $b + \frac{1}{2}$. What is the maximum likelihood estimator for $b$ based on a random sample of $n$ observations of $X$?

**7.** What is the maximum likelihood estimator for the parameter $\lambda$ of an exponential random variable for a sample of size $n$?

**8.** A random sample of 30 electron tubes was placed in operation and the time to failure for each was recorded. If you assume that the recorded times were such that $\sum x_i = 32{,}916$ hours, what is the maximum likelihood estimate for the parameter of the exponential distribution of tube lives?

**9.** Suppose that $X$ is a binomial random variable with parameters $n$ (known) and $p$. What is the maximum likelihood estimator for $p$ based on a random sample of $N$ observations of $X$?

### 7.3. Properties of Estimators

In the last two sections we have seen two different, generally applicable methods for constructing estimators of unknown parameters. In many cases the two methods generate the same estimator, but in many important problems they do not lead to the same estimator. When we are faced with the choice of two or more estimators for the same parameter, it becomes important to develop criteria for comparing them; if we were able to develop some scale of "goodness" of estimators, then we would always want to use the estimator that was best for the given problem. Unfortunately there is no universally appropriate scale of goodness which can be used in comparing estimators of the same parameter.

As we have seen, the estimator of an unknown parameter is a statistic, a random variable whose value can be observed on the basis of a sample. Since an estimator $\hat{\Gamma}$ is always a random variable, the particular value it takes on varies from one sample to another; thus we certainly would not expect to have the estimate $\hat{\gamma}$ equal to the true unknown value $\gamma$ for every sample we take. If we are considering two estimators, $\tilde{\Gamma}$ and $\hat{\Gamma}$, for the same parameter $\gamma$, we can in theory derive the probability laws of $\tilde{\Gamma}$ and of $\hat{\Gamma}$ from the probability law of the random sample. Comparison of $\tilde{\Gamma}$ and $\hat{\Gamma}$ then reduces to a comparison of their two respective probability laws in some way. If, for example, $\tilde{\Gamma}$ were uniformly distributed from $\gamma - \frac{1}{4}$ to $\gamma + \frac{1}{4}$ while $\hat{\Gamma}$ was uniformly distributed on the interval from $\gamma - \frac{1}{16}$ to $\gamma + \frac{1}{16}$, we would obviously prefer $\hat{\Gamma}$ as an estimator of $\gamma$ (and thus would use $\hat{\gamma}$ rather than $\tilde{\gamma}$ as our estimate for any particular sample). Unfortunately the comparison of probability laws of estimators is generally not this straightforward; in most cases the probability density functions (or probability functions) of various estimators are of widely divergent types and there is then a variety of ways in which they could be compared. Perhaps the most obvious property to investigate of the probability law of an estimator $\Gamma$ would be its mean value. As the following definition states, if the mean value of $\Gamma$ is $\gamma$, then $\Gamma$ is unbiased; otherwise it is a biased estimator.

DEFINITION 7.3.1. An estimator $\Gamma$ of an unknown parameter $\gamma$ is *unbiased* if $E[\Gamma] = \gamma$.

Thus an unbiased estimator is a random variable whose expected value is the parameter being estimated; if we were to take samples of size $n$ repeatedly and for each compute the observed value of $\Gamma$ (the estimate for that sample outcome), then the average of these observed values would be $\gamma$, the parameter being estimated.

*Example 7.3.1.* We saw in Chapter 6, Section 3, that $E[\bar{X}] = \mu$, no matter what the probability law from which we were sampling. Thus $\bar{X}$ is an unbiased estimator of $\mu$ in a normal density, of the parameter $p$ for a Bernoulli random variable, and of the parameter $\lambda$ for a Poisson random variable (since in each of these cases the parameter mentioned is the mean of the random variable being sampled).

The property of unbiasedness, while generally desirable for an estimator, is not the only criterion that should be employed when comparing estimators. In all cases we can find a large number of unbiased estimators of a parameter; thus we should have at least one additional criterion for judging which is best among all unbiased estimators.

If an estimator is unbiased, then the average of the values it takes on over an infinite number of samples is the parameter estimated; however, this statement does not imply that the estimator has very high probability of lying close to the unknown parameter for any given sample. Clearly, if $\Gamma_1$ and $\Gamma_2$ are two different unbiased estimators of $\gamma$ for the same sample, and $\Gamma_1$ had probability .9 of being within 1 unit of $\gamma$ while $\Gamma_2$ had probability .5 of being within 1 unit of $\gamma$, then we would prefer to use the observed value of $\Gamma_1$ as our estimate rather than the observed value of $\Gamma_2$. This additional idea of choosing the unbiased estimator that has high probability of lying close to the parameter leads one to consider the variances of the unbiased estimators. The unbiased estimator that has smaller variance (about the true value of the parameter) then would have the larger probability of being close to the unknown parameter. Generally in the literature this comparison of variances of estimators is described as comparisons of efficiency of estimators (the more efficient estimator has smaller variance). Accordingly, we have the following definition.

DEFINITION 7.3.2. If $\Gamma_1$ and $\Gamma_2$ are both unbiased estimators of $\gamma$ for the same sample, then $\Gamma_1$ is *more efficient* than $\Gamma_2$ if

$$V(\Gamma_1) < V(\Gamma_2)$$

($V(\Gamma_1)$ is the variance of $\Gamma_1$).

The following example compares two unbiased estimators.

*Example 7.3.2.* Suppose that $X$ is a Poisson random variable with parameter $\lambda$ and $X_1, X_2, \ldots, X_n$ is a random sample of $X$. Then $\Lambda_1 = \bar{X}$ and $\Lambda_2 = \frac{1}{2}(X_1 + X_2)$ are both unbiased estimators of $\lambda$ and

$$V(\Lambda_1) = \frac{\lambda}{n}, \qquad V(\Lambda_2) = \frac{\lambda}{2}.$$

Thus, if $n > 2$, $\Lambda_1$ is a more efficient estimator of $\lambda$ than $\Lambda_2$ since

$$\frac{\lambda}{n} < \frac{\lambda}{2}.$$

DEFINITION 7.3.3. $\Gamma$ is the *best linear unbiased estimator* for $\gamma$ if: (1) $\Gamma$ is a linear function of $X_1, X_2, \ldots, X_n$, (2) $E[\Gamma] = \gamma$ for all $\gamma$, and (3) among all estimators $\Gamma^*$ satisfying (1) and (2), $V(\Gamma) \leq V(\Gamma^*)$.

A best linear unbiased estimator then is one that has both the property of unbiasedness and the smallest possible variance within a certain class of estimators. (Thus it could also be called most efficient within the class of linear unbiased estimators.)

***Theorem 7.3.1.*** $X$ is a random variable with mean $\mu$ and variance $\sigma^2$. If $X_1, X_2, \ldots, X_n$ is a random sample of $X$, $\bar{X}$ is the best linear unbiased estimator of $\mu$.

*Proof:* Let M be a linear unbiased estimator of $\mu$. Then

$$\mathrm{M} = \sum_{i=1}^{n} a_i X_i + b$$

and

$$E[\mathrm{M}] = \mu \sum a_i + b.$$

Since M must be unbiased, we must have $\mu \sum a_i + b = \mu$, no matter what value $\mu$ may have. This implies then that $b = 0$ and $\sum a_i = 1$. The variance of M is

$$\begin{aligned} V(\mathrm{M}) &= V(\sum a_i X_i) \\ &= \sigma^2 \sum a_i^2. \end{aligned}$$

Finding the best linear unbiased estimator then reduces to finding $a_1, a_2, \ldots, a_n$ such that $\sum a_i = 1$ and $\sum a_i^2$ is as small as possible. Now, $\sum a_i = 1$

implies that

$$a_1 = 1 - \sum_2^n a_i$$

and thus we want to minimize

$$Q = \left[1 - \sum_2^n a_i\right]^2 + \sum_2^n a_i^2.$$

This minimization is accomplished by setting

$$\frac{\partial Q}{\partial a_2}, \frac{\partial Q}{\partial a_3}, \ldots, \frac{\partial Q}{\partial a_n}$$

each equal to zero and solving the resulting equations for $a_2^*, a_3^*, \ldots, a_n^*$ $\left(\text{then } a_1^* = 1 - \sum_2^n a_i^*\right)$. We find

$$\frac{\partial Q}{\partial a_j} = 2\left[1 - \sum_2^n a_i\right](-1) + 2a_j, \qquad j = 2, 3, \ldots, n$$

and thus

$$a_j^* = \left[1 - \sum_2^n a_i^*\right] = a_1^*, \qquad j = 2, 3, \ldots, n.$$

But then

$$\sum_1^n a_i^* = na_1^* = 1$$

and thus

$$a_j^* = \frac{1}{n}, \qquad j = 1, 2, \ldots, n$$

are the values that minimize $\sum_1^n a_i^2$ subject to the condition that $\sum_1^n a_i = 1$. Thus,

$$\mathrm{M} = \sum \frac{1}{n} X_i = \bar{X}$$

is the linear unbiased estimator of $\mu$ with the smallest variance. It can, in fact, be shown that $\bar{X}$ has the smallest variance among all unbiased estimators of $\mu$, regardless of whether they are linear functions of $X_1, X_2, \ldots, X_n$. ◀

Note that the preceding theorem holds true without assuming a specific form for the distribution function for $X$ (other than the fact that $\mu$ and $\sigma^2$ are assumed to exist). Thus the sample mean is the best linear unbiased estimator for $p$ in the Bernoulli probability function, for $\lambda$ in the Poisson probability function, and for $\mu$ in the normal density function.

Let us make note of one further property which estimators may possess, that of consistency.

DEFINITION 7.3.4. $\Theta$ is a *consistent* estimator of $\theta$ if

$$\lim_{n\to\infty} P[|\Theta - \theta| > \varepsilon] = 0$$

or, equivalently, if

$$\lim_{n\to\infty} P[|\Theta - \theta| < \varepsilon] = 1, \qquad \text{for any } \varepsilon > 0$$

($n$ is the sample size).

Note first of all that an estimator $\Theta$ is consistent if a sequence of probabilities (evaluated from the probability or density function of $\Theta$) converges to zero (or 1) as the sample size increases. Thus consistency, strictly speaking, has to do only with the limiting behavior of an estimator as the sample size increases without limit and does not imply that the observed value of $\Theta$ is necessarily close to $\theta$ for any specific size of sample $n$. The consistency of an estimator is called an asymptotic or large sample property of the estimator; presumably if $n$ is fixed at some "large" value, then

$$P[|\Theta - \theta| > \varepsilon]$$

is quite small. What sample size is "large" for one estimator is not necessarily large for another in this sense. In particular, if we are only going to have a relatively small sample available, and only one at that, it would seem quite immaterial whether we use a consistent estimator or not.

The following theorem, which we shall not prove, gives a relatively easy check on whether a particular estimator is consistent.

***Theorem** 7.3.2.* If

$$\lim_{n\to\infty} E[\Gamma] = \gamma$$

and

$$\lim_{n\to\infty} V(\Gamma) = 0,$$

then $\Gamma$ is a consistent estimator of $\gamma$ ($n$ is the sample size).

Notice that if $\Gamma$ is a random variable whose expected value gets closer and closer to $\gamma$ as $n$ increases, and whose variance shrinks to zero as $n$ increases, it will be a consistent estimator of $\gamma$.

*Example 7.3.3.* Suppose that $X$ is a random variable with mean $\mu$ and variance $\sigma^2$. If $X_1, X_2, \ldots, X_n$ is a random sample of $X$, then we know that

$$E[\bar{X}] = \mu, \qquad V(\bar{X}) = \frac{\sigma^2}{n}.$$

Thus

$$\lim_{n\to\infty} E[\bar{X}] = \mu, \qquad \lim_{n\to\infty} V(\bar{X}) = \lim_{n\to\infty} \frac{\sigma^2}{n} = 0$$

and $\bar{X}$ is a consistent estimator of $\mu$.

*Example 7.3.4.* If $X$ is a normal random variable with mean $\mu$ and variance $\sigma^2$, and $X_1, X_2, \ldots, X_n$ is a random sample of $X$, then $[(n-1)/\sigma^2]S^2$ is a $\chi^2$ random variable with $n - 1$ degrees of freedom (see Theorem 6.3.5). Furthermore, the mean of a $\chi^2$ random variable is its degrees of freedom and the variance of a $\chi^2$ random variable is twice its degrees of freedom. Thus

$$E\left[\frac{n-1}{\sigma^2} S^2\right] = n - 1$$

$$V\left(\frac{n-1}{\sigma^2} S^2\right) = 2(n - 1);$$

that is

$$E[S^2] = \sigma^2$$

$$V(S^2) = \frac{2\sigma^4}{n-1}$$

and thus $S^2$ is a consistent estimator of $\sigma^2$.

Let us now make note of some general results regarding maximum likelihood estimators of parameters. The first of these is generally called the *invariant property* of maximum likelihood estimators. Suppose $\hat{\Gamma}$ is the maximum likelihood estimator of a parameter $\gamma$. If for some reason it is desired to estimate a function of $\gamma$, say $\theta = h(\gamma)$, rather than $\gamma$ itself, we could revert to the likelihood function $L(\gamma)$ and express it in terms of $\theta$ rather than $\gamma$ (by substituting the values of $\theta$ corresponding to those of $\gamma$ through the function $\theta = h(\gamma)$). Then the maximum likelihood estimator of $\theta$, $\hat{\Theta}$, would be determined by the value that maximizes $L$. The invariant property of maximum likelihood estimators says that this maximizing value $\hat{\theta}$ of $L$ will yield the estimator $\hat{\Theta} = h(\hat{\Gamma})$; thus there is no need actually to express $L$ in terms of $\theta$ to get the estimator if we already have $\hat{\Gamma}$.

It can also be shown that maximum likelihood estimators are generally consistent and, for large sample sizes, approximately normally distributed. This latter result is quite strong and says that we actually know the density function (approximately) for the maximum likelihood estimators for large sample sizes. This result is given, without proof, in the following theorem.

***Theorem 7.3.3.*** If $X$ is a random variable with probability function $p_X$ or density function $f_X$ which depends on an unknown parameter $\gamma$ and $\hat{\Gamma}$ is the maximum likelihood estimator based on a random sample of size $n$ of $X$, then $\hat{\Gamma}$ is approximately normally distributed for large $n$ with mean $\gamma$ and variance

$$\left\{nE\left[\frac{d}{d\gamma}\ln p_X(X)\right]^2\right\}^{-1}$$

or

$$\left\{nE\left[\frac{d}{d\gamma}\ln f_X(X)\right]^2\right\}^{-1},$$

depending on whether $X$ is discrete or continuous.

This result again is one where "large" must remain undefined; for some probability laws $n = 50$ would undoubtedly be large enough for sufficient accuracy while for others the sample size might have to be 500 to attain the same accuracy.

*Example 7.3.5.* Suppose that the time to failure $X$ of a certain type of electron tube is an exponential random variable with parameter $\theta$. Thus, $E[X] = 1/\theta$. If we take a random sample of $n$ observations of $X$, the maximum likelihood estimator of $\theta$ is

$$\hat{\Theta} = \frac{1}{\bar{X}}$$

(see problem 7, Exercise 7.2). By the invariance principle the maximum likelihood estimator of $E[X] = 1/\theta$ is $1/\hat{\Theta} = \bar{X}$, the sample mean.

*Example 7.3.6.* Suppose that $X_1, X_2, \ldots, X_n$ is a random sample of a geometric random variable with parameter $p$. Then the maximum likelihood estimator of $p$ is

$$\hat{P} = \frac{1}{\bar{X}}$$

(see problem 5, Exercise 7.2). For large $n$, then, Theorem 7.3.3 says that $1/\bar{X}$ is approximately normally distributed with mean $p$ and variance $p^2(1-p)/n$. This large sample variance is derived as follows: Since $X$ is a geometric random variable,

$$p_X(X) = (1-p)^{X-1}p$$

$$\frac{d}{dp}\ln p_X(X) = \frac{1}{1-p}\left[\frac{1}{p} - X\right]$$

$$E\left[\frac{d}{dp}\ln p_X(X)\right]^2 = \frac{1}{(1-p)^2}E\left[\left(X - \frac{1}{p}\right)^2\right]$$

$$= \frac{1}{(1-p)^2}\frac{1-p}{p^2} = \frac{1}{p^2(1-p)}$$

and thus

$$\left\{nE\left[\frac{d}{dp}\ln p_X(X)\right]^2\right\}^{-1} = \frac{p^2(1-p)}{n}.$$

**EXERCISE 7.3.**

**1.** Suppose that $X_1, X_2, \ldots, X_6$ is a random sample of a normal random variable with mean $\mu$ and variance $\sigma^2$. Determine $C$ such that

$$C[(X_1 - X_2)^2 + (X_3 - X_4)^2 + (X_5 - X_6)^2]$$

is an unbiased estimator of $\sigma^2$.

**2.** Compute the variance of the estimator of $\sigma^2$ given in problem 1. Is it more efficient than the sample variance $S^2$?

**3.** Suppose that the time it takes a distance runner to run a mile is a normal random variable with mean $\mu$ and variance $\sigma^2$. Suppose he will run $n_1$ mile races in May and $n_2$ mile races in June. Let $\overline{X}_1$ be the mean value of his times in May and $\overline{X}_2$ be the mean value of his times in June. Then $a\overline{X}_1 + (1 - a)\overline{X}_2$ will be an unbiased estimator of $\mu$ for any value of $a$. Find the value of $a$ that minimizes the variance of $a\overline{X}_1 + (1 - a)\overline{X}_2$, assuming $\overline{X}_1$ and $\overline{X}_2$ are independent.

**4.** $X_1, X_2, \ldots, X_n$ is a random sample of a normal random variable with mean $\mu$ and variance $\sigma^2$ ($n$ is an even number). Is

$$\frac{1}{n}\sum_{i=1}^{n/2}(X_{2_i} - X_{2_{i-1}})^2$$

a consistent estimator for $\sigma^2$?

**5.** On the basis of his experience, an automobile dealer feels that the number of new car sales he makes per working day is a Poisson random variable with parameter $\lambda$. He examines his records for the past year (which consisted of 310 working days) and finds that the total number of cars he sold was 279. Compute the maximum likelihood estimate of the probability that he will sell no cars on his next working day.

**6.** Assume that the lengths of lives of American males are normally distributed with mean $\mu$ and variance $\sigma^2$. A random sample of 10,000 American males' mortality histories yielded the estimate $\bar{x} = 72.1$ years, $s^2 = 144$ years$^2$. Compute the maximum likelihood estimate of the probability that an American male will live to be 50 or older and the estimate that he will not live to be 90.

**7.** Derive the asymptotic probability density function for the maximum likelihood estimator of the parameter $\lambda$ of the exponential random variable.

**8.** Assume that $X$ is uniform on the interval $(0, \gamma)$. Based on a random sample of $n$ observations, the maximum likelihood estimator is $\hat{\Gamma} = \max(X_1, X_2, \ldots, X_n)$ while the method of moments estimator is $\tilde{\Gamma} = 2\overline{X}$. Compute $E[\hat{\Gamma}]$ and $E[\tilde{\Gamma}]$ and the variances of the two estimators. Which would you prefer? Are they consistent? (The reader might like to review the density function of the maximum in Section 6.3.)

**9.** Suppose that $T$ is an exponential random variable with parameter $\lambda$. Given a random sample of $n$ observations of $T$, construct an unbiased estimator of $1/\lambda$, using the minimum sample value. Would you prefer this estimator or $\bar{T}$ for estimating $1/\lambda$?

## 7.4. Confidence Intervals

So far in this chapter we have considered the problem of finding: (1) a point estimator $\Gamma$ of an unknown parameter $\gamma$, given a random sample of observations of a random variable, and (2) criteria for comparing two or more estimators of the same parameter. By point estimator we are simply referring to the fact that, after the sampling has been done and the observed value of the estimator computed, our end product is this single number, a point on the real line, which we feel is a good guess for the unknown true value of the parameter. If we are using a good estimator, according to some criteria, then we feel that the estimate we give should probably be close to the unknown true value. However, the single number itself does not include any indication of how high the probability might be that the estimator has taken on a value close to the unknown true value. The methods of confidence intervals are meant to give both an idea of the actual numerical value the parameter may have and also an indication of how confident we may be, on the basis of the sample, that we have given a correct indication of the possible numerical value of the parameter. Let us now define a confidence interval for an unknown parameter.

DEFINITION 7.4.1. Suppose that $X$ is a random variable whose probability law depends on an unknown parameter $\theta$. Given a random sample of $X$, $X_1, X_2, \ldots, X_n$, the two statistics $L_1$ and $L_2$ form a 100 $(1 - \alpha)\%$ *confidence interval* for $\theta$ if

$$P(L_1 \leq \theta \leq L_2) \geq 1 - \alpha,$$

no matter what the unknown value of $\theta$.

Note first of all that the interval $(L_1, L_2)$ might be termed a random interval, since both of its end points are random variables and hence will vary from one sample to another. Once we have taken the sample, we can compute the observed values for $L_1$ and $L_2$ and then can say that we are 100 $(1 - \alpha)\%$ confident that the interval bracketed by their observed values contains the unknown value of $\theta$.

*Example 7.4.1.* Suppose that the number of ounces of beer that a machine puts into a bottle is a normal random variable with unknown mean $\mu$ and known standard deviation of $\frac{1}{2}$ ounce. If we randomly select $n$ bottles filled by this machine and let $X_1, X_2, \ldots, X_n$, respectively, be the number of ounces they contain, then $\bar{X}$ is a normal random variable with mean $\mu$ and standard deviation $1/2\sqrt{n}$. From a standard normal table we know that

$$P(-1.96 \leq Z \leq 1.96) = .95,$$

where $Z$ is a standard normal random variable. But, for our sample,

$$\frac{\bar{X} - \mu}{1/2\sqrt{n}}$$

is a standard normal random variable and thus

$$P\left(-1.96 \leq \frac{\bar{X} - \mu}{1/2\sqrt{n}} \leq 1.96\right) = .95.$$

The inequality,

$$-1.96 \leq \frac{\bar{X} - \mu}{1/2\sqrt{n}} \leq 1.96,$$

is equivalent to

$$\bar{X} - 1.96\frac{1}{2\sqrt{n}} \leq \mu \leq \bar{X} + 1.96\frac{1}{2\sqrt{n}},$$

since any time the first statement is true so is the second and vice versa; thus their probabilities must be equal and we have

$$P\left(\bar{X} - 1.96\frac{1}{2\sqrt{n}} \leq \mu \leq \bar{X} + 1.96\frac{1}{2\sqrt{n}}\right) = .95.$$

The two statistics

$$L_1 = \bar{X} - 1.96\frac{1}{2\sqrt{n}}$$

$$L_2 = \bar{X} + 1.96\frac{1}{2\sqrt{n}}$$

then form a 95% confidence interval for $\mu$. If we actually took a sample of 75 bottles filled by this machine and found $\bar{x} = 12.1$, the observed values of $L_1$ and $L_2$, the 95% confidence interval for $\mu$, would be 11.99 and 12.21, respectively. We then say that we are 95% confident that the interval (11.99, 12.21) includes $\mu$.

It is important to note the sort of statement that we can make regarding the observed values of a confidence interval. If we were repeatedly to take samples of 75 bottles from the output of this machine, and compute the observed values of $L_1$ and $L_2$ for each, we would find that 95% of these observed intervals actually did include $\mu$ while 5% did not. Given the results of a single sample, such as that quoted above, we do not know whether the particular interval we have computed brackets $\mu$ or not. However, 95% of the intervals computed in this fashion bracket $\mu$, which is the basis for saying we are 95% confident that we have bracketed $\mu$. The following theorem shows how to construct a 100 $(1 - \alpha)$% confidence interval, with $0 < \alpha < 1$, for the mean value of a normal random variable whose standard deviation $\sigma$ is known.

***Theorem 7.4.1.*** If $X_1, X_2, \ldots, X_n$ is a random sample of a normal random variable with unknown mean $\mu$ and known variance $\sigma^2$, then the two statistics

$$L_1 = \bar{X} - z_{1-\alpha/2}\frac{\sigma}{\sqrt{n}}$$

$$L_2 = \bar{X} + z_{1-\alpha/2}\frac{\sigma}{\sqrt{n}},$$

where

$$P(Z \le z_{1-\alpha/2}) = 1 - \frac{\alpha}{2},$$

constitute a 100 $(1 - \alpha)$% confidence interval for $\mu$. ($Z$ is a standard normal random variable.)

*Proof:* Given that $X_1, X_2, \ldots, X_n$ is a random sample of a normal random variable, we know that $\bar{X}$ is normal with mean $\mu$ and variance $\sigma^2/n$. Thus

$$Z = \frac{\bar{X} - \mu}{\sigma/\sqrt{n}}$$

is a standard normal random variable and we can find $z_{1-\alpha/2}$ from a standard normal table ($z_{1-\alpha/2}$ is simply the 100 $(1 - \alpha/2)$-th percentile of the standard normal distribution). Since the standard normal density is symmetric about the line $z = 0$,

$$P(-z_{1-\alpha/2} \le Z \le z_{1-\alpha/2}) = 1 - \alpha$$

and, in particular,

$$P\left(-z_{1-\alpha/2} \le \frac{\bar{X} - \mu}{\sigma/\sqrt{n}} \le z_{1-\alpha/2}\right) = 1 - \alpha.$$

But

$$-z_{1-\alpha/2} \le \frac{\bar{X} - \mu}{\sigma/\sqrt{n}} \le z_{1-\alpha/2}$$

is equivalent to

$$\bar{X} - z_{1-\alpha/2}\frac{\sigma}{\sqrt{n}} \le \mu \le \bar{X} + z_{1-\alpha/2}\frac{\sigma}{\sqrt{n}}.$$

Thus $P(L_1 \le \mu \le L_2) = 1 - \alpha$, where $L_1$ and $L_2$ are as defined above. ◀

*Example 7.4.2.* Suppose that Mr. Jones knows that the length of time it takes him to drive from his home to his business is a normal random variable with a standard deviation of 2 minutes. Over a period of a month, which includes 22 working days, he records the actual time it takes him to make this trip and finds the average of these times to be 28.5 minutes. By checking the standard normal tables, we find $z_{.9} = 1.28$, $z_{.95} = 1.64$, and thus the observed value of the 80% confidence interval for $\mu$ is given by

$$\left(28.5 - \frac{1.28(2)}{\sqrt{22}}, \quad 28.5 + \frac{1.28(2)}{\sqrt{22}}\right) = (27.95, 29.05);$$

the observed value of the 90% confidence interval for $\mu$ is

$$\left(28.5 - \frac{1.64(2)}{\sqrt{22}}, \quad 28.5 + \frac{1.64(2)}{\sqrt{22}}\right) = (27.80, 29.20).$$

Note in this example that length of the observed value of the 90% confidence interval is longer than the observed length of the 80% confidence interval; this will in general be the case since we need to take a longer interval to be more confident of bracketing $\mu$.

*Example 7.4.3.* Notice that if we have a random sample of size $n$ of a normal random variable with unknown mean $\mu$ and known variance $\sigma^2$, the length of the 100 $(1 - \alpha)$% confidence interval for $\mu$ is a constant. In fact,

$$L_2 - L_1 = 2z_{1-\alpha/2}\frac{\sigma}{\sqrt{n}}.$$

and we could select a sample size $n$ to give us any desired preselected length of interval for $\mu$. For example, if $\sigma = 2$, the length of the 95% confidence interval for $\mu$ would be

$$\frac{2(1.96)2}{\sqrt{n}} = \frac{7.84}{\sqrt{n}}$$

Thus if we choose $n = 10$ the 95% confidence interval for $\mu$ will be 2.48 units long, whereas if we choose $n = 100$ it will be only .78 unit long.

Occasionally we might want to construct a *one-sided confidence interval* in which we simply observe a one-sided interval with at least probability $1 - \alpha$ of enclosing an unknown parameter. The following example is of this type.

*Example 7.4.4.* A manufacturer of light bulbs, through his past experience, is quite sure that the lengths of lives of bulbs he manufactures are normally distributed with mean $\mu$ and standard deviation of 50 hours. He has an idea that the mean $\mu$ is close to 1500 hours, but for advertising purposes he would like to find a number of hours that he is quite sure $\mu$ will exceed. He decides that a lower 99% confidence interval will serve this purpose. (That is, if he can find a statistic, $L_1$, such that

$$P(L_1 \leq \mu) = .99,$$

then the observed value of $L_1$ will fill his need.) Assume that he takes a random sample of light bulbs, whose lives will be $X_1, X_2, \ldots, X_n$ hours. Then, as we know,

$$\frac{\bar{X} - \mu}{50/\sqrt{n}}$$

is a standard normal random variable; from the standard normal tables we can find $z_{1-\alpha}$ such that

$$P\left[\frac{\bar{X} - \mu}{50/\sqrt{n}} \leq z_{1-\alpha}\right] = 1 - \alpha.$$

But this statement is equivalent to

$$P\left[\bar{X} - \frac{50}{\sqrt{n}} z_{1-\alpha} \leq \mu\right] = 1 - \alpha.$$

Thus,

$$L_1 = \bar{X} - \frac{50}{\sqrt{n}} z_{1-\alpha}$$

will specify a lower confidence interval for $\mu$.

It has been necessary, when deriving the confidence interval for $\mu$ in each case mentioned thus far, that the variance $\sigma^2$ of the normal random variable be known. In almost all cases this quantity would not be known but luckily it is still possible to construct a confidence interval for $\mu$, using what is known as the $t$-distribution. The following theorem, which will not be proved, shows how the $t$-distribution occurs.

***Theorem 7.4.2.*** If $Z$ is a standard normal random variable and $V$ is a $\chi^2$ random variable with $n$ degrees of freedom, then if $Z$ and $V$ are independent,

$$T = \frac{Z}{\sqrt{V/n}}$$

has density

$$f_T(t) = \frac{\Gamma((n+1)/2)}{\Gamma(n/2)\sqrt{n\pi}}\left[1 + \frac{t^2}{n}\right]^{-(n+1)/2}.$$

This is known as the $t$-distribution with $n$ degrees of freedom.

Figure 7.3 gives a picture of the $t$-density function, for $n = 5$ degrees of freedom, as well as a plot of a standard normal density. Note that the two are extremely similar in shape but that the normal includes more density close to zero. It can be shown that as $n$, the degrees of freedom, becomes large without limit the $t$-distribution approaches the standard normal density function.

Table E in Appendix 5 presents selected percentiles of the $t$-distribution.

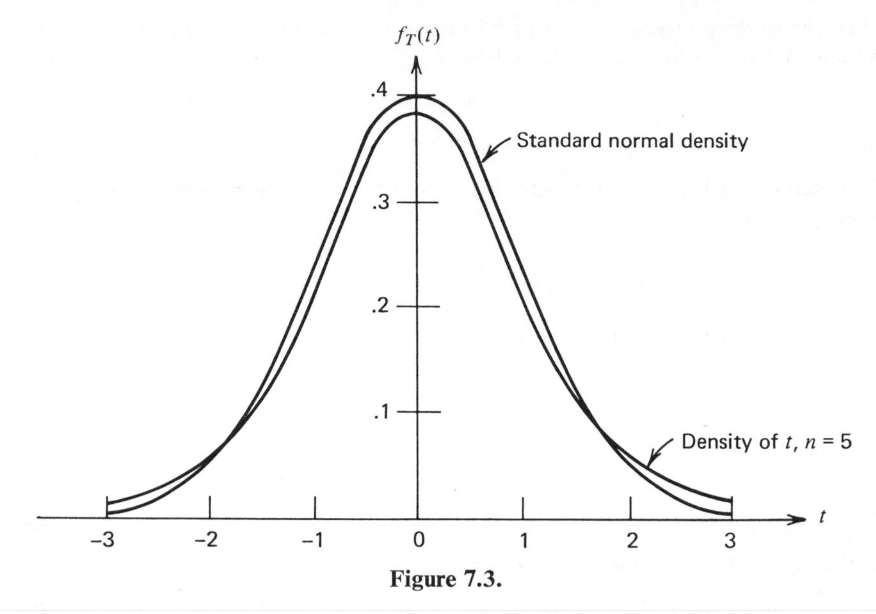


**Figure 7.3.**

The 100 $(1 - \gamma)$ percentile, $t_{1-\gamma}$, is the number such that $F_T(t_{1-\gamma}) = 1 - \gamma$ (see Figure 7.4). We see then, for a $t$ random variable with 5 degrees of freedom, that $t_{.95} = 2.015$, $t_{.99} = 3.365$. For a $t$ random variable with 20 degrees of freedom, $t_{.90} = 1.325$, $t_{.975} = 2.086$.

***Theorem 7.4.3.*** Suppose that $X$ is a normal random variable with mean $\mu$ and variance $\sigma^2$. Let $X_1, X_2, \ldots, X_n$ be a random sample of $n$ observations

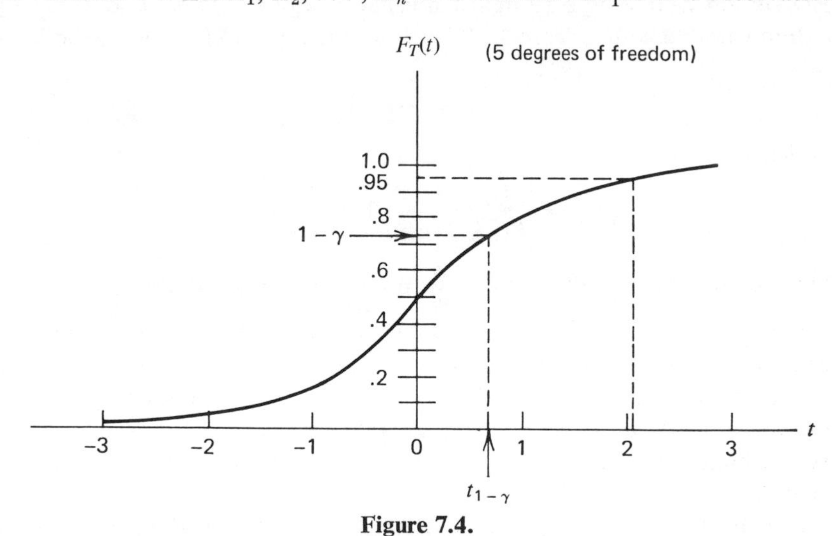


**Figure 7.4.**

of $X$. Then

$$\frac{\bar{X} - \mu}{S/\sqrt{n}}$$

has the $t$-distribution with $n - 1$ degrees of freedom. ($S$ is the sample standard deviation.)

*Proof:* We know that

$$\frac{\bar{X} - \mu}{\sigma/\sqrt{n}}$$

is a standard normal random variable, that

$$\frac{(n-1)S^2}{\sigma^2}$$

is a $\chi^2$ random variable with $n - 1$ degrees of freedom, and that $\bar{X}$ and $S^2$ are independent random variables (Theorem 6.3.5). Then, by Theorem 7.4.2,

$$\frac{\bar{X} - \mu}{\sigma/\sqrt{n}} \Big/ \sqrt{\frac{(n-1)S^2}{\sigma^2(n-1)}} = \frac{\bar{X} - \mu}{S/\sqrt{n}}$$

has the $t$-distribution with $n - 1$ degrees of freedom (the number of degrees of freedom of the $\chi^2$ random variable in the denominator). ◀

As we shall see, the role of the $t$-distribution, in confidence intervals for $\mu$ with $\sigma^2$ unknown, is identical with the role of the standard normal distribution in computing confidence intervals for $\mu$ when $\sigma$ is known. The following theorem derives this result.

***Theorem 7.4.4.*** Suppose that $X$ is a normal random variable with mean $\mu$ and variance $\sigma^2$, both of which are unknown. Let $X_1, X_2, \ldots, X_n$ be a random sample of $X$; then

$$L_1 = \bar{X} - t_{1-\alpha/2}\frac{S}{\sqrt{n}}$$

and

$$L_2 = \bar{X} + t_{1-\alpha/2}\frac{S}{\sqrt{n}}$$

constitute a $100\,(1 - \alpha)\%$ confidence interval for $\mu$ ($t_{1-\alpha/2}$ is the 100 $(1 - \alpha/2)$ percentile of the $t$-distribution with $n - 1$ degrees of freedom).

*Proof:* We saw in Theorem 7.4.3 that

$$\frac{\bar{X} - \mu}{S/\sqrt{n}}$$

is a $t$ random variable with $n - 1$ degrees of freedom. Then

$$P\left[-t_{1-\alpha/2} \leq \frac{\bar{X} - \mu}{S/\sqrt{n}} \leq t_{1-\alpha/2}\right] = 1 - \alpha,$$

where $t_{1-\alpha/2}$ is the 100 $(1 - \alpha/2)$ percentile of the $t$-distribution with $n - 1$ degrees of freedom. But, just as in Theorem 7.4.1,

$$-t_{1-\alpha/2} \leq \frac{\bar{X} - \mu}{S/\sqrt{n}} \leq t_{1-\alpha/2}$$

is equivalent to

$$\bar{X} - t_{1-\alpha/2}\frac{S}{\sqrt{n}} \leq \mu \leq \bar{X} + t_{1-\alpha/2}\frac{S}{\sqrt{n}}.$$

Thus,

$$P[L_1 \leq \mu \leq L_2] = 1 - \alpha$$

where

$$L_1 = \bar{X} - t_{1-\alpha/2}\frac{S}{\sqrt{n}}$$

$$L_2 = \bar{X} + t_{1-\alpha/2}\frac{S}{\sqrt{n}}.$$ ◀

The interpretation of a confidence interval for $\mu$ with $\sigma^2$ unknown is identical to that of a confidence interval for $\mu$ with $\sigma^2$ known. Again, in 95% of all samples taken, the observed values of the confidence interval will bracket the unknown value of $\mu$.

*Example 7.4.5.* On a bright sunny day in May, a track man put the shot 6 times; the distances thrown were 58 feet, 69 feet, 62 feet, 55 feet, 64 feet, and 65 feet. We might assume that the distance he will put the shot on a day such as this is a normal random variable $X$ with unknown mean $\mu$ and unknown variance $\sigma^2$. If we look at these 6 values as being the observed values of a random sample of size 6 of the random variable $X$, we can compute the observed value of a 90% confidence interval for $\mu$, the unknown average distance he would put the shot, as follows. We find

$$\sum_{i=1}^{6} x_i = 373, \qquad \sum_{i=1}^{6} x_i^2 = 23{,}315$$

and thus $\bar{x} = 62.2$ feet, $s = 5.04$; then the observed value of the 90% confidence interval is from $62.2 - 2.015(2.06) = 58.1$ feet to $62.2 + (2.015)(2.06) = 66.3$ feet. We are 90% sure that the interval (58.1 feet, 66.3 feet) brackets the unknown value of $\mu$.

*Example 7.4.6.* If $X_1, X_2, \ldots, X_n$ is a random sample of a normal random variable $X$ with unknown mean $\mu$ and unknown variance $\sigma^2$, note that the length of a 100 $(1 - \alpha)\%$ confidence interval for $\mu$ is a random variable. In fact

$$L_2 - L_1 = 2t_{1-\alpha/2}\frac{S}{\sqrt{n}}$$

and in this case there is no sample size $n$ which would assure us with probability 1 that we would get a given length for the interval. In fact, without knowledge of $\sigma^2$, we cannot even say we have high probability of getting a certain length or less; for

$$P(L_2 - L_1 \leq g) = P\left(2t_{1-\alpha/2}\frac{S}{\sqrt{n}} \leq g\right)$$

$$= P\left(\frac{(n-1)S^2}{\sigma^2} \leq \frac{n(n-1)g^2}{4\sigma^2 t^2_{1-\alpha/2}}\right).$$

Recall that

$$\frac{(n-1)S^2}{\sigma^2}$$

is a $\chi^2$ random variable with $n - 1$ degrees of freedom, so its probability law is known; however, the value it should not exceed is a function of $\sigma^2$ and hence cannot be evaluated without knowing $\sigma^2$. It is possible to construct a table of values for this probability, as a function of $n$ and $\sigma^2$, which then can be used to investigate questions regarding $n$ and the length of the interval for $\mu$.

We have discussed two methods of constructing confidence intervals for the mean $\mu$ of a normal random variable, depending on whether or not the variance $\sigma^2$ is known. Frequently it is desired to derive a confidence interval for the variance $\sigma^2$ of a normal random variable. The following theorem illustrates how this may be done.

***Theorem 7.4.5.*** If $X_1, X_2, \ldots, X_n$ is a random sample of a normal random variable with unknown mean $\mu$ and variance $\sigma^2$, then

$$L_1 = \frac{\sum (X_i - \bar{X})^2}{\chi^2_{1-\alpha/2}}$$

and

$$L_2 = \frac{\sum (X_i - \bar{X})^2}{\chi^2_{\alpha/2}}$$

form a 100 $(1 - \alpha)\%$ confidence interval for $\sigma^2$. ($\chi^2_{\alpha/2}$ and $\chi^2_{1-\alpha/2}$ are the 100 $\alpha/2$ and 100 $(1 - \alpha/2)$ percentiles, respectively, of the $\chi^2$ distribution with $n - 1$ degrees of freedom.)

*Proof:* From Theorem 6.3.5, we know that

$$\frac{\sum (X_i - \bar{X})^2}{\sigma^2}$$

is a $\chi^2$ random variable with $n - 1$ degrees of freedom. Thus

$$P\left[\chi^2_{\alpha/2} \leq \frac{\sum (X_i - \bar{X})^2}{\sigma^2} \leq \chi^2_{1-\alpha/2}\right] = 1 - \alpha,$$

where $\chi^2_{\alpha/2}$ and $\chi^2_{1-\alpha/2}$ are the 100 $\alpha/2$ and 100 $(1 - \alpha/2)$ percentiles of the $\chi^2$ distribution with $n - 1$ degrees of freedom. The inequality

$$\chi^2_{\alpha/2} \leq \frac{\sum (X_i - \bar{X})^2}{\sigma^2} \leq \chi^2_{1-\alpha/2}$$

is equivalent to

$$\frac{\sum (X_i - \bar{X})^2}{\chi^2_{1-\alpha/2}} \leq \sigma^2 \leq \frac{\sum (X_i - \bar{X})^2}{\chi^2_{\alpha/2}}$$

and thus the result is established. ◄

*Example 7.4.7.* Assume that a machine is used to clip metal stock into pieces (which may be used to make bolts, nuts, face plates, etc.) and that the natural variability of the machine causes the length of a clipped piece to be a normal random variable with unknown mean $\mu$ and unknown variance $\sigma^2$. If the variance $\sigma^2$ appears to be excessively large, then the machine will have to be reset. Assume that we have selected a random sample of 25 pieces clipped by this machine and find that $\sum x_i = 101.1$, $\sum x_i^2 = 412.75$ ($x_i$'s are measured in inches). Then $\sum (x_i - \bar{x})^2 = 3.90$ and a 90% confidence interval for $\sigma^2$ would extend from $\frac{3.90}{36.4} = .11$ to $\frac{9.9}{13.8} = .28$. If this range of values makes it appear that $\sigma^2$ is excessively large—the machine is producing too large a proportion of pieces too short and too long—we would want to stop production and reset the machine.

The following theorem shows how to construct a 100 $(1 - \alpha)$%, one-sided confidence interval for the parameter of an exponential random variable. It has frequently been mentioned that the time to failure of electronic components of various sorts is often an exponential random variable with parameter $\gamma$. Recall that the expected value of an exponential random variable (the expected time to failure in this case) is $1/\gamma$. If we want to be assured that the expected time to failure of an exponential random variable is at least some number of hours, we would want to say that $\gamma$ is no larger than a certain number. Thus the usual practical problem that occurs is to derive a one-sided, upper confidence interval for $\gamma$ (which would imply a one-sided, lower confidence interval for $1/\gamma$). Theorem 7.4.6 derives such a confidence interval.

***Theorem 7.4.6.*** If $X_1, X_2, \ldots, X_n$ is a random sample of an exponential random variable $X$ with parameter $\gamma$, then

$$L_2 = \frac{\chi^2_{1-\alpha}}{2n\bar{X}}$$

is an upper 100 $(1 - \alpha)\%$ confidence interval for $\gamma$ where $\chi^2_{1-\alpha}$ is the 100 $(1 - \alpha)$ percentile of the $\chi^2$ distribution with $2n$ degrees of freedom.

*Proof:* If $X_1, X_2, \ldots, X_n$ is a random sample of an exponential random variable with parameter $\gamma$, then $n\bar{X} = \sum X_i$ has moment generating function

$$m_{n\bar{X}}(t) = \left[\frac{\gamma}{\gamma - t}\right]^n = \frac{1}{(1 - (t/\gamma))^n}.$$

(See Theorem 5.4.2.) Then $2\gamma n\bar{X}$ has moment generating function

$$m_{2\gamma n\bar{X}}(t) = \frac{1}{(1 - 2t)^n};$$

thus $2\gamma n\bar{X}$ is a $\chi^2$ random variable with $2n$ degrees of freedom (see Section 3.5). We then know that

$$P(2\gamma n\bar{X} \le \chi^2_{1-\alpha}) = 1 - \alpha$$

where $\chi^2_{1-\alpha}$ is the 100 $(1 - \alpha)$ percentile of the $\chi^2$ distribution with $2n$ degrees of freedom. Thus,

$$P\left(\gamma \le \frac{\chi^2_{1-\alpha}}{2n\bar{X}}\right) = 1 - \alpha,$$

as was to be proved. ◀

*Example 7.4.8.* Assume that the time to failure of a certain type of electron tube is an exponential random variable with parameter $\gamma$. We randomly select 12 of these tubes, put them on test until all fail (independently), and observe that the sum of their times to failure is

$$\sum_{i=1}^{12} x_i = 11{,}196 \text{ hours.}$$

Then the observed value of a 90% upper confidence interval for $\gamma$ is from

$$\frac{33.2}{2(11{,}196)} = .0015$$

on down. $\Big($Alternatively the observed value of a lower 90% confidence interval for $1/\gamma$, the expected time to failure, is from $\dfrac{1}{.0015} = 674.5$ on up.$\Big)$

To this point we have only considered the problem of constructing various confidence intervals for the parameters of certain continuous random variables. Let us now discuss the construction of a 100 $(1 - \alpha)\%$ confidence interval for $p$, the parameter of a Bernoulli random variable. If $X_1, X_2, \ldots, X_n$ is a random sample of a Bernoulli random variable with parameter $p$, we know that $\bar{X}$ is approximately normally distributed with mean $p$ and variance $p(1 - p)/n$ for $n$ large. Thus

$$P\left[\frac{|\bar{X} - p|}{\sqrt{p(1 - p)/n}} \le z_{1-\alpha/2}\right] \doteq 1 - \alpha$$

where $z_{1-\alpha/2}$ is the 100 $(1 - \alpha/2)$ percentile of the standard normal distribution function. This inequality is equivalent to

$$(\bar{X} - p)^2 \le z^2 \frac{p(1 - p)}{n}$$

where the subscript has been dropped from $z_{1-\alpha/2}$. Placing all terms on the left-hand side of the inequality, we have

$$p^2\left(1 + \frac{z^2}{n}\right) - p\left(2\bar{X} + \frac{z^2}{n}\right) + \bar{X}^2 \le 0;$$

note that the left-hand side is simply the equation of a parabola, such as appears in Figure 7.5. We thus want to find the two numbers, $p_1$ and $p_2$, defining the interval for which the parabola is negative. These two numbers, for a given sample value of $\bar{X}$, will be the observed value of the confidence

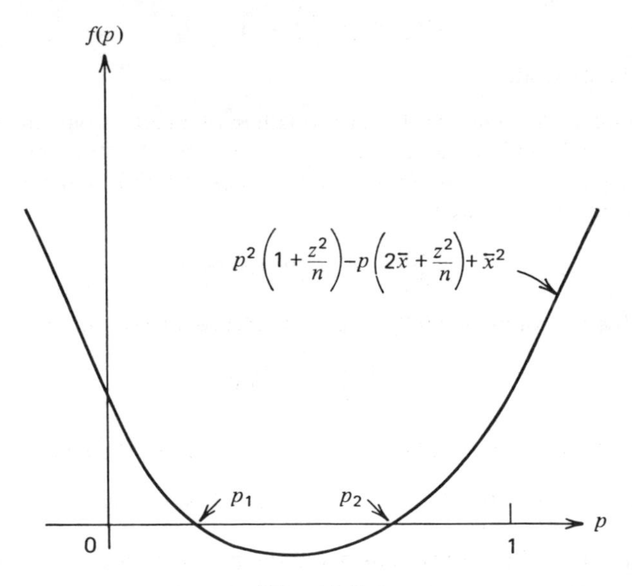


**Figure 7.5.**

interval for $p$. Using the formula for the roots of a quadratic equation, we find

$$P_1 = \frac{\bar{X} + z^2/2n - (z/\sqrt{n})\sqrt{\bar{X}(1 - \bar{X}) + z^2/4n}}{1 + z^2/n}$$

$$P_2 = \frac{\bar{X} + z^2/2n + (z/\sqrt{n})\sqrt{\bar{X}(1 - \bar{X}) + z^2/4n}}{1 + z^2/n}.$$

Assuming that $n$ is large enough for the normal approximation to be sufficiently accurate, we have

$$P(P_1 \leq p \leq P_2) = 1 - \alpha;$$

thus the two statistics, $P_1$ and $P_2$, form an approximate 100 $(1 - \alpha)\%$ confidence interval for $p$. Especially if $n$ is quite large compared to $z^2$, we might just as well set the terms in $z^2/n$ equal to zero; the two limits then become

$$P_1 = \bar{X} - \frac{z}{\sqrt{n}}\sqrt{\bar{X}(1 - \bar{X})}$$

$$P_2 = \bar{X} + \frac{z}{\sqrt{n}}\sqrt{\bar{X}(1 - \bar{X})}.$$

Note that these are the two statistics we would use for a confidence interval for the mean $\mu$ of a normal random variable whose variance is $\bar{X}(1 - \bar{X})$. (This is in fact the maximum likelihood estimator of the variance of a Bernoulli random variable.)

*Example 7.4.9.* Suppose that a professional baseball player has 400 times at bat in a certain year and that he gets 140 hits in this year. We might assume that he has a constant (unknown) probability $p$ of getting a hit each time he is at bat and that his 400 times at bat constitute a random sample of size 400 of a Bernoulli random variable with parameter $p$. Then, to construct an approximate 95% confidence interval for $p$, we find that $z_{.975} = 1.96$ and $z^2/n = .01$. Since $z^2/n$ is quite small, we might use the simplified confidence interval computations. The observed values of these simplified statistics are .303 and .397, respectively. Thus we are (approximately) 95% sure that the interval (.303, .397) covers the true value of $p$. (The observed values of the original $P_1$ and $P_2$, not assuming $z^2/n$ is zero, are .305 and .398, respectively.)

**EXERCISE 7.4.**

**1.** Assume that the top speed attainable by a certain design racing car is a normal random variable $X$ with unknown mean $\mu$ and standard deviation known to be 10 miles per hour. A random sample of 10 cars, built according to this design, was selected and each car was tested. The sum and the sum of squares of the top speeds attained (in miles per hour) were

$$\sum_{i=1}^{10} x_i = 1652, \qquad \sum_{i=1}^{10} x_i^2 = 273{,}765.$$

Compute the observed value of a 95% confidence interval for $\mu$.

**2.** For the data given in problem 1, assume that $\sigma$ is not known and compute the observed value of a 95% confidence interval for $\mu$.

**3.** Find a statistic $L_2$ such that

$$P(\sigma^2 \le L_2) = 1 - \alpha,$$

given a random sample of a size $n$ of a normal random variable with mean $\mu$ and variance $\sigma^2$.

**4.** Since $\sigma$ is the natural scale factor for the normal density function, it is frequently of use to have a confidence interval for $\sigma$ rather than $\sigma^2$. Derive a 100 $(1 - \alpha)$% confidence interval for $\sigma$, based on a random sample of $n$ observations of a normal random variable.

**5.** Suppose that the monthly income of a certain family is a normal random variable $X$ with mean $\mu_X$ and variance $\sigma^2$ and that the monthly expenses of the same family is a normal variable $Y$ with mean $\mu_Y$ and variance $\sigma^2$ (note that both random variables have the same variance). Construct a 100 $(1 - \alpha)$% confidence interval for $\mu_X - \mu_Y$, assuming $\sigma^2$ is known, based on two independent random samples of $X$ and $Y$, both of size $n$.

**6.** Construct a 100 $(1 - \alpha)$% confidence interval for $\mu_X - \mu_Y$, as in problem 5, assuming $\sigma^2$ is unknown.

**7.** The reliability of an electronic component for $t$ hours is simply the probability that its time to failure $T$ exceeds $t$ hours. Assume that the time to failure of a certain type of component is an exponential random variable with parameter $\gamma$ and that $T_1, T_2, \ldots, T_n$ is a random sample of $T$. Based on the result of Theorem 7.4.6, construct a 100 $(1 - \alpha)$% confidence interval for the reliability of a component of this type for 1000 hours.

**8.** How large a sample $n$ should we take in order that the 95% confidence interval for the mean of a normal random variable will be $\frac{1}{2}$ unit long, given that $\sigma^2 = 16$? Given that $\sigma^2 = 100$?

**9.** Given a random sample of 10 observations of a normal random variable with unknown mean $\mu$ and unknown variance $\sigma^2$, what is the probability that the length of the 90% confidence interval for $\mu$ exceeds 1.17 units, given that $\sigma^2$ is 1? Given that $\sigma^2$ is 2?

**10.** Assuming that we have a large random sample of a Poisson random variable $X$ with parameter $\lambda$, use the normal approximation to the distribution of $\bar{X}$ to derive an approximate 100 $(1 - \alpha)$% confidence interval for $\lambda$ similar to the binomial derivation discussed above.

**11.** Compute the expected length of a 100 $(1 - \alpha)\%$ confidence interval for $\mu$, based on a random sample of $n$ observations of a normal random variable $X$ with unknown variance $\sigma^2$. (**Hint:** The length is a multiple of the square root of a $\chi^2$ random variable. Review Theorem 3.5.3.)

**12.** Given $X_1, X_2, \ldots, X_n$ is a random sample of a normal random variable $X$ with *known* mean $\mu$ and unknown variance $\sigma^2$, derive a 100 $(1 - \alpha)\%$ confidence interval for $\sigma^2$. There is a better interval than the one given in Theorem 7.4.5 for this case. (In what sense is it better?)

**13.** For any sample size $n$, the ratio of the lengths of the 100 $(1 - \alpha_1)\%$ and the 100 $(1 - \alpha_2)\%$ confidence intervals for $\mu$ are fixed, whether $\sigma^2$ is known or not. Determine the value of this fixed ratio if $\sigma^2$ is known and if $\sigma^2$ is unknown.

# 8

# Statistical Inference: Tests of Hypotheses

In this chapter we shall study the classical theory of tests of hypotheses. A hypothesis, first of all, is simply a statement about the probability law of a random variable (which can be sampled). A test of a hypothesis, then, amounts to sampling the random variable whose probability law is referred to in the hypothesis and, on the basis of the sample, deciding to accept or reject the stated hypothesis.

In general, if our sample results seem consistent with the hypothesis, we would like to accept the stated hypothesis; if however the sample results do not seem consistent with the stated hypothesis, we would like to reject it. Usually there are many different ways of making the decision to accept or reject a hypothesis on the basis of a sample; we shall examine some of the methodology which has been developed for judging the best way (in a sense) of making the decision.

## 8.1. Hypotheses and Tests

As was just mentioned, for our purposes a hypothesis will always be a statement about the probability law of a random variable. There are two distinct sorts of statements which can be made about the probability law of a random variable. The first of these is to assume that the form of the distribution function $F_X$ for $X$ is known and the hypothesis then merely refers to the value or values of the parameters of $F_X$. The second type of hypothesis is a statement regarding the form of $F_X$ itself. An example of the first type of hypothesis would be a case in which we know, or are willing to assume, that a random variable $X$ is binomial and the hypothesis is a statement regarding the value of $p$. An example of the second type would be a case in which our hypothesis is that $X$ is a binomial random variable. For reasons that will become apparent as we proceed, it is advantageous to treat these two types of hypotheses separately. At first we shall concern ourselves with only hypotheses of the first type, those in which the form of the density or probability function is assumed known and we are interested in accepting or rejecting statements about the values of the parameters.

If our hypothesis gives exact values for all unknown parameters of the assumed probability law, we shall call it a *simple* hypothesis; otherwise the hypothesis is *composite*. The following example illustrates this difference.

*Example 8.1.1.* Suppose that a random variable $X$ is binomial with parameter $n = 20$, and we hypothesize that $p = \frac{1}{4}$ or $p = \frac{1}{2}$ or $p = .9$. Each of these is an example of a simple hypothesis. However, the statements

$$0 \le p \le .1, \qquad p < \tfrac{1}{4}, \qquad .49 < p < .51$$

are each composite hypotheses. Or, if $Y$ is a normal random variable, $\mu = 2$, $\sigma = 1$ is a simple hypothesis while $\mu = 2$ is a composite hypothesis if $\sigma$ is unknown (since $\sigma$ is unspecified).

As has been mentioned, we shall decide to accept or reject a hypothesis on the basis of a random sample of the random variable concerned. Thus we are, in effect, partitioning our sample space of all possible outcomes for the experiment into two disjoint pieces, those outcomes for which we decide to accept the hypothesis versus those outcomes for which we decide to reject the hypothesis. We would like to find that way of partitioning the sample space which is good in some sense.

DEFINITION 8.1.1. A *test* of a hypothesis is a partitioning of the sample space into two parts, called the rejection region (or critical region) and the acceptance region.

If we compare two different ways of partitioning the sample space, then we say we are comparing two tests (of the same hypothesis).

Tests are generally compared on the basis of the probabilities of errors which might be committed. A little reflection will reveal that there are two types of errors which might be committed. We might reject the hypothesis when it is true or we might accept the hypothesis when it is false.

DEFINITION 8.1.2. If we reject the hypothesis when it is true, we commit a *type I error*. If we accept the hypothesis when it is false, we commit a *type II error*.

Then, if we could find a test for a hypothesis which simultaneously minimized the probabilities of both of these types of errors, it would clearly be the test we would want to use. Unfortunately, it is not in general possible to do this, as a simple example will illustrate.

*Example 8.1.2.* Suppose, that we have a brand new coin and wish to test the hypothesis that the probability $p$ of getting a head when this coin is flipped is equal to $\frac{1}{2}$. We might decide to flip the coin 100 times, in which case we would have a random sample of 100 observations of a Bernoulli random variable. We could decide to accept the hypothesis no matter what happens when the coin is flipped 100 times (thus the rejection region is null and the whole sample space is the acceptance region). This partition yields $P$(type I error) equal to zero, clearly the smallest possible value it might have. Or we might decide to reject the hypothesis no matter what happens on the 100 flips (now the acceptance region is null), in which case $P$(type II error) equals zero. Thus, we could always get a test which makes the probability of one of the two errors equal to zero.

The way in which different tests are compared, therefore, is to fix the probability of type I error at $\alpha$ (.05 and .01 are common values) and then choose the test that gives the smallest probability of type II error. If we recall that accepting the hypothesis when it is false is called a type II error, it is evident that it is necessary to assume some value of the parameter not specified by the hypothesis before the probability of a type II error can be calculated. That is, we must specify what we assume the true value of the parameter to be before we can do the calculation; we must state what it is we do accept as the value of the parameter. To make this comparison of probabilities of type II errors more concrete (really to specify those values of the parameter which should be relevant to the comparison of tests), we shall always state two hypotheses, one of which is to be accepted on the basis of the sample. The hypothesis which we want to test is called the null hypothesis; the range of parameter values pertinent to the comparison of probabilities of type II errors specifies the alternative hypothesis. We shall use $H_0$ to indicate the null hypothesis and $H_1$ to indicate the alternative hypothesis. Thus, in Example 8.1.2 above, we would write $H_0{:}p = \frac{1}{2}$ and, depending on which alternatives we might want to consider, we would write $H_1{:}p \neq \frac{1}{2}$ or $H_1{:}p > \frac{1}{2}$ or $H_1{:}p = \frac{3}{4}$, etc.

*Example 8.1.3.* Suppose that the lengths of lives of light bulbs made by a certain company are, through past experience, assumed to be normally distributed with $\mu = 1500$ hours and $\sigma = 100$ hours. Someone in the research division of this company invents a new process for making bulbs which preliminary research shows may have a mean lifetime of more than 1500 hours with a standard deviation of 100 hours (again it would appear plausible that the distribution of lifetimes for bulbs made the new way is normal in form). It is apparent that the new process would cost the same amount per bulb once installed. The company feels that if this new method does produce bulbs with longer average lifetimes, they should adopt it since it would lead to increased consumer goodwill, possibly a bigger share of the light bulb market, etc.

It is assumed, then, that lifetimes of bulbs made the new way are normally distributed with unknown mean $\mu$ and known standard deviation of 100 hours. The company wants to test $H_0:\mu = 1500$ versus $H_1:\mu > 1500$. A group of 100 bulbs are made the new way (and it is assumed that their lifetimes constitute a random sample of size 100 from the distribution of all possible lifetimes for this process). It is decided that if the sample mean exceeds 1525 hours, then $H_0$ would be rejected ($H_1$ accepted) and the new process would be adopted. Otherwise $H_0$ would be accepted and they would continue to make bulbs in the old way. Thus the partition of the sample space is into those outcomes where $\bar{X} > 1525$ (rejection region) versus those for which $\bar{X} \leq 1525$ (acceptance region). The probability of a type I error using this test then is

$$P(\bar{X} > 1525 \mid \mu = 1500) = P\left(\frac{\bar{X} - 1500}{10} > \frac{1525 - 1500}{10} \,\middle|\, \mu = 1500\right)$$

$$= P(Z > 2.5) = .0062,$$

where $Z$ is a standard normal random variable.

In the preceding example, $H_0$ specified that the new process was the same as the old (in terms of average lifetime), while $H_1$ then consisted of the possible pertinent alternative values for $\mu$. Note that rejecting $H_0$ when it is true (type I error) then would consist of adopting the new process when in fact it was no better than the old; this would be an error which would cause the company needless expense. We noted that its probability of occurrence is quite small.

As we shall see, the usual methods for testing hypotheses allow easy control of the probability of type I error. Thus, in practical examples, $H_0$ and $H_1$ are selected so that the more important or more serious error is the type I; its probability of occurrence can then be easily made as small as one would like. In Example 8.1.3 it was deemed more important not to adopt the new process when it was no better than the old rather than to fail to adopt it when in fact it was better.

We made no mention of the probability of type II error in Example 8.1.3. Since $H_1$ is composite, the probability of type II error varies with the particular value of $\mu$ specified by $H_1$ that we want to assume is true. The operating characteristic ($OC$) curve describes how the probability of type II error varies with $\mu$.

DEFINITION 8.1.3. When testing hypotheses about a parameter $\mu$, the probability of accepting $H_0$ is a function of $\mu$, denoted by $C(\mu)$, and is called the $OC$ curve of the test. $Q(\mu) = 1 - C(\mu)$ is called the *power function* of the test.

*Example 8.1.4.* In Example 8.1.3, $H_0$ is accepted if $\bar{X} \leq 1525$. Thus

$$\begin{aligned} C(\mu) &= P(\bar{X} \leq 1525) \\ &= P\left(\frac{\bar{X} - \mu}{10} \leq \frac{1525 - \mu}{10}\right) \\ &= N_Z\left(\frac{1525 - \mu}{10}\right), \end{aligned}$$

where $N_Z$ is the standard normal distribution function. Since

$$\frac{d}{d\mu} N_Z\left(\frac{1525 - \mu}{10}\right) = -\frac{1}{10} n_Z\left(\frac{1525 - \mu}{10}\right),$$

we know that $C(\mu)$ is a monotonically decreasing function of $\mu$. Figure 8.1 gives a sketch of this curve. The power function for the test then is

$$\begin{aligned} Q(\mu) &= 1 - C(\mu) \\ &= 1 - N_Z\left(\frac{1525 - \mu}{10}\right) \end{aligned}$$

and is sketched in Figure 8.2. Since the power curve, for a given value of $\mu$ specified by $H_1$, gives the probability of making no error, we would clearly prefer the test that has the biggest power, preferably for all $\mu$ specified by $H_1$.

*Example 8.1.5.* Joe and Hugh are in the habit of flipping coins daily to see which of them will buy their coffee. Suppose that Joe is of a suspicious nature and would like to test formally the hypothesis that his probability of winning, $p$, is $\frac{1}{2}$ in playing

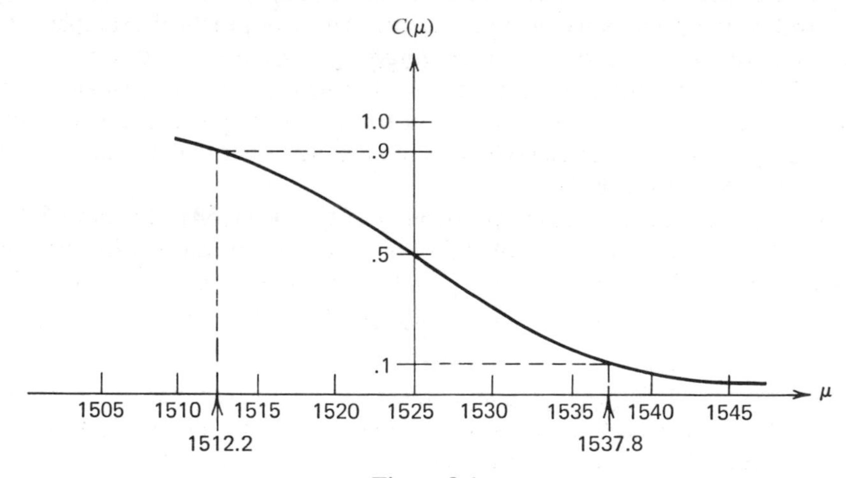


**Figure 8.1.**

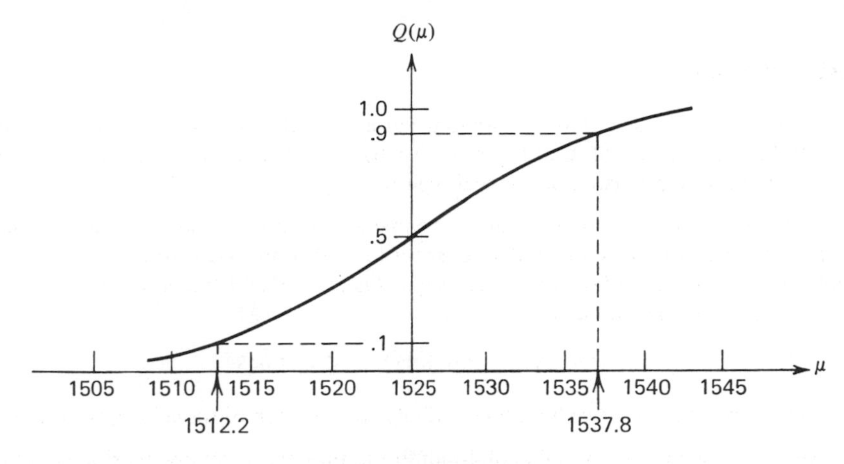


**Figure 8.2.**

this game versus the alternative that $p < \frac{1}{2}$. He decides to keep track of the outcomes the next 10 times that he and Hugh gamble and to reject $H_0: p = \frac{1}{2}$ if he does not win at least 2 times out of the 10. Then, formally, he is defining

$$\begin{aligned} X_i &= 1, \quad \text{if he wins on the } i\text{-th game} \\ &= 0, \quad \text{if he loses on the } i\text{-th game,} \end{aligned}$$

for $i = 1, 2, \ldots, 10$; thus $X_1, X_2, \ldots, X_{10}$ is a random sample of 10 observations of a Bernoulli random variable $X$ with parameter $p$. His critical region then is defined by

$$\sum_{i=1}^{10} X_i \leq 1.$$

$$Y = \sum_{i=1}^{10} X_i$$

is a binomial random variable with parameters 10 and $p$; his probability of type I error then is

$$P[Y \leq 1 \mid p = \tfrac{1}{2}] = (\tfrac{1}{2})^{10} + \frac{10}{2^{10}} = .01.$$

The $OC$ curve for this test then is

$$\begin{aligned} C(p) &= 1 - P[Y \leq 1 \mid p] \\ &= 1 - p^{10} - 10p^9(1 - p) \end{aligned}$$

and we find, for example, that

$$C(.4) = .95, \qquad C(.3) = .85, \qquad C(.2) = .63, \qquad C(.1) = .27$$

(see Table A in Appendix 5).

## EXERCISE 8.1.

**1.** Suppose that $X$ is a Poisson random variable with parameter $\lambda$. We take a sample $X_1, X_2, \ldots, X_n$ of $X$ and want to test $H_0: \lambda = 2$ versus $H_1: \lambda > 2$. What do you feel would be a reasonable critical region?

**2.** Suppose that $X$ is a normal random variable with unknown mean $\mu$ and variance 9. If we take a random sample of 16 observations of $X$ and want to test $H_0: \mu = 2$, show that each of the following critical regions is of size .05 (that is, the probability of type I error is .05 for each).

(a) $\bar{X} > 3.47$ (b) $\bar{X} < .53$ (c) $1.952 < \bar{X} < 2.048$

For what type of alternative hypothesis, if any, would each of these be appropriate?

**3.** We might assume that weights of 4-month-old pigs are normally distributed with standard deviation $\sigma = 10$ pounds. A standard ration is known to produce an average weight of 120 pounds. If we desire to check whether a new ration is equally good, we might take a sample of 25 pigs, feed them the new ration, check their weights at age 4 months, and test $H_0: \mu = 120$ versus $H_1: \mu < 120$ using the critical region $\bar{X} < 116.08$. Compute the probability of type I error for this test and derive the power function. What is the power of the test if $\mu = 115$ pounds? (That is, what is $Q(115)$?)

**4.** Suppose that the number of ounces that a bottling machine puts in a coke bottle is a normal random variable with mean $\mu$ and variance $\sigma^2$. If $\sigma^2$ is too large, many bottles will be filled to overflowing; for the setting on the machine controlling the mean, a $\sigma^2$ not exceeding $\frac{1}{4}$ is tolerable. Suppose we take a sample of 20 bottles filled by the machine and want to test $H_0: \sigma^2 = \frac{1}{4}$ versus $H_1: \sigma^2 > \frac{1}{4}$. We decide to reject $H_0$ if

$$\sum_{i=1}^{20} (X_i - \bar{X})^2 > 7.536.$$

What is the probability of type I error for this test?

**5.** Assume that the time to failure (in hours) of a certain type of vacuum tube is an exponential random variable with parameter $\lambda$. We put 10 of these tubes on test and want to test the hypothesis $H_0: \lambda = .001$ versus $H_1: \lambda < .001$. We decide to wait only until the first tube fails and will reject $H_0$ if the time to this first failure is no larger than 5.129 hours. Show that the probability of type I error for this test is .05. (Thus we are concerned with the minimum sample value of a sample of size 10 of an exponential random variable; the reader might like to review Example 6.3.4 to see the density function for this minimum value.)

**6.** Assume that $X$ is a uniform random variable on the interval from $-\theta$ to $\theta$. We observe one value of $X$ and want to test $H_0: \theta = 1$ versus $H_1: \theta > 1$; we decide to reject $H_0$ if the sample value exceeds .99. Compute the probability of type I error for this test and draw the power function.

## 8.2. Simple Hypotheses

It will be recalled from Section 8.1 that a simple hypothesis is one that completely specifies the density or probability function for a random variable. Suppose that $X$ is a random variable with density $f_X$ indexed by a single parameter $\theta$. Then, if we wanted to test $H_0:\theta = 1$ versus $H_1:\theta = 5$, we would say that we were testing a simple hypothesis versus a simple alternative, since each of the two hypotheses completely specifies the density $f_X$. Note that if we are testing a simple $H_0$ versus a simple $H_1$, then both the probability of a type I error and the probability of a type II error are well defined; since each hypothesis specifies a single value for the unknown parameter, we can quote unique probabilities for both type I and type II errors. These two probabilities will be denoted by $\alpha$ and $\beta$, respectively.

It was also mentioned in Section 8.1 that selecting a test for a hypothesis was equivalent to simply selecting a critical region; if we could find some method of constructing a critical region which was good in some sense, this method would yield what we might call a good test of the hypothesis.

In the special case in which we are testing a simple hypothesis against a simple alternative, we would clearly like to use the critical region that minimizes $\beta$, the probability of a type II error, among all regions with a probability of type I error equal to $\alpha$. The Neyman-Pearson lemma (Theorem 8.2.1) shows how we can always construct such a region.

***Theorem 8.2.1.*** (Neyman-Pearson) Assume that $X$ is a random variable whose distribution function is indexed by the parameter $\theta$. $X_1, X_2, \ldots, X_n$ is a random sample of $X$; $L(\theta)$ is the likelihood function of the sample. We desire to test $H_0:\theta = \theta_0$ versus $H_1:\theta = \theta_1$ where $\theta_0$ and $\theta_1$ are two specific values for $\theta$. Let $R$ (the critical region) be the subset of $S$ such that

$$\frac{L(\theta_1)}{L(\theta_0)} \geq k.$$

Then $\bar{R}$ will consist of those points such that

$$\frac{L(\theta_1)}{L(\theta_0)} < k$$

where $k$ is chosen so that the probability of type I error is $\alpha$. Then the region $R$ has the smallest possible probability of type II error among all critical regions of size $\alpha$ (probability of type I error). $R$ is called the most powerful critical region for the given $\alpha$.

*Proof:* To indicate the proof of this theorem (in the continuous case), we shall use $\int_A L(\theta_0)$ and $\int_A L(\theta_1)$ as shorthand for the integrals of the sample density function over a subset $A \subset S$, assuming $\theta_0$ and $\theta_1$ as the true values of the parameter $\theta$, respectively. Let $R$ be the region defined above; then

$$\int_R L(\theta_0) = \alpha$$

and we want to show that

$$\int_R L(\theta_1) \geq \int_{R^*} L(\theta_1)$$

where $R^*$ is any other region satisfying

$$\int_{R^*} L(\theta_0) = \alpha.$$

Clearly,

$$R = (R \cap R^*) \cup (R \cap \bar{R}^*)$$
$$R^* = (R^* \cap R) \cup (R^* \cap \bar{R});$$

thus

$$\int_{R^*} L(\theta_1) = \int_{R^* \cap R} L(\theta_1) + \int_{R^* \cap \bar{R}} L(\theta_1)$$

and

$$\int_R L(\theta_1) = \int_{R \cap R^*} L(\theta_1) + \int_{R \cap \bar{R}^*} L(\theta_1).$$

Then

$$\int_{R^*} L(\theta_1) - \int_R L(\theta_1) = \int_{R^* \cap \bar{R}} L(\theta_1) - \int_{R \cap \bar{R}^*} L(\theta_1).$$

Now, by definition, $L(\theta_1) \geq kL(\theta_0)$ for points in $R$ and $L(\theta_1) < kL(\theta_0)$ for points in $\bar{R}$. Thus

$$\int_{R^* \cap \bar{R}} L(\theta_1) < k \int_{R^* \cap \bar{R}} L(\theta_0)$$

and

$$\int_{R \cap \bar{R}^*} L(\theta_1) \geq k \int_{R \cap \bar{R}^*} L(\theta_0);$$

then we have

$$\int_{R^* \cap \bar{R}} L(\theta_1) - \int_{R \cap \bar{R}^*} L(\theta_1) \leq k\left(\int_{R^* \cap \bar{R}} L(\theta_0) - \int_{R \cap \bar{R}^*} L(\theta_0)\right).$$

But

$$\begin{aligned}\int_{R^* \cap \bar{R}} L(\theta_0) - \int_{R \cap \bar{R}^*} L(\theta_0) &= \int_{R^* \cap \bar{R}} L(\theta_0) + \int_{R^* \cap R} L(\theta_0) \\ &\quad - \int_{R^* \cap R} L(\theta_0) - \int_{R \cap \bar{R}^*} L(\theta_0) \\ &= \int_{R^*} L(\theta_0) - \int_R L(\theta_0) = \alpha - \alpha = 0.\end{aligned}$$

Thus

$$\int_{R^*} L(\theta_1) - \int_R L(\theta_1) \le k(0) = 0$$

and the result is established. The same sort of reasoning is appropriate in the discrete case, using summation rather than integration. ◀

This theorem then gives a method which will guarantee, among all possible critical regions with probability of type I error equal to $\alpha$, the smallest possible value for the probability of type II error. Let us apply this theorem to some specific examples and see the form of this best critical region.

*Example 8.2.1.* Suppose that the time to failure of light bulbs made a certain way is a normal random variable with mean $\mu$ and standard deviation $\sigma = 100$. We take a sample of 25 of these bulbs and want to test $H_0: \mu = 1500$ versus $H_1: \mu = 1525$. The likelihood function for the sample then is

$$L(\mu) = \left(\frac{1}{2\pi}\right)^{25/2} \left(\frac{1}{100}\right)^{25} e^{-\Sigma(x_i-\mu)^2/2(100)^2};$$

the critical region specified by the Neyman-Pearson fundamental lemma is derived as follows:

$$\frac{L(\mu_1)}{L(\mu_0)} = e^{-[\Sigma(x_i-1525)^2 - \Sigma(x_i-1500)^2]/2(100)^2}$$

$$= e^{.0125\bar{x} - .75875}.$$

Now $e^{.0125\bar{x} - .75875} \ge k$ is equivalent to

$$\bar{x} \ge 80 \ln k + 60.7 = C;$$

that is, if we determine a constant $C$ such that the probability that $\bar{X}$ exceeds $C$ is $\alpha$ when $H_0$ is true, and then reject $H_0$ if $\bar{X} > C$, we have the test of $H_0$ versus $H_1$ of size $\alpha$ which has the smallest possible value of $\beta$. If $\mu = 1500$, then $(\bar{X} - 1500)/20$ is a standard normal random variable and

$$P(\bar{X} > C) = 1 - N_Z\left(\frac{C - 1500}{20}\right).$$

If we want the probability of type I error to be $\alpha$, then $C$ is chosen so that

$$N_Z\left(\frac{C - 1500}{20}\right) = 1 - \alpha;$$

that is

$$\frac{C - 1500}{20} = z_{1-\alpha}$$

(the 100 $(1 - \alpha)$ percentile of the standard normal distribution) and

$$C = 1500 + 20z_{1-\alpha}.$$

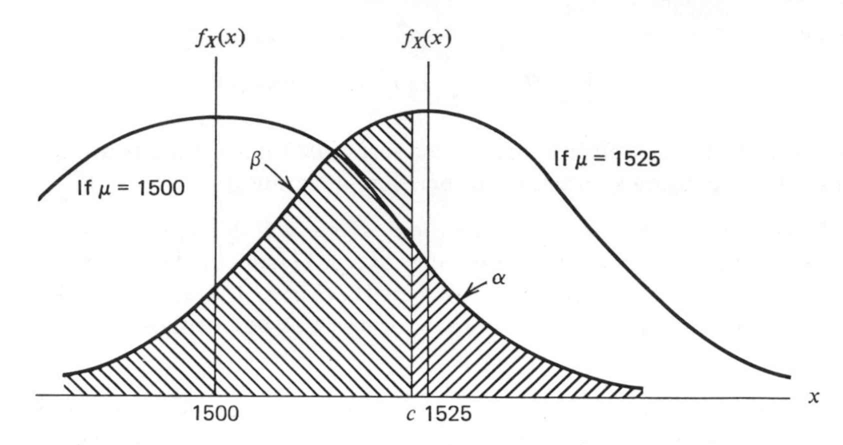


**Figure 8.3.**

Thus, if we want $\alpha = .05$, we should reject $H_0$ if $\bar{X} > 1531.1$; if we want $\alpha = .01$, we should reject $H_0$ if $\bar{X} > 1546.6$. In either of these two cases for the given $\alpha$, we have the smallest possible value for the probability of a type II error. If $\mu =$ 1525, then $(\bar{X} - 1525)/20$ is a standard normal random variable and

$$P[\bar{X} \leq 1500 + 20z_{1-\alpha}] = N_Z\left(\frac{-25 + 20z_{1-\alpha}}{20}\right) = \beta.$$

Thus for $\alpha = .05$, $z_{1-\alpha} = 1.645$ and

$$\beta = N_Z(-1.25 + 1.645) = N_Z(.395) = .6535.$$

Whereas, if $\alpha = .01$, $z_{1-\alpha} = 2.33$ and

$$\beta = N_Z(-1.25 + 2.33) = N_Z(1.08) = .8599.$$

Figure 8.3 illustrates the probabilities of occurrence of these two types of errors.

It is to be stressed that the critical region derived in Example 8.2.1 has the smallest $\beta$ for a given $\alpha$. We could simply always reject $H_0$, regardless of what happens on the sample, in which case we would never commit a type II error and we would have $\beta = 0$ (we would also have $\alpha = 1$). The proof of the following theorem is asked for in problem 3 below.

***Theorem 8.2.2.*** Assume that $X_1, X_2, \ldots, X_n$ is a random sample of a normal random variable $X$ with mean $\mu$ and known variance $\sigma^2$. Then the most powerful (the one with smallest $\beta$) critical region of size $\alpha$ for testing $H_0: \mu = \mu_0$ versus $H_1: \mu = \mu_1$ (where $\mu_1 > \mu_0$) is specified by

$$\bar{x} \geq \mu_0 + \frac{\sigma}{\sqrt{n}} z_{1-\alpha};$$

if $\mu_1 < \mu_0$, then the most powerful critical region of size $\alpha$ is specified by

$$\bar{x} \leq \mu_0 - \frac{\sigma}{\sqrt{n}} z_{1-\alpha}.$$

Let us now consider an example concerning a test of a hypothesis about the variance of a normal random variable whose mean is known. (It is necessary that we assume that the mean $\mu$ is known so that we may have a simple hypothesis and a simple alternative.)

*Example 8.2.2.* A manufacturer of nuts and bolts makes nuts whose interior diameter is, on the average, $\frac{1}{4}$ inch. In order that a nut made by the manufacturer will fit one of his bolts, it is necessary that the interior diameter is neither much bigger than $\frac{1}{4}$ inch nor much smaller than $\frac{1}{4}$ inch. If we assume that the interior diameters of all nuts (of this size) made by this manufacturer are normally distributed with mean $\frac{1}{4}$ inch and unknown standard deviation $\sigma$, we might take a random sample of 36 nuts that he has made and test the hypothesis $H_0: \sigma = .01$ versus $H_1: \sigma = .02$. Presumably, $\sigma = .02$ would lead to a great number of nuts being too large or too small, whereas $\sigma = .01$ would give essentially all usable nuts. Let us derive the most powerful test for this case. The likelihood of the sample is

$$L(\sigma^2) = \left(\frac{1}{2\pi\sigma^2}\right)^{18} e^{-\Sigma(x_i - 1/4)^2/2\sigma^2}$$

thus

$$\frac{L(\sigma_1^2)}{L(\sigma_0^2)} = \left(\frac{1}{2}\right)^{36} e^{3750\Sigma(x_i - 1/4)^2}$$

and

$$(\tfrac{1}{2})^{36} e^{3750\Sigma(x_i - 1/4)^2} \geq k$$

is equivalent to

$$\sum \left(x_i - \frac{1}{4}\right)^2 \geq \frac{\ln k}{3750} - \frac{\ln (\frac{1}{2})^{36}}{3750} = C.$$

Thus, if we can find a value of $C$ such that

$$P[\sum (X_i - \tfrac{1}{4})^2 \geq C] = \alpha,$$

given that $H_0$ is true, then rejecting $H_0$ if

$$\sum (X_i - \tfrac{1}{4})^2 \geq C$$

is the most powerful test of size $\alpha$. We saw in Chapter 6 that if $X_1, X_2, \ldots, X_{36}$ are independent normal random variables, each with mean $\frac{1}{4}$ and variance $(.01)^2$, then

$$\frac{\sum (X_i - \frac{1}{4})^2}{(.01)^2}$$

is a $\chi^2$ random variable with 36 degrees of freedom. Thus

$$P\left[\frac{\sum (X_i - \frac{1}{4})^2}{(.01)^2} > \chi^2_{1-\alpha}\right] = \alpha$$

where $\chi^2_{1-\alpha}$ is the 100 $(1 - \alpha)$ percentile of the $\chi^2$ distribution with 36 degrees of freedom. Thus, if we take $C = (.01)^2\chi^2_{1-\alpha}$ and reject $H_0$ whenever $\sum (X_i - \frac{1}{4})^2 > C$, we have the most powerful test of $H_0$ versus $H_1$. Then, for $\alpha = .05$, we find $\chi^2_{.95} = 50.75$ and we would reject $H_0$ whenever $\sum (X_i - \frac{1}{4})^2 > .005075$; in so doing we have the smallest possible probability of type II error among all tests of size .05.

In each of the two examples studied so far the critical region is entirely specified by the sample value of a statistic; for the more commonly used tests this is generally the case. Such a statistic, whose sample value indicates acceptance or rejection of the null hypothesis, is frequently referred to as a *test statistic*. The test of the hypothesis reduces to merely observing where the sample value of the statistic lies. Theorem 8.2.3 shows the form of the most powerful critical region for tests of simple hypotheses about the standard deviation of a normal random variable with known mean; its proof is asked for in problem 4 below.

***Theorem 8.2.3.*** Assume that $X_1, X_2, \ldots, X_n$ is a random sample of a normal random variable $X$ with known mean $\mu$ and unknown variance $\sigma^2$. Then the most powerful critical region of size $\alpha$ for testing $H_0: \sigma = \sigma_0$ versus $H_1: \sigma = \sigma_1 (\sigma_1 > \sigma_0)$ is specified by

$$\sum (x_i - \mu)^2 > \sigma_0{}^2\chi^2_{1-\alpha}$$

where $\chi^2_{1-\alpha}$ is the 100 $(1 - \alpha)$ percentile of the $\chi^2$ distribution with $n$ degrees of freedom. If $\sigma_1 < \sigma_0$, the most powerful critical region of size $\alpha$ is specified by

$$\sum (x_i - \mu)^2 < \sigma_0{}^2\chi_\alpha{}^2.$$

Let us now consider testing a simple hypothesis versus a simple alternative for the parameter of a discrete variable.

*Example 8.2.3.* Assume that 10 years ago traffic fatality data for accidents occurring in San Francisco were examined and it was found that the number of traffic fatalities within the city limits, per day, was apparently a Poisson random variable with parameter $\lambda = .1$. Due to the increased congestion of the present day, it is suspected that the rate is now .15 per day rather than .1. To investigate this point, assume that we have available the records for the number of traffic fatalities for $n$ recent days and want to test $H_0: \lambda = .1$ versus $H_1: \lambda = .15$. We shall use the Neyman-Pearson fundamental lemma to derive the most powerful test. The likelihood function for the sample is

$$L(\lambda) = \frac{\lambda^{\Sigma x_i}}{\Pi x_i!} e^{-n\lambda};$$

thus

$$\frac{L(\lambda_1)}{L(\lambda_0)} = \frac{L(.15)}{L(.1)} = (1.5)^{\Sigma x_i} e^{-.05n}$$

and

$$\frac{L(\lambda_1)}{L(\lambda_0)} \geq k$$

is equivalent to

$$\sum x_i \geq \frac{\ln k + .05n}{\ln (1.5)} = C.$$

Thus, to use the most powerful critical region of size $\alpha$, we would have to determine $C$ such that

$$P(\sum X_i > C) = \alpha,$$

given $\lambda = .1$, and then would reject $H_0$ if we found that $\sum x_i > C$. We found in Chapter 5 that $\sum X_i$ is a Poisson random variable with parameter $n\lambda$ if $X_1, X_2, \ldots, X_n$ are independent Poisson random variables each with parameter $\lambda$. Thus, if we choose $C$ equal to the 100 $(1 - \alpha)$ percentile of the Poisson distribution with parameter $n/10$, we have the most powerful test. Since the Poisson distribution is discrete, we do not have the complete freedom for choices of $\alpha$ that the continuous cases afford. Note in particular that if $n = 20$, then $n/10 = 2$, and from the Poisson tables we find that

$$P(\sum X_i > 4) = .0527, \qquad P(\sum X_i > 5) = .0166;$$

thus there is no constant $C$ such that we have $\alpha$ equal to exactly .05 (or .04 or .03 or .02). This is a minor point, but one which the reader should be aware of. As $n$ gets larger, of course, there are more available values for $\alpha$.

It may be evident, especially in the case of testing a simple hypothesis versus a simple alternative, that the choice of which statement is $H_0$ and which is $H_1$ is arbitrary. However, the following general sort of reasoning is frequently applied when choosing $H_0$. The Neyman-Pearson fundamental lemma gives a method for finding the most powerful test for arbitrary $\alpha$, the probability of type I error. Thus, depending on how much of a gamble we would like to take, we can make the probability of rejecting $H_0$ when it is true equal to .1 or .05 or .0001. This causes people to state as $H_0$ something that they believe the sample will indicate to be false; then in rejecting $H_0$ on the basis of the sample, they can make a concrete probability statement about whether they are wrong in so rejecting it and the value of that probability is within their control (since it is the probability of type I error). If we were to state as $H_0$ something that we believe the sample would indicate to be true, then in accepting $H_0$ on the basis of the sample we could only make a type II error and its probability would not be within our control (for a given $\alpha$); most people feel that this is a weak position to occupy.

Let us try to make the preceding discussion more concrete on the basis of Example 8.2.3. There it was suspected that the rate of occurrence of traffic fatalities was now .15 per day rather than .1; thus it was felt that the actual sample data would support $\lambda = .15$ as a conclusion and this was stated as

$H_1$. We can then select $\alpha = .01$ if we like and, if the sample results bear out our belief that $\lambda = .15$ (and we reject $H_0$), we know that there is only 1 chance in 100 that we have reached a wrong conclusion. If, on the other hand, we interchange the two hypotheses with $\alpha = .01$ and again assume that the sample results are consistent with $\lambda = .15$ (now we would accept $H_0$), the probability that we are wrong in accepting $H_0$ is $\beta$, a quantity quite out of our control in the usual method of constructing tests. We could, of course, manipulate $\alpha$ in such a way to set $\beta$ at a level that we like, but it is much more straightforward simply to state our hypotheses (as $H_0$ and $H_1$) in a way that the error we want to control is $\alpha$.

When a rather complete tabulation of the distribution function of the test statistic is available, it is possible to find a sample size $n$ that will set both $\alpha$ and $\beta$ at desired levels. In such a case it is quite immaterial which of the two hypotheses is called $H_0$. The following example illustrates this technique.

*Example 8.2.4.* The reader will recall Example 8.2.1 in which we assumed that the time to failure, $X$, of light bulbs made a certain way was a normal random variable with $\sigma = 100$. We wanted to test $H_0{:}\,\mu = 1500$ versus $H_1{:}\,\mu = 1525$. We found in that example that the most powerful critical region was of the form $\bar{x} > C$. Suppose now that we want to find $n$ and $C$ such that $\alpha = .01$ and $\beta = .05$. Then

$$\alpha = .01 = N_Z\left(\frac{C - 1500}{100/\sqrt{n}}\right)$$

which implies that

$$\frac{C - 1500}{100/\sqrt{n}} = 2.33,$$

and

$$\beta = .05 = N_Z\left(\frac{C - 1525}{100/\sqrt{n}}\right)$$

which implies that

$$\frac{C - 1525}{100/\sqrt{n}} = -1.64$$

Thus we have two equations in two unknowns ($C$ and $n$) which we can solve simultaneously. Their solution is $C = 1514.67$, $n = 252.8$; since we cannot take a fractional sample size we should take a sample of 253 bulbs and then reject $H_0$ if $\bar{X} > 1514.67$. In so doing our actual probability of type I error is slightly less than .01 and our actual probability of type II error is slightly less than .05.

The following example illustrates the sort of statement that can be made about the power of the Neyman-Pearson critical region when testing a simple $H_0$ against a one-sided alternative $H_1$.

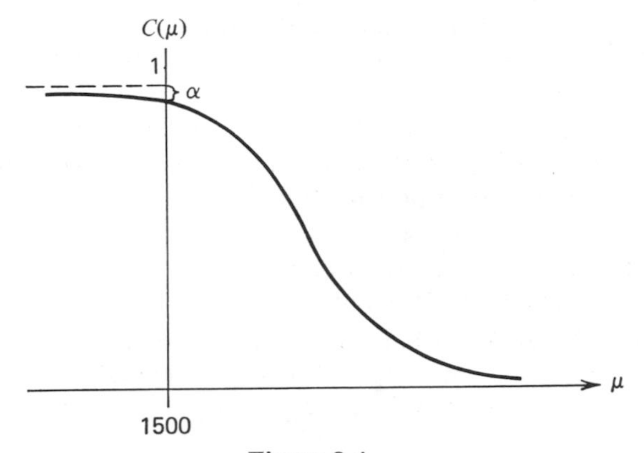


**Figure 8.4.**

*Example 8.2.5.* Let us again return to the light bulb example of 8.2.1. Suppose that we want to test the hypothesis $H_0: \mu = 1500$ versus the composite one-sided alternative $H_1: \mu > 1500$ (the alternative is called one sided since it simply states that $\mu$ takes on one of the values in the interval to the right of 1500, those values on one side of 1500). We take a sample of $n = 100$ bulbs and shall reject $H_0$ if $\bar{X} > 1516.4$ (thus $\alpha = .05$); we are using the Neyman-Pearson critical region, the one which is most powerful against the simple alternative that $\mu = \mu_1$ where $\mu_1$ is any value greater than 1500. Thus, if we were to plot the *OC* curve for this test (see Figure 8.4), we know that, among all tests with $\alpha = .05$, the value of $C(\mu_1)$ for this test is the smallest possible. That is, if we were to consider any other critical region for $H_0: \mu = 1500$ versus $H_1: \mu > 1500$, with $\alpha = .05$, its *OC* curve cannot be below the *OC* curve for the Neyman-Pearson critical region for any $\mu > 1500$. Such a test is called uniformly most powerful and is clearly a desirable sort of test to use. For many common testing problems uniformly most powerful tests do exist, while for others they can be shown not to exist.

## EXERCISE 8.2.

**1.** Suppose that $X$ is a Bernoulli random variable with parameter $p$. We take a random sample of 4 observations of $X$ and want to test $H_0: p = \frac{1}{4}$ versus $H_1: p = \frac{3}{4}$. If we reject $H_0$ only if we get 4 successes in the sample, compute the values of $\alpha$ and $\beta$.

**2.** Given that $X$ is a uniform random variable on the interval $(0, \theta)$, we might test $H_0: \theta = 1$ versus the alternative $H_1: \theta_1 = 2$ by taking a sample of 2 observations of $X$ and rejecting $H_0$ if $\bar{X} > .99$. Compute $\alpha$ and $\beta$ for this test.

**3.** Prove Theorem 8.2.2.

**4.** Prove Theorem 8.2.4.

**5.** Compute the probability of type II error (as a function of $C$) for the test derived in Example 8.2.2.

**6.** If $X$ is a normal random variable with mean 0, derive the uniformly most powerful test of $H_0: \sigma^2 = 2$ versus $H_1: \sigma^2 > 2$ of size $\alpha$, based on a random sample of $n$ observations of $X$.

**7.** Derive the form of the Neyman-Pearson critical region for testing $H_0: \lambda = \lambda_0$ versus $H_1: \lambda = \lambda_1 (\lambda_1 > \lambda_0)$, based on a random sample of $n$ observations of an exponential random variable with parameter $\lambda$.

**8.** Assume that the probability of getting a head on a single flip of a given coin is $p$ and derive the most powerful critical region for testing $H_0: p = \frac{1}{2}$ versus $H_1: p = \frac{3}{4}$, based on a sample of $n$ flips of the coin.

**9.** Assume that $X$ and $Y$ are independent normal random variables, each with variance 1. Given a random sample of $n$ observations of each, derive the most powerful critical region of size $\alpha$ for testing $H_0: \mu_X - \mu_Y = 0$ versus

$$H_1: \mu_X - \mu_Y = 1.$$

### 8.3. Composite Hypotheses

In Section 8.2 we were especially concerned with tests of simple hypotheses versus simple alternatives; in such situations the Neyman-Pearson fundamental lemma can be used to find the most powerful test of size $\alpha$. If $H_0$ or $H_1$ or both are composite hypotheses, then unfortunately no such theorem exists. However, there is a general procedure that can be followed to construct a critical region. It is similar to the procedure used in the Neyman-Pearson lemma and generally leads to a good test.

Let us first make note of the fact that if $H_0$ is composite, then we do not have a unique value for the probability of type I error. That is, the probability of rejecting $H_0$ when it is true is generally dependent on which of the several possible values of the parameter stated in $H_0$ is assumed to be the true one. In such situations it is usual to let $\alpha$ denote the largest value of the probability of type I error for all values of the parameter specified by $H_0$. Then, for all possible tests with the same $\alpha$, we would prefer the one with the lowest $OC$ curve for all values specified by $H_1$, if such a test exists.

The likelihood ratio test criterion is a method which can be employed to derive a critical region for testing $H_0$ versus $H_1$, even if one or both are composite; it generally leads to a good test in the sense of giving small probabilities of type II error for a given $\alpha$. It is defined as follows.

DEFINITION 8.3.1. Suppose that $X_1, X_2, \ldots, X_n$ is a random sample of a random variable $X$ with a probability law indexed by $\theta_1, \theta_2, \ldots, \theta_r$. To test $H_0$ versus $H_1$, where either one or both may be composite hypotheses, let $L(\hat{\omega})$ be the maximum value of the likelihood function of the sample where the parameters $\theta_1, \theta_2, \ldots, \theta_r$ are restricted to values specified by $H_0$, and let $L(\hat{\Omega})$ be the maximum value of the likelihood function of the sample where the parameters may take on any value specified by the union of $H_0$ and $H_1$. Then the critical region $R$ consists of those sample outcomes such that

$$\lambda = \frac{L(\hat{\omega})}{L(\hat{\Omega})} < A,$$

where $A$ is chosen so that $P(\text{type I error}) = \alpha$.

Notice that the likelihood ratio defined above, $L(\hat{\omega})/L(\hat{\Omega})$, will never be greater than 1 since it is the ratio of two maximum values of the likelihood function for the same sample and the denominator allows a range for the parameters which includes the range for the numerator. Thus the constant $A$ will always be less than 1 and can be determined from the probability law of the likelihood ratio. In many common cases the probability law of the ratio is quite straightforward, while in others the exact distribution is quite cumbersome to compute. Luckily, the approximate probability distribution of the ratio is easily determined, for large samples, by the result given in Theorem 8.3.4. Let us apply the likelihood ratio test criterion to a case in which we want to test a simple hypothesis about the mean of a normal random variable against a composite alternative with $\sigma^2$ known.

*Example 8.3.1.* Suppose that we own a large racing stable and are contemplating the purchase of a 1-year-old stallion. We assume that the length of time that it will take this stallion to run a mile practice race is a normal random variable $X$ with mean $\mu$ and standard deviation of 2 seconds. Then, based upon our experience with horses of this type and on the basis of $n$ practice runs, we want to test $H_0: \mu = 100$ versus $H_1: \mu < 100$, where our measurements are made in seconds. If we reject $H_0$, then we would be interested in purchasing the horse; otherwise we would not. We choose a probability of type I error of $\alpha$ and then derive the likelihood ratio test criterion as follows. Clearly, the likelihood function for the sample values is

$$\begin{aligned} L(\mu) &= \left(\frac{1}{8\pi}\right)^{n/2} e^{-1/8\Sigma(x_i-\mu)^2} \\ &= \left(\frac{1}{8\pi}\right)^{n/2} e^{-\Sigma(x_i-\bar{x})^2/8-n(\bar{x}-\mu)^2/8} \end{aligned}$$

and, since $H_0$ specifies a unique value for $\mu$,

$$L(\hat{\omega}) = \max_{H_0} L(\mu) = L(100).$$

For any outcome for the experiment such that $\bar{x} < 100$, we clearly have

$$L(\hat{\Omega}) = \max_{H_0 \cup H_1} L(\mu) = L(\bar{x})$$

whereas, if $\bar{x} \geq 100$, then

$$L(\hat{\Omega}) = \max_{H_0 \cup H_1} L(\mu) = L(100).$$

Thus, if

$$\bar{x} \geq 100, \qquad \lambda = \frac{L(\hat{\omega})}{L(\hat{\Omega})} = \frac{L(100)}{L(100)} = 1,$$

which simply indicates that any sample outcome with $\bar{x} \geq 100$ will not be in the critical region. For $\bar{x} < 100$,

$$\lambda = \frac{L(\hat{\omega})}{L(\hat{\Omega})} = \frac{L(100)}{L(\bar{x})} = e^{-n(\bar{x}-100)^2/8},$$

a quantity which always is less than 1. Then the critical region $R$ is described by

$$e^{-n(\bar{x}-100)^2/8} < A$$

which is equivalent to

$$\bar{x} < 100 - \sqrt{-\frac{8}{n} \ln A} = C$$

where $C$ (or $A$) should be chosen such that the probability of type I error is $\alpha$. Since $\bar{X}$ is normal with mean 100 and variance $4/n$ if $H_0$ is true, we should set $C$ equal to $100 - 2z_{1-\alpha}/\sqrt{n}$, where $z_{1-\alpha}$ is the 100 $(1 - \alpha)$ percentile of the standard normal (since we want

$$N_Z\left(\frac{C - 100}{2/\sqrt{n}}\right) = \alpha\Bigg).$$

Thus, if we have available $n = 4$ practice times for this horse, and we want $\alpha = .05$, we should reject $H_0$ (and purchase the horse) only if the average of his 4 times is less than 98.36 seconds. It is important to note that in this example the likelihood ratio test criterion has led to the uniformly most powerful test for $H_0$ versus $H_1$ (see Theorem 8.2.2).

The following theorem describes the likelihood ratio test criterion critical region for several common hypotheses about the mean of a normal random variable whose variance is known.

***Theorem 8.3.1.*** $X_1, X_2, \ldots, X_n$ is a random sample of a normal random variable $X$ whose variance $\sigma^2$ is known. Then the likelihood ratio test criterion critical region $R$ for a test of size $\alpha$ of $H_0$ versus $H_1$ is specified as follows for the stated $H_0$ and $H_1$.

| Test | $H_0$ | $H_1$ | $R$ |
|---|---|---|---|
| 1. | $\mu = \mu_0$ | $\mu > \mu_0$ | $\bar{x} > \mu_0 + \frac{\sigma}{\sqrt{n}} z_{1-\alpha}$ |
| 2. | $\mu = \mu_0$ | $\mu < \mu_0$ | $\bar{x} < \mu_0 - \frac{\sigma}{\sqrt{n}} z_{1-\alpha}$ |
| 3. | $\mu = \mu_0$ | $\mu \neq \mu_0$ | $\lvert\bar{x} - \mu_0\rvert > \frac{\sigma}{\sqrt{n}} z_{1-\alpha/2}$ |

Each of the first two tests listed is the uniformly most powerful test of size $\alpha$ for the stated $H_0$ and $H_1$ while the last test is not.

Problem 2 below asks for proof of the fact that the critical regions are as listed.

That test 3 given above is not the uniformly most powerful test for $H_0$ versus $H_1$ can be seen as follows. Suppose that we wanted to test $H_0: \mu = \mu_0$ versus $H_1: \mu \neq \mu_0$ with $\alpha = .05$. Then the likelihood ratio test criterion says that we should reject $H_0$ if

$$|\bar{x} - \mu_0| > \frac{\sigma}{\sqrt{n}} z_{.975}.$$

Now, if we assume as correct any alternative value for $\mu$, say $\mu_1$, such that $\mu_1 > \mu_0$, we know that

$$P\left[|\bar{x} - \mu_0| > \frac{\sigma}{\sqrt{n}} z_{.975}\right] \leq P\left[\bar{x} > \mu_0 + \frac{\sigma}{\sqrt{n}} z_{.95}\right]$$

since the one-sided test with $\alpha = .05$ is the uniformly most powerful test for all $\mu_1 > \mu_0$. Similarly, if we assume the true mean to be any value $\mu_2 < \mu_0$, then

$$P\left[|\bar{x} - \mu_0| > \frac{\sigma}{\sqrt{n}} z_{.975}\right] \leq P\left[\bar{x} < \mu_0 - \frac{\sigma}{\sqrt{n}} z_{.95}\right]$$

since the one-sided test (on the other side) with $\alpha = .05$ is uniformly most powerful for any $\mu_2 < \mu_0$. In words, when choosing our critical region so that we reject $H_0$ if $\bar{x}$ is either too large or too small with $\alpha$ fixed, we cannot have as high a probability of rejecting $H_0$, given $\mu \neq \mu_0$, as we would if we had chosen the appropriate one-sided critical region for the same $\alpha$. Figure 8.5 shows the power curve for the two-sided test versus the two one-sided tests. We would clearly not want to use either one-sided test if our alternative were $\mu \neq \mu_0$, since values of $\mu$ on the other side of $\mu_0$ lead to very large probabilities of type II error; yet we cannot make the uniformly most powerful claim for values of $\mu$ on either side of $\mu_0$ for the two-sided test.

It is generally more realistic to assume that the variance of the normal random variable is unknown when testing hypotheses about its mean. The following example derives the likelihood ratio test criterion for this important case.

*Example 8.3.2.* We might return to the race horse example (8.3.1) and see how we might test $H_0: \mu = 100$ versus $H_1: \mu < 100$, if $\sigma^2$ is not assumed to be known. With both $\mu$ and $\sigma^2$ unknown the likelihood function for the sample is a function of two arguments and thus the maximization of $L$ in $H_0 \cup H_1$ must be accomplished

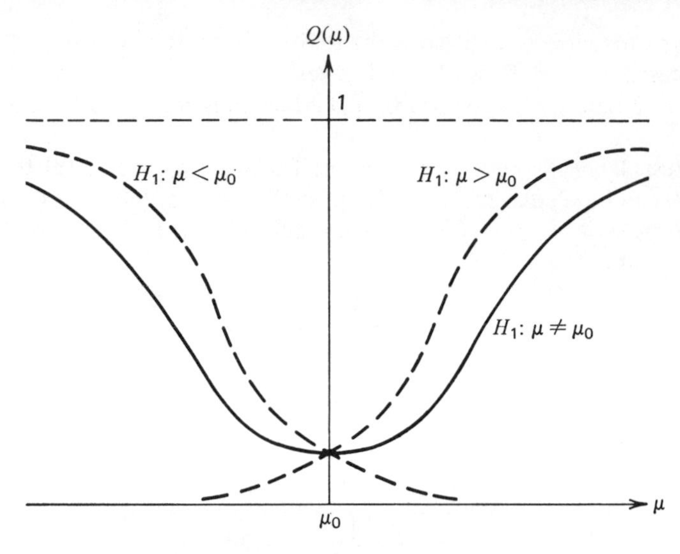


**Figure 8.5.**

with respect to both $\mu$ and $\sigma^2$. The likelihood function is

$$L(\mu, \sigma^2) = \left(\frac{1}{2\pi\sigma^2}\right)^{n/2} e^{-\Sigma(x_i-\mu)^2/2\sigma^2}$$

and, for values of the parameters specified by $H_0$, we would clearly maximize $L$ by setting

$$\mu = 100, \qquad \sigma^2 = \frac{1}{n}\sum(x_i - 100)^2.$$

Thus,

$$L(\hat{\omega}) = \left(\frac{n}{2\pi\sum(x_i - 100)^2}\right)^{n/2} e^{-n/2}.$$

In the parameter space specified by $H_0 \cup H_1$ the situation is very similar to that discussed in Example 8.3.1; for sample outcomes with $\bar{x} \leq 100$ the maximizing values of $\mu$ and $\sigma^2$ are $\bar{x}$ and $(1/n)\sum(x_i - \bar{x})^2$. Thus, if $\bar{x} \leq 100$,

$$L(\Omega) = \left[\frac{n}{2\pi\sum(x_i - \bar{x})^2}\right]^{n/2} e^{-n/2};$$

for sample results with $\bar{x} > 100$, the maximizing values for $\mu$ and $\sigma^2$ are 100 and $(1/n)\sum(x_i - 100)^2$, respectively, and thus $L(\hat{\Omega}) = L(\hat{\omega})$. This again merely states that any sample result with $\bar{x} > 100$ will not fall in the critical region, since we then would have

$$\lambda = \frac{L(\hat{\omega})}{L(\hat{\Omega})} = 1.$$

Thus, for points in the critical region, $\bar{x} \leq 100$ and, moreover, $\lambda < A$ implies

$$\frac{L(\hat{\omega})}{L(\hat{\Omega})} = \left[\frac{\sum (x_i - \bar{x})^2}{\sum (x_i - 100)^2}\right]^{n/2} < A.$$

The identity

$$\sum (x_i - 100)^2 = \sum (x_i - \bar{x})^2 + n(\bar{x} - 100)^2$$

may be substituted in the denominator of this ratio. Then, after dividing both numerator and denominator by the numerator, the points in the critical region must satisfy

$$\left[\frac{1}{1 + \dfrac{1}{n-1}\dfrac{n(\bar{x} - 100)^2}{\sigma^2}\Big/\dfrac{\sum (x_i - \bar{x})^2}{\sigma^2(n-1)}}\right]^{n/2} > A$$

or

$$\frac{1}{1 + t^2/(n-1)} > A^{2/n}$$

where

$$t^2 = \frac{n(\bar{x} - 100)^2}{\sigma^2}\Big/\frac{\sum (x_i - \bar{x})^2}{\sigma^2(n-1)}.$$

Note that

$$t = \frac{\sqrt{n}(\bar{x} - 100)}{\sigma}\Big/\sqrt{\frac{\sum (x_i - \bar{x})^2}{\sigma^2(n-1)}}$$

is the observed value of a $t$-random variable with $n - 1$ degrees of freedom if $H_0$ is true (see Theorem 7.4.2). The critical region then, when $\bar{x} \leq 100$, is described by

$$t^2 > \frac{(n-1)(1 - A^{2/n})}{A^{2/n}}$$

or

$$t < -\sqrt{\frac{(n-1)(1 - A^{2/n})}{A^{2/n}}} = C.$$

(Why do we take only the negative radical?) Thus, if we choose $C = -t_{1-\alpha}$, the 100 $(1 - \alpha)$ percentile of the $t$-distribution, and reject $H_0$ only if

$$\frac{(\bar{X} - 100)}{S/\sqrt{n}} < -t_{1-\alpha}$$

we will have $P(\text{type I error}) = \alpha$. Thus, if we had a sample of times for this horse, we would base our decision on whether to reject $H_0$ on the observed value of a $t$ random variable rather than a standard normal.

The following theorem gives the likelihood ratio test criterion critical region for tests about the mean of a normal random variable whose variance is unknown.

***Theorem 8.3.2.*** $X_1, X_2, \ldots, X_n$ is a random sample of a normal random variable with unknown mean $\mu$ and unknown variance $\sigma^2$. Then the likelihood ratio test criterion critical region $R$ for a test of size $\alpha$ of $H_0$ versus $H_1$ is specified as follows for the stated $H_0$ and $H_1$.

| Test | $H_0$ | $H_1$ | $R$ |
|---|---|---|---|
| 1. | $\mu = \mu_0$ | $\mu > \mu_0$ | $\bar{x} > \mu_0 + \frac{S}{\sqrt{n}} t_{1-\alpha}$ |
| 2. | $\mu = \mu_0$ | $\mu < \mu_0$ | $\bar{x} < \mu_0 - \frac{S}{\sqrt{n}} t_{1-\alpha}$ |
| 3. | $\mu = \mu_0$ | $\mu \neq \mu_0$ | $\left\lvert \frac{\bar{x} - \mu_0}{S/\sqrt{n}} \right\rvert > t_{1-\alpha/2}$ |

Problem 3 below asks for the proof of this theorem.

Let us now turn our attention to tests about the variance of a normal random variable whose mean $\mu$ is unknown. The following example derives the likelihood ratio test criterion critical region for a test of this sort.

*Example 8.3.3.* Some types of machines must be periodically adjusted or reset for the performance of their job; frequently the need for such adjustments is indicated by the increased variability of the product made by the machine. As a specific example, imagine that a creamery uses a machine to cut $\frac{1}{4}$-pound sections from large slabs of butter; typically the "true" weight of a section cut by the machine is a normal random variable $X$ with mean $\mu$ and variance $\sigma^2$. (We assume that the large slab of butter is uniformly mixed and thus the variability in the sections cut is mainly due to slippage of the machine.) Presumably the true mean $\mu$ is close to, if not exactly equal to, $\frac{1}{4}$ pound. If we imagine that the standard deviation $\sigma$ should be no larger than .005 pound in order that the machine has a high proportion of sections whose weights are close to $\mu$, we might be interested in testing $H_0\colon \sigma^2 = (.005)^2$ versus $H_1\colon \sigma^2 > (.005)^2$ without making any statement about $\mu$. (Presumably, if we reject $H_0$, we should adjust the machine since we would then be getting too large a variability in the weights of the sections cut by the machine.) Suppose we selected a random sample of sections cut by the machine; then we derive the likelihood ratio test criterion critical region as follows.

$$L(\mu, \sigma^2) = \left(\frac{1}{2\pi\sigma^2}\right)^{n/2} e^{-\Sigma(x_i-\mu)^2/2\sigma^2}.$$

Among values of the parameters given by $H_0$, we clearly have $\bar{x}$ and $(.005)^2$ as the maximizing values for $\mu$ and $\sigma^2$. Thus,

$$L(\hat{\omega}) = L(\bar{x}, (.005)^2) = \left[\frac{1}{2\pi(.005)^2}\right]^{n/2} e^{-\Sigma(x_i-\bar{x})^2/2(.005)^2}$$

$H_0 \cup H$ specifies that we must have $\sigma^2 \geq (.005)^2$; no matter what the value of $\sigma^2$

the maximizing value for $\mu$ is $\bar{x}$. If

$$\frac{\sum (x_i - \bar{x})^2}{n} < (.005)^2,$$

then the maximizing value for $\sigma^2$ is $(.005)^2$; if

$$\frac{1}{n}\sum (x_i - \bar{x})^2 \geq (.005)^2,$$

then the maximizing value of $\sigma^2$ is

$$\frac{1}{n}\sum (x_i - \bar{x})^2.$$

Thus,

$$\begin{aligned} L(\hat{\Omega}) &= L(\hat{\omega}), && \text{if } \frac{1}{n}\sum (x_i - \bar{x})^2 < (.005)^2 \\ &= \left[\frac{n}{2\pi \sum (x_i - \bar{x})^2}\right]^{n/2} e^{-n/2}, && \text{if } \frac{1}{n}\sum (x_i - \bar{x})^2 \geq (.005)^2, \end{aligned}$$

and we have

$$\begin{aligned} \lambda = \frac{L(\hat{\omega})}{L(\hat{\Omega})} &= 1, && \text{if } \frac{1}{n}\sum (x_i - \bar{x})^2 < (.005)^2 \\ &= \left[\frac{\sum (x_i - \bar{x})^2}{n(.005)^2}\right]^{n/2} e^{n/2 - \Sigma (x_i - \bar{x})^2/2(.005)^2}, && \text{if } \frac{1}{n}\sum (x_i - \bar{x})^2 \geq (.005)^2. \end{aligned}$$

Figure 8.6 gives the graph of $\lambda$ versus

$$\frac{\sum (x_i - \bar{x})^2}{n(.005)^2}.$$

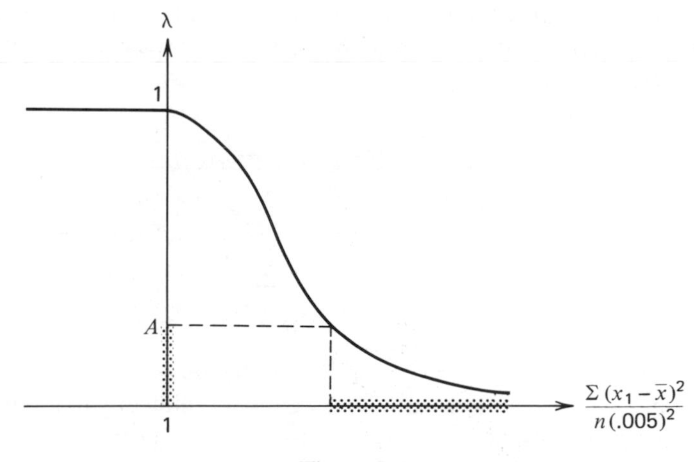


**Figure 8.6.**

Clearly, if $\lambda < A$, then

$$\frac{\sum (x_i - \bar{x})^2}{n(.005)^2} > C.$$

Thus, choosing $A$ so that $P(\text{type I error}) = \alpha$ is equivalent to choosing $C$ such that

$$P\left[\frac{\sum (X_i - \bar{X})^2}{(.005)^2} > Cn\right] = \alpha;$$

but, if $H_0$ is true, then

$$\frac{\sum (X_i - \bar{X})^2}{(.005)^2}$$

is a $\chi^2$ random variable with $n - 1$ degrees of freedom and thus

$$C = \frac{1}{n}\chi^2_{1-\alpha}$$

where $\chi^2_{1-\alpha}$ is the 100 $(1 - \alpha)$ percentile of the $\chi^2$ distribution with $n - 1$ degrees of freedom. If, for example, we randomly selected 10 sections of butter cut by the machine and wanted to test $H_0\colon \sigma^2 = (.005)^2$ versus $H_1\colon \sigma^2 > (.005)^2$ with $\alpha = .05$, then $\chi^2_{.95} = 16.9$ (9 degrees of freedom) and we should reject $H_0$ if

$$\sum (x_i - \bar{x})^2 > \frac{16.9}{10}(.005)^2 = .00004225,$$

where $x_1, x_2, \ldots, x_{10}$ are the weights of the 10 selected sections.

Theorem 8.3.3 gives the likelihood ratio test criterion critical regions for testing hypotheses about the variance of a normal random variable with unknown mean.

***Theorem 8.3.3.*** $X_1, X_2, \ldots, X_n$ is a random sample of a normal random variable $X$ whose mean $\mu$ is unknown. Then the likelihood ratio test criterion critical region $R$ for a test of size $\alpha$ of $H_0$ versus $H_1$ is specified as follows for the stated $H_0$ and $H_1$.

| Test | $H_0$ | $H_1$ | $R$ |
|---|---|---|---|
| 1. | $\sigma^2 = \sigma_0^2$ | $\sigma^2 > \sigma_0^2$ | $\sum (x_i - \bar{x})^2 > \sigma_0^2\chi^2_{1-\alpha}$ |
| 2. | $\sigma^2 = \sigma_0^2$ | $\sigma^2 < \sigma_0^2$ | $\sum (x_i - \bar{x})^2 < \sigma_0^2\chi_\alpha^2$ |
| 3. | $\sigma^2 = \sigma_0^2$ | $\sigma^2 \neq \sigma_0^2$ | $\sum (x_i - \bar{x})^2 < \sigma_0^2\chi^2_{\alpha/2}$ |
| | | and | $\sum (x_i - \bar{x})^2 > \sigma_0^2\chi^2_{1-\alpha/2}$ |

Problem 4 below asks for the proof of this theorem.

In each of the examples treated so far, we had no difficulty in finding the $A$ such that the probability of type I error was $\alpha$. In each case we were able to manipulate the inequality describing the critical region to get an inequality for a statistic whose distribution was known if $H_0$ were true. This known distribution then can effectively be used to choose $A$. In many problems, however, the exact distribution for $\lambda$ may be very difficult to obtain and it may not be possible to manipulate the necessary inequality to derive an equivalent statement for a statistic whose distribution is known, if $H_0$ is true. In cases such as this, the following theorem, which we shall not prove, is of use in describing the critical region.

***Theorem 8.3.4.*** Suppose that $X$ is a random variable whose distribution function satisfies certain regularity conditions and is indexed by $k$ parameters, $\theta_1, \theta_2, \ldots, \theta_k$, and $X_1, X_2, \ldots, X_n$ is a random sample of $X$. For

$$H_0:\theta_1 = \theta_1^*, \theta_2 = \theta_2^*, \ldots, \theta_r = \theta_r^*, \qquad \text{where } r \leq k,$$

versus the alternative $H_1:H_0$ is false, define

$$\lambda = \frac{L(\hat{\omega})}{L(\hat{\Omega})} = g(x_1, x_2, \ldots, x_n)$$

in the usual way. Then, if we define $\Lambda = g(X_1, X_2, \ldots, X_n)$, $-2 \ln \Lambda$ is approximately a $\chi^2$ random variable with $r$ degrees of freedom, if $H_0$ is true, for large values of $n$. (Thus, if $Y = -2 \ln \Lambda$,

$$\lim_{n \to \infty} F_Y(t) = F_{\chi^2}(t)$$

where $F_{\chi^2}(t)$ is the distribution function of a $\chi^2$ random variable with $r$ degrees of freedom.)

Thus, we can use the $\chi^2$ table to determine the value of $A$ if we have a "large" sample size $n$. Since

$$P[\Lambda < A] = P[-2 \ln \Lambda > -2 \ln A],$$

we simply choose $-2 \ln A = \chi^2_{1-\alpha}$ or $A = e^{-\chi^2_{1-\alpha}/2}$ where $\chi^2_{1-\alpha}$ is the 100 $(1 - \alpha)$ percentile of the $\chi^2$ distribution with $r$ degrees of freedom.

*Example 8.3.4.* Suppose that we would like to test the hypothesis that the probability that a particular experienced dart player will hit the bull's-eye of a target, on a single toss, is .9. Thus, if we have him toss a dart at the bull's-eye $n$ times, we will have a random sample of $n$ observations of a Bernoulli random variable with parameter $p$ and would like to test $H_0:p = .9$ versus $H_1:p \neq .9$ with $\alpha = .01$, for example. Then the likelihood function for the sample is

$$L(p) = p^{\Sigma x}(1 - p)^{n - \Sigma x_i}$$

and

$$L(\hat{\omega}) = L(.9) = (.9)^{\Sigma x_i}(.1)^{n - \Sigma x_i};$$

we also know that the maximum likelihood estimator of $p$ is $\bar{X}$, so

$$L(\hat{\Omega}) = L(\bar{x}) = \bar{x}^{\Sigma x_i}(1 - \bar{x})^{n-\Sigma x_i}.$$

Then

$$\lambda = \frac{L(\hat{\omega})}{L(\hat{\Omega})} = \left(\frac{.9}{\bar{x}}\right)^{\Sigma x_i}\left(\frac{.1}{1 - \bar{x}}\right)^{n-\Sigma x_i},$$

and we would reject $H_0$ only if $-2 \ln \lambda > \chi^2_{.99} = 6.635$ (for 1 degree of freedom). If, for example, he threw 200 darts at the bull's-eye, 160 of which hit it, then $\bar{x} = .8$,

$$\lambda = (\tfrac{9}{8})^{160}(\tfrac{1}{2})^{40}, \qquad -2 \ln \lambda = 17.88,$$

so we would reject $H_0$. The (approximate) probability that we are wrong in so rejecting is .01.

The reader should note the close tie between confidence intervals for parameters derived at the end of Chapter 7 and the tests of hypotheses versus composite alternatives discussed in this section. For a given sample of size $n$, let $T$ be the set of numbers specified by the 100 $(1 - \alpha)\%$ two-sided confidence interval for $\mu$, the mean of a normal random variable, with $\sigma^2$ known; thus

$$T = \left\{y : \bar{x} - z_{1-(\alpha/2)} \frac{\sigma}{\sqrt{n}} \le y \le \bar{x} + z_{1-(\alpha/2)} \frac{\sigma}{\sqrt{n}}\right\}.$$

Then, for exactly the same observed sample values, the hypothesis $H_0 : \mu = \mu_0$ versus $H_1 : \mu \ne \mu_0$ will be accepted if and only if $\mu_0 \in T$ with probability of type I error equal to $\alpha$. Similarly, let $S$ be the one-sided 100 $(1 - \alpha)\%$ lower confidence interval for $\mu$ for a sample of size $n$ of a normal random variable with known variance $\sigma^2$; that is

$$S = \left\{y : \bar{x} - z_{1-\alpha} \frac{\sigma}{\sqrt{n}} \le y\right\}.$$

Then, for the same observed sample values, the hypothesis $H_0 : \mu = \mu_0$ versus $H_1 : \mu < \mu_0$ will be accepted if and only if $\mu_0 \in S$ where $\alpha$ is the probability of type I error. The same relation exists for hypotheses about $\mu$ with $\sigma^2$ unknown and for hypotheses about $\sigma^2$ whether or not $\mu$ is known.

## EXERCISE 8.3.

**1.** Assume that the annual rainfall at a certain recording station is a normal random variable with mean $\mu$ and standard deviation 2 inches. The rainfall recorded (in inches) in each of 5 years was 18.6, 20.4, 17.3, 15.1, and 22.6. Test the hypothesis that $\mu = 21$ versus the alternative $\mu < 21$ with $\alpha = .1$.

**2.** Prove Theorem 8.3.1.

**3.** Prove Theorem 8.3.2.

**4.** Prove Theorem 8.3.3.

**5.** A producer of frozen fish is being investigated by the Bureau of Fair Trades. Each package of fish which this producer markets carries the claim that it contains 12 ounces of fish; a complaint has been registered that this claim is not true. The Bureau acquires 100 packages of fish marketed by this company and, letting $x_i$ be the observed weight (in ounces) of the $i$-th package, $i = 1, 2, \ldots, 100$, they find

$$\sum x_i = 1150, \qquad \sum x_i^2 = 13{,}249.75.$$

It would seem reasonable to assume that the true weights of packages that they market are normally distributed with mean $\mu$ and variance $\sigma^2$, neither of which is known. With $\alpha = .01$, would the Bureau accept or reject $H_0: \mu = 12$ versus $H: \mu < 12$, based on this sample?

**6.** In deciding whether a certain type of plant would be appropriate for hedges, it is of some importance that individual plants exhibit small variability in the amounts they will grow in a year (at the same age). Specifically, we might assume that the growth made by a plant of a specific type and age (for given climatic conditions) is a normal random variable with mean $\mu$ and variance $\sigma^2$. Then, to decide whether the plant would be appropriate for hedges, we might like to test $H_0: \sigma^2 = \frac{1}{4}$ versus $H_1: \sigma^2 < \frac{1}{4}$ with $\alpha = .05$ (measurements made in feet). Suppose that we record the growth of 5 plants of this type for 1 year and find them to be 1.9, 1.1, 2.7, 1.6, and 2.0 feet. Should we accept $H_0$?

**7.** Derive the likelihood ratio test criterion critical region for testing the hypothesis that $\lambda = \lambda_0$ versus the alternative that $\lambda > \lambda_0$ where $\lambda$ is the parameter of an exponential random variable. Assume that you have available a random sample of $n$ observations of $X$ and that you want your probability of type I error to be $\alpha$.

**8.** The time to failure of a certain vacuum tube is known to be an exponential random variable with parameter $\lambda$ when used in a particular type of circuit. A thousand of these tubes are placed in operation and the sum of their times to failure is 109,652 hours. With $\alpha = .05$, would you reject $H_0: \lambda = .008$ versus $H_1: \lambda > .008$ on the basis of this sample?

**9.** The number of accidents in an industrial plant per month is assumed to be a Poisson random variable with parameter $\lambda$. Given a sample of the number of accidents per month in a given year (thus $n = 12$), derive the likelihood ratio test criterion for testing $H_0: \lambda = \lambda_0$ versus $H_1: \lambda < \lambda_0$ for a fixed $\alpha$. Suppose that a given industry finds that they had a total of 15 accidents last year. With $\alpha = .1$, should they reject $H_0: \lambda = 2$ versus $H_1: \lambda < 2$? (Use the result of Theorem 8.3.4.)

**10.** The number of times that an electric light switch can be turned on and off before it fails is a geometric random variable $X$ with parameter $p$ (thus $X$ is the number of turnons until failure). Given a random sample of 10 switches, and that the sum of the number of turnons to failure (for all 10) was 15,169, would you reject $H_0: p = .00005$ versus $H_1: p > .00005$ with $\alpha = .05$? (Use Theorem 8.3.4.)

## 8.4. Goodness-of-Fit Tests

We have discussed ways of testing what might be called parametric hypotheses about a random variable; that is, we have assumed that the probability law of the random variable was known and discussed testing hypotheses that stated values for the parameters of the probability law. In many practical problems it is desired to test a hypothesis that specifies the probability law for the random variable being sampled versus the alternative that the probability law is not of the stated type. The likelihood ratio methodology discussed in the last section is then not appropriate, since the likelihood function is not well defined, assuming the alternative is true. If we assume that the probability law is not of the form specified by the hypothesis, then it could be any one from a large class of possible functions and we would find it very difficult to maximize the likelihood function over this class of possible alternatives. The fact that the class of alternatives is so large also makes it difficult to make definitive comparative statements about different possible tests of the hypothesis (in terms of probability of type II error). Thus, we shall follow the usual convention and simply not attempt to evaluate the probability of type II error when our hypothesis specifies the distributional form for the random variable we have sampled.

One method of testing such distributional hypotheses that we shall discuss is the $\chi^2$ goodness-of-fit test. All such tests rest upon the multinomial random variable and the following theorem, which we shall present with no proof. (The multinomial random variable was discussed in Section 5.4.)

***Theorem 8.4.1.*** If $(X_1, X_2, \ldots, X_k)$ is a multinomial random variable with parameters $n, p_1, p_2, \ldots, p_k$, then the distribution function of the random variable

$$U = \sum_{i=1}^{k} \frac{(X_i - np_i)^2}{np_i}$$

approaches the $\chi^2$ distribution function with $k - 1$ degrees of freedom as $n \to \infty$. That is,

$$\lim_{n\to\infty} F_U(t) = F_{\chi^2}(t), \qquad \text{for any } t,$$

where $F_{\chi^2}(t)$ is the $\chi^2$ distribution function with $k - 1$ degrees of freedom.

To make the result just quoted seem a little more plausible, we might examine the case of the multinomial random variable with $k = 2$. Then each of our $n$ trials falls in one of two classes. If we call the first class success, then $X_1$ is simply the number of successes and $X_2 = n - X_1$ is the number of failures in the $n$ trials, where $p_1$ is the probability of a success on a single trial and $p_2 = 1 - p_1$ is the probability of a failure. Then

$$U = \frac{(X_1 - np_1)^2}{np_1} + \frac{(X_2 - np_2)^2}{np_2}$$

$$= \frac{(X_1 - np_1)^2}{np_1} + \frac{(n - X_1 - n(1 - p_1))^2}{n(1 - p_1)}$$

$$= \frac{(X_1 - np_1)^2(1 - p_1) + (-X_1 + np_1)^2 p_1}{np_1(1 - p_1)}$$

$$= \frac{(X_1 - np_1)^2}{np_1(1 - p_1)} = V^2,$$

for example. But

$$V = \frac{X_1 - np_1}{\sqrt{np_1(1 - p_1)}}$$

is approximately a standard normal random variable for large $n$, and thus $V^2$ is approximately the square of a standard normal random variable for large $n$. Thus, the distribution of $V^2$ will be a $\chi^2$ with 1 degree of freedom, for large $n$, which is just what Theorem 8.4.1 claims. The proof of Theorem 8.4.1 for arbitrary $k$ would proceed in quite an analogous fashion. Essentially the $k$ terms that are summed together can be shown to be the same as a sum of squares of $k - 1$ independent variables, each of which tends to be the square of a standard normal random variable as $n$ increases. Because of Theorem 6.3.3, the sum has the $\chi^2$ distribution with $k - 1$ degrees of freedom.

To use Theorem 8.4.1, then, to test the hypothesis that $(X_1, X_2, \ldots, X_k)$ is a multinomial random variable with specified parameters $p_1, p_2, \ldots, p_k$, we would take our sample, compute

$$u = \sum_{i=1}^{k} \frac{(x_i - np_i)^2}{np_i},$$

and reject the hypothesis if $u > \chi^2_{1-\alpha}$, the 100 $(1 - \alpha)$ percentile of the $\chi^2$ distribution with $k - 1$ degrees of freedom. If the hypothesis is true, then the probability we would reject is $\alpha$; that is, $\alpha$ is our probability of type I error. Many authors agree that the $\chi^2$ approximation is quite good for the distribution of $U$ so long as $np_i \geq 5$ for $i = 1, 2, \ldots, k$. If necessary, some of the classes may be combined to satisfy this requirement. Some of the following examples illustrate this procedure.

*Example 8.4.1.* Suppose that we are given a particular die and want to test the hypothesis that it is fair, with $\alpha = .05$, based on a sample of 100 rolls. Then, the hypothesis might be written

$$H_0: p_i = \tfrac{1}{6}, \qquad i = 1, 2, \ldots, 6,$$

and we could let $X_i$, $i = 1, 2, \ldots, 6$ be the number of rolls on which we observe face $i$. Then, suppose we observe $x_1 = 20$, $x_2 = 24$, $x_3 = 15$, $x_4 = 21$, $x_5 = 12$, and $x_6 = 8$; the observed value of $u$ is

$$\sum_{i=1}^{6} \frac{(x_i - 16\frac{1}{3})^2}{16\frac{1}{3}} = 10.94.$$

Since the 95-th percentile of the $\chi^2$ distribution with 5 degrees of freedom is 11.070, we would accept $H_0$ and conclude that this die is fair. (Note that if we had chosen $\alpha = .1$, we would have rejected $H_0$; the probability that we would be wrong in such a conclusion would be .1.)

By making use of the following theorem, the multinomial test just described can be applied to testing the hypothesis that the distribution function $F_Y$ of a random variable $Y$ has any specified form.

***Theorem 8.4.2.*** Assume that $Y$ has distribution function $F_Y$ (either discrete or continuous) and that $Y_1, Y_2, \ldots, Y_n$ is a random sample of $Y$. Then, if we define

$$I_1 = \{y : y \le a_1\}, \qquad I_2 = \{y : a_1 < y \le a_2\}, \ldots,$$
$$I_{k-1} = \{y : a_{k-2} < y \le a_{k-1}\}, \qquad I_k = \{y : a_{k-1} < y\},$$

and let $X_i$ be the number of sample values that fall in $I_i$, $i = 1, 2, \ldots, k$, $(X_1, X_2, \ldots, X_k)$ is the multinomial random variable with parameters $n$, $p_1, p_2, \ldots, p_k$, where $n$ is the sample size (for the sample of $Y$) and

$$p_i = P[Y \text{ falls in } I_i], \qquad i = 1, 2, \ldots, k.$$

*Proof:* Clearly, the first sample value, $Y_1$, will fall in $I_1$ or $I_2$ or $\cdots$ or $I_k$; the probability that it falls in $I_i$ is

$$p_i = P[Y_1 \text{ falls in } I_i] = P[Y \text{ falls in } I_i]$$

since $F_{Y_1}(t) = F_Y(t)$. Thus, this first sample of $Y$ can be thought of as being a multinomial trial with $k$ possible outcomes whose probabilities of occurrence are $p_1, p_2, \ldots, p_k$. The second sample value, $Y_2$, can be thought of as a second, independent, multinomial trial with the same probabilities of occurrence; the same is true for each of the succeeding $n - 2$ sample values. Since $Y_1, Y_2, \ldots, Y_n$ are independent, the trials they generate are independent and thus $(X_1, X_2, \ldots, X_k)$ is the multinomial random variable with parameters $n, p_1, \ldots, p_k$, where $X_i$ is the total number of $Y_j$'s that fall in $I_i$, $i = 1, 2, \ldots, k$. ◀

Then, no matter what the form of $F_Y$, we can test the hypothesis that $Y_1, Y_2, \ldots, Y_n$ each has distribution function $F_Y$ by testing the hypothesis that the multinomial random variable $(X_1, X_2, \ldots, X_k)$ generated by $Y_1, Y_2, \ldots, Y_n$ has the appropriate parameters $p_1, p_2, \ldots, p_k$. If we reject this hypothesis, then $F_Y$ is apparently not appropriate, compared with the sample; if we accept this hypothesis, then $F_Y$ would appear to be consistent with the sample values.

*Example 8.4.2.* The number of misprints per page in a printed book is frequently taken to be a Poisson random variable, since they are presumably independent and, if the whole book were done by the same typesetter, the rate should be constant from the first page to the last. A misprint count was made for 100 pages of a recent novel with the numbers of misprints found as given below.

| Number of Misprints | Number of Pages |
|---|---|
| 0 | 65 |
| 1 | 25 |
| 2 | 8 |
| 3 | 2 |
| total | 100 |

We will test the hypothesis that $Y$, the number of misprints per page, is a Poisson random variable with parameter $\lambda = .4$ per page with $\alpha = .1$. Since $Y$ is discrete, the natural intervals to use in defining the multinomial random variable would be ones centered around the values the variable may take on. Thus, we define

$$I_1 = \{y : y \le \tfrac{1}{2}\}, \qquad I_2 = \{y : \tfrac{1}{2} < y \le 1\tfrac{1}{2}\},$$
$$I_3 = \{y : 1\tfrac{1}{2} < y \le 2\tfrac{1}{2}\}, \qquad I_4 = \{y : 2\tfrac{1}{2} < y\}.$$

Then we define $X_1$ to be the number of 0's we observe, $X_2$ the number of 1's, $X_3$ the number of 2's, and $X_4$ the number of 3's (or higher values). By referring to the Poisson table, we find that $p_1 = .6703$, $p_2 = .2681$, $p_3 = .0536$, and $p_4 = .0080$. Then we note that $np_4 = .80$ which is not at least 5 so we combine the two classes for 2 and for 3 misprints per page. Thus, $(X_1, X_2, X_3)$, where $X_3$ now is the number of times we observe 2 or more misprints, is the multinomial random variable with parameters 100, .6703, .2681, and .0616, if the hypothesis is true. The observed value of the test statistic is

$$u = \frac{(65 - 67.03)^2}{67.03} + \frac{(25 - 26.81)^2}{26.81} + \frac{(10 - 6.16)^2}{6.16} = 2.577.$$

Since the 90-th percentile of the $\chi^2$ distribution with 2 degrees of freedom is 4.605, we accept the hypothesis.

In making goodness-of-fit tests for a discrete random variable, such as in the example above, the intervals $I_1, I_2, \ldots, I_k$ that the data are summarized into are actually determined by the discrete values that the variable can take on. In continuous cases, though, the construction of $I_1, I_2, \ldots, I_k$ is generally much more arbitrary. In such cases, there is general agreement that the lengths of the finite intervals used should be equal and that the number used should be as large as possible consistent with the requirement that $np_i \geq 5$, $i = 1, 2, \ldots, k$.

*Example 8.4.3.* The times to failure of a thousand electron tubes were as given in the following table.

| Time to Failure | Number |
|---|---|
| $t \leq 150$ | 543 |
| $150 < t \leq 300$ | 258 |
| $300 < t \leq 450$ | 120 |
| $450 < t \leq 600$ | 48 |
| $600 < t \leq 750$ | 20 |
| $750 < t$ | 11 |

The manufacturer claims that lives of tubes of this type are exponentially distributed with $\lambda = .005$. Should we reject his claim, with $\alpha = .01$? Defining $I_1$ through $I_6$ in the same way as the above data are summarized, the numbers of failure times in these 6 intervals should then be a multinomial random variable with parameters 1000, $p_1, p_2, \ldots, p_6$, where

$$p_1 = \int_0^{150} f_Y(t)\, dt = \int_0^{150} .005e^{-.005t}\, dt = .5277,$$

$$p_2 = \int_{150}^{300} f_Y(t)\, dt = \int_{150}^{300} .005e^{-.005t}\, dt = .2492,$$

etc. The remaining probabilities are $p_3 = .1177$, $p_4 = .0556$, $p_5 = .0263$, $p_6 = .0235$. Then, the observed value of $u$ is

$$\frac{(543 - 527.7)^2}{527.7} + \frac{(258 - 249.2)^2}{249.2} + \cdots + \frac{(11 - 23.5)^2}{23.5} = 9.997;$$

the 99-th percentile of the $\chi^2$ distribution with 5 degrees of freedom is 15.1. Thus we would accept his claim.

In the two goodness-of-fit cases examined above, the original hypothesis completely specified the distribution function of the random variable $Y$ which had been sampled; the specification was complete down to the value of the parameter of the probability law. Such complete specification was necessary, of course, in order that we could compute the values of $p_1, p_2, \ldots, p_k$. In many cases it is desirable to be able to test the hypothesis that the distribution function has a specified form without having to also specify the values of the parameters of the distribution function. Thus, we might simply like to test the hypothesis that a random variable $Y$ follows the Poisson probability law, rather than that $Y$ follows the Poisson probability law with parameter $\lambda = .4$; or we might like to test that $Y$ is exponential, rather than that $Y$ is exponential with parameter $\lambda = .005$. Fortunately, a slight adaptation of the test already discussed is all that is necessary to test these less restrictive hypotheses which do not specify the values of the parameters. The following theorem, stated without proof, gives the necessary methodology.

***Theorem 8.4.3.*** Assume that $Y$ has distribution function $F_Y$ which satisfies certain regularity conditions and which is indexed by $r$ unknown parameters, $\theta_1, \theta_2, \ldots, \theta_r$, and that $Y_1, Y_2, \ldots, Y_n$ is a random sample of $Y$. Let $\hat{\Theta}_1, \hat{\Theta}_2, \ldots, \hat{\Theta}_r$ be the maximum likelihood estimators for $\theta_1, \theta_2, \ldots, \theta_r$, respectively, and define $I_1, I_2, \ldots, I_k$ and $X_1, X_2, \ldots, X_k$ as in Theorem 8.4.2. Then, if we define

$$\hat{P}_i = P[Y \text{ falls in } I_i], \qquad i = 1, 2, \ldots, k$$

where $\hat{\Theta}_1, \hat{\Theta}_2, \ldots, \hat{\Theta}_r$ replace $\theta_1, \theta_2, \ldots, \theta_r$ in $F_Y$, the distribution of the statistic

$$V = \sum_{i=1}^{k} \frac{(X_i - n\hat{P}_i)^2}{n\hat{P}_i}$$

tends to that of a $\chi^2$ random variable with $k - r - 1$ degrees of freedom as $n$ gets large.

Thus, if we want to test the hypothesis that $F_Y$ has a specified form, saying nothing about the values of the parameters, we simply use the data to estimate the parameters and then proceed as we have in previous examples. That is, we would compute the observed value of $V$, in this case, and reject $H_0$ if this observed value exceeds the 100 $(1 - \alpha)$ percentile of the $\chi^2$ distribution with $k - r - 1$ degrees of freedom. Note that we have "lost" 1 degree of freedom for every parameter that we had to estimate.

*Example 8.4.4.* A manufacturer of yardsticks has a certain production process which turns out yardsticks. He feels that $Y$, the "true" length of one of his yardsticks, should be a normal random variable. To test the hypothesis that $Y$ is normal, without specifying $\mu$ and $\sigma$, he randomly selects 300 yardsticks and measures their actual lengths. These 300 measurements are summarized in the table at the top of page 286.

He may test the hypothesis that $Y$, the true length of a yardstick made this way, is a normal random variable with $\alpha = .10$ by using the result given as Theorem 8.4.3. Thus, he would first want to use the sample to compute the observed values of the maximum likelihood estimators for $\mu$ and $\sigma^2$; then he would use these

| True Length $y$ | Frequency $x$ |
|---|---|
| $35.7 < y \leq 35.8$ | 10 |
| $35.8 < y \leq 35.9$ | 30 |
| $35.9 < y \leq 36.0$ | 104 |
| $36.0 < y \leq 36.1$ | 97 |
| $36.1 < y \leq 36.2$ | 51 |
| $36.2 < y \leq 36.3$ | 8 |

quantities to compute the observed values of the estimators of the probabilities of individual sample values falling into the six classes used to summarize the data above. We find

$$\bar{y}=36.008, \qquad \frac{n-1}{n}s^2 = .012291, \qquad \sqrt{\frac{n-1}{n}s^2} = .1109.$$

Thus, to compute $\hat{p}_1, \hat{p}_2, \ldots, \hat{p}_6$, we simply compute the probability that a normal random variable with mean 36.008 and standard deviation .1109 would fall in each of the six intervals. Using the standard normal table, we find

$$\hat{p}_1 = .0301, \quad \hat{p}_2 = .1359, \quad \hat{p}_3 = .3061,$$
$$\hat{p}_4 = .3246, \quad \hat{p}_5 = .1615, \quad \hat{p}_6 = .0418.$$

Thus, if the hypothesis were true the numbers we expect in each of these six classes are 9.03, 40.77, 91.83, 97.38, 48.45, and 12.54, respectively, and we find

$$v = \sum_{i=1}^{6} \frac{(x_i - 300\hat{p}_i)^2}{300\hat{p}_i} = 6.341.$$

The 90-th percentile of the $\chi^2$ distribution with 3 degrees of freedom is 6.25 so the hypothesis is rejected. The probability this conclusion is incorrect is .1, since there is a 10% chance that we would observe this large (or larger) a value of $V$ if the hypothesis were true.

*Example 8.4.5.* Let us test the hypothesis that the number of suicides per month in Manhattan, $Y$, is a Poisson random variable with $\alpha = .05$, based upon the following observed numbers of suicides per month for a 5-year span.

| Number of Suicides $y$ | Frequency $x$ |
|---|---|
| 0 | 33 |
| 1 | 17 |
| 2 | 7 |
| 3 | 3 |

We must first estimate $\lambda$, assuming the hypothesis is true; thus we find $\bar{y} = .667$. Then the estimated probabilities of observing 0, 1, 2, and 3 suicides per month are

$$\hat{p}_1 = .5314, \quad \hat{p}_2 = .3415, \quad \hat{p}_3 = .1181, \quad \hat{p}_4 = .0090$$

and the expected frequencies would be 31.884, 20.490, 7.086, and .54, respectively. Since the expected frequency for $Y = 3$ is less than 5, we combine both the observed frequencies and these expected frequencies for $Y = 3$ with those for $Y = 2$. Then,

$$\hat{p}_1 = .5314, \qquad \hat{p}_2 = .3415, \qquad \hat{p}_3 = .1271$$

and we find

$$v = \sum_{i=1}^{3} \frac{(x_i - 60\hat{p}_i)^2}{60\hat{p}_i} = 1.372.$$

The 95-th percentile of the $\chi^2$ distribution with 1 degree of freedom is 3.84 so we accept the hypothesis that $Y$ is a Poisson random variable.

## EXERCISE 8.4.

**1.** A college track man put the shot 100 times in practice in a week. The distances he threw it (measured in feet) are recorded in the following table.

| Distance $y$ | Frequency $x$ |
|---|---|
| $y \le 61$ | 12 |
| $61 < y \le 63$ | 20 |
| $63 < y \le 65$ | 40 |
| $65 < y \le 67$ | 25 |
| $y > 67$ | 3 |

With $\alpha = .01$, test the hypothesis that the distance, $Y$, which he can put the shot is a normal random variable with $\mu = 63$ feet, $\sigma = 2$ feet.

**2.** In each of 100 games a baseball player had 4 times at bat. If the number of hits he got were as follows, test the hypothesis that the number of hits he will get in 4 times at bat is a binomial random variable with $n = 4$ and $p = .3$. (Use $\alpha = .05$.)

| Number of Hits $y$ | Frequency (Number of Days) $x$ |
|---|---|
| 0 | 18 |
| 1 | 45 |
| 2 | 29 |
| 3 | 6 |
| 4 | 2 |

**3.** It would seem reasonable to assume that the final digit of a randomly selected telephone number is equally likely to be 0 or 1 or 2 or . . . or 9. Select a page at random from your telephone book and count the number of 0's, 1's, 2's, . . . , 9's that occur as last digits. Then test the hypothesis that the last digit is equally likely to be 0 or 1 or . . . or 9, with $\alpha = .1$. (You should get a sample of size $n = 280$ or so since this is roughly the number of phone numbers listed on a single page.)

**4.** A pair of dice was rolled 500 times. The sums that occurred were as recorded in the following table. Test the hypothesis that the dice were fair, with $\alpha = .05$

| Sum | Frequency |
|---|---|
| 2, 3, or 4 | 74 |
| 5 or 6 | 120 |
| 7 | 83 |
| 8 or 9 | 135 |
| 10, 11, or 12 | 88 |

**5.** If cars are arriving at a supermarket consistent with a Poisson process, the length of time between successive arrivals is an exponential random variable. Arrival times for all autos in a 2-hour period were recorded and the times between arrivals (in minutes) were as summarized below.

| Time Between Arrivals $t$ | Frequency |
|---|---|
| $t \leq 1$ | 40 |
| $1 < t \leq 2$ | 29 |
| $2 < t \leq 3$ | 15 |
| $t > 3$ | 8 |
| Total | 102 |

Test the hypothesis that this time distribution is consistent with an exponential distribution for time between arrivals, with $\alpha = .1$.

**6.** A hundred students took the same I.Q. test. The scores they made were as summarized below.

| Score $x$ | Frequency |
|---|---|
| $x \leq 90$ | 8 |
| $90 < x \leq 110$ | 38 |
| $110 < x \leq 130$ | 45 |
| $x > 130$ | 9 |

We might assume that these 100 scores are a random sample of the scores that would be made by all possible people that could take this exam. Test the hypothesis that the scores made by the (conceptually infinite) population would be normally distributed, with $\alpha = .05$.

**7.** In a recent 72-hour holiday period in the United States there was a total of 290 fatal auto accidents. The number of fatal accidents, per hour, during this period was as follows.

| Number per Hour | Number of Hours |
|---|---|
| 0 or 1 | 5 |
| 2 | 11 |
| 3 | 15 |
| 4 | 14 |
| 5 | 12 |
| 6 | 8 |
| 7 or more | 6 |

Test the hypothesis that the number of accidents per hour, during such a holiday weekend, is a Poisson random variable.

### 8.5. Contingency Tables

Many times the elements of a sample may be categorized according to two or more criteria and it then is of interest to know whether the methods of classification are independent. For example, suppose that there are exactly two candidates in a presidential election; everyone that casts a vote may be categorized according to the person he votes for and, quite separately, according to party registration (for example, Republican, Democrat, or independent). Presumably these two modes of classification may be dependent upon each other. Or we might classify each person in a sample from a given racial group according to eye color and also according to hair color and then ask whether these two modes of classification are independent. Or we might examine state health statistics and find the number of boys and the number of girls born in the state in each month of a given year and ask whether these two modes of classification are independent (that is, does the proportion of boys born seem to be the same for all months).

In each of the cases just mentioned, the elements of the sample can be classified according to each of two different types of categories. Thus, suppose we take a sample of size $n$ and have two ways of categorizing each element of the sample. Assume that the first method of classification has $r$ levels and the second method has $c$ levels. Then the numbers of sample elements falling into each possible combination of classes from the two methods can be conveniently summarized in tabular form as shown in Table 8.1.

Thus, $x_{ij}$ is the observed number of sample elements falling in the $i$-th level of classification 1 *and* the $j$-th level of classification 2, $i = 1, 2, \ldots, r$, $j = 1, 2, \ldots, c$. Note that we are using a dot instead of a subscript to denote

**Table 8.1**

| Level of Classification 1 | Level of Classification 2 | | | | | |
|---|---|---|---|---|---|---|
| | 1 | 2 | 3 | ... | $c$ | Total |
| 1 | $x_{11}$ | $x_{12}$ | $x_{13}$ | ... | $x_{1c}$ | $x_{1.}$ |
| 2 | $x_{21}$ | $x_{22}$ | $x_{23}$ | ... | $x_{2c}$ | $x_{2.}$ |
| 3 | $x_{31}$ | $x_{32}$ | $x_{33}$ | ... | $x_{3c}$ | $x_{3.}$ |
| . | . | . | . | . | . | |
| . | . | . | . | . | . | |
| . | . | . | . | . | . | |
| $r$ | $x_{r1}$ | $x_{r2}$ | $x_{r3}$ | ... | $x_{rc}$ | $x_{r.}$ |
| Total | $x_{.1}$ | $x_{.2}$ | $x_{.3}$ | | $x_{.c}$ | $n$ |

summation over that subscript; thus

$$x_{1.} = \sum_{j=1}^{c} x_{1j}, \qquad x_{4.} = \sum_{j=1}^{c} x_{4j}, \qquad x_{.3} = \sum_{i=1}^{r} x_{i3},$$

etc. Since our total sample size is $n$, then

$$\sum_i \sum_j x_{ij} = \sum_i x_{i.} = \sum_j x_{.j} = n.$$

As was mentioned, a frequent question that is asked concerning such samples is whether or not the methods of classification are independent. We are then asking whether the probability that a sample element falls in the $i$-th level of category 1 is the same for each level of category 2. If so, the two methods of categorizing are independent and, if not, they are not independent. Let us make the same statement in the context of the presidential election example; we are asking whether the probability a person will vote for candidate A is the same, regardless of whether the voter is registered as a Republican, Democrat, or independent. (If the answer to this is yes, then it would appear that party registration does not have much effect on the outcome of a presidential election.) Thus, the question of independence could also be phrased: Are the probabilities of occurrence of the various levels of classification 1 contingent on the levels of classification 2? As we shall see the hypothesis of no contingency (independence) can be tested using a summarization such as given in Table 8.1. Thus, such a table as 8.1 is commonly called a contingency table; it simply summarizes the number of sample elements that fall in the $i$-th level of category 1 for each of the $c$ levels of category 2, $i = 1, 2, \ldots, r$.

Testing the hypothesis of independence in a contingency table is simply a special case of the result given in Theorem 8.4.3. Suppose that we have an infinite population, each element of which falls in exactly one level of classification 1 and in exactly one level of classification 2, where the two modes of classification have a total of $r$ and $c$ levels, respectively. Let $p_{ij}$ be the probability that a randomly selected element falls in the $i$-th level of classification 1 and the $j$-th level of classification 2. Then $\sum_i \sum_j p_{ij} = 1$. If we randomly select a sample of $n$ elements from the population and define

$$X_{ij} = \text{number of sample elements in } i\text{-th level of classification } 1 \text{ and } j\text{-th level of classification } 2$$

for $i = 1, 2, \ldots, r$ and $j = 1, 2, \ldots, c$, then the $rc$ random variables $X_{ij}$ are multinomial with parameters $n$, $p_{ij}$, $i = 1, 2, \ldots, r$, $j = 1, 2, \ldots, c$. Furthermore, if the two methods of classification are independent, then the probability that a randomly selected element falls in class $ij$ is the product of the probability that it falls in class $i$ times the probability it falls in class $j$. That is, we would have

$$p_{ij} = w_i s_j, \qquad i = 1, 2, \ldots, r, \qquad j = 1, 2, \ldots, c$$

where $w_i$ is the probability that a randomly selected element falls in the $i$-th level of classification and $S_j$ is the probability it falls in the $j$-th level of classification 2. Then, assuming independence of the two methods of classification (that is, assuming $p_{ij} = w_i s_j$), the maximum likelihood estimator of $w_i$ is

$$\hat{W}_i = \frac{X_{i.}}{n} = \frac{1}{n} \sum_j X_{ij}, \qquad i = 1, 2, \ldots, r$$

since $X_{1.}, X_{2.}, \ldots, X_{r.}$ would be multinomial with parameters $n$, $w_1, w_2, \ldots, w_r$. Similarly, the maximum likelihood estimator of $s_j$ is

$$\hat{S}_j = \frac{X_{.j}}{n}, \qquad j = 1, 2, \ldots, c.$$

Then, by Theorem 8.4.3,

$$V = \sum_i \sum_j \frac{(X_{ij} - n\hat{W}_i\hat{S}_j)^2}{n\hat{W}_i\hat{S}_j}$$

is approximately a $\chi^2$ random variable with $(r - 1)(c - 1)$ degrees of freedom for large $n$. (We have $(r - 1)(c - 1)$ degrees of freedom because initially the data can be classified into $k = rc$ classes. If all the $p_{ij}$'s were specified, then we would have $k - 1 = rc - 1$ degrees of freedom. To estimate the $p_{ij}$'s, we actually only need to estimate $w_1, w_2, \ldots, w_{r-1}$ and $s_1, s_2, \ldots, s_{c-1}$; since we lose 1 degree of freedom for each parameter estimated we end up with $rc - 1 - (r - 1) - (c - 1) = (r - 1)(c - 1)$

**Table 8.2**

| | Died of Lung Cancer | All Other Causes | Total |
|---|---|---|---|
| Smokers | 348 | 3152 | 3500 |
| Nonsmokers | 82 | 1418 | 1500 |
| Total | 430 | 4570 | 5000 |

degrees of freedom.) The use of this statistic in testing the hypothesis of independence is identical with that in the examples given in Section 8.4.

*Example 8.5.1.* Assume that the medical records of 5000 people (all now deceased) were available for examination. For each person we know whether or not death was caused by lung cancer and whether or not the individual was a smoker. Table 8.2 might result.

If these records were a random sample from some population, say that of the United States, then we might test the hypothesis that smoking and death by lung cancer are independent, with $\alpha = .01$, by using the $V$ statistic mentioned above. We can label the smoking versus nonsmoking as classification 1 and the lung cancer versus other causes as classification 2. Then we have

$$\hat{w}_1 = \tfrac{3500}{5000} = .7, \qquad \hat{w}_2 = \tfrac{1500}{5000} = .3,$$

$$\hat{s}_1 = \tfrac{430}{5000} = .086, \qquad \hat{s}_2 = \tfrac{4570}{5000} = .914.$$

The observed value of $v$ is

$$v = \frac{[348 - 5000(.7)(.086)]^2}{5000(.7)(.086)} + \cdots + \frac{[1418 - 5000(.3)(.914)]^2}{5000(.3)(.914)} = 26.764;$$

the 99-th percentile of the $\chi^2$ distribution with 1 degree of freedom is 6.63. Thus we would reject the hypothesis of independence.

At this point a remark may be in order concerning the difference between causality and independence. In Example 8.5.1 the hypothetical data led to rejection of the hypothesis of independence between smoking and lung cancer (although the data were hypothetical, actual records are very similar). The fact that the two appear to be dependent is not equivalent to saying that one causes the other. Both classifications may in fact be related to some third factor that causes these two to act as they do. For example, when tracking two satellites in essentially the same orbit about the earth, we would observe a marked association between their relative positions about the earth at various times, yet it is not fact that one causes the other to behave as it does. The same is true for a test of independence in a contingency table; the test indicates whether or not there appears to be an association between two variables or modes of classification, but this association is not necessarily causal in nature.

*Example 8.5.2.* The captain of a sport fishing boat, who daily takes tourists out rock fishing, is interested in whether the same type of fish is caught with the same frequency throughout the summer months. The only types of fish caught in this area are ling cod, blue fish, and yellowtails. He records the total catch for each of four months of the three types of fish; see Table 8.3.

**Table 8.3**

| | Number of | | | |
|---|---|---|---|---|
| | Ling Cod | Blue Fish | Yellowtail | Total |
| June | 315 | 1347 | 620 | 2282 |
| July | 270 | 1250 | 514 | 2034 |
| Aug. | 295 | 1480 | 710 | 2485 |
| Sept. | 246 | 1200 | 494 | 1940 |
| Total | 1126 | 5277 | 2338 | 8741 |

Testing the independence of the two categories is equivalent to testing that the relative proportions of the three types of fish caught are the same in each of the four months (the frequencies don't change). To test the hypothesis of independence, with $\alpha = .05$, we compute

$$\hat{w}_1 = \tfrac{2282}{8741} = .2611, \qquad \hat{w}_2 = .2327, \qquad \hat{w}_3 = .2843, \qquad \hat{w}_4 = .2219$$

$$\hat{s}_1 = \tfrac{1126}{8741} = .1288, \qquad \hat{s}_2 = .6037, \qquad \hat{s}_3 = .2675.$$

Then we find

$$v = \sum_i \sum_j \frac{(x_{ij} - 8741\,\hat{w}_i\hat{s}_j)^2}{8741\,\hat{w}_i\hat{s}_j} = 11.897.$$

Since the 95-th percentile of the $\chi^2$ distribution with 6 degrees of freedom is 12.6, we would accept the hypothesis of independence. Note that if we had chosen $\alpha = .10$ rather than .05, we would reject the hypothesis.

Contingency tables can be generalized to three or more dimensions and the same procedure, in essence, is applicable. In a 3-dimensional contingency table, each of the $n$ elements of the sample would be categorized according to each of three different modes of classification. There are actually four different independence hypotheses that could be tested; we might test that all three are completely independent or that any single mode of classification is independent of the other two (without specifying that these two are independent of each other).

Let us make the foregoing a little plainer by introducing some exact notation. Suppose that we select a sample of $n$ elements; let $X_{ijk}$ be the number of these in the $i$-th level of classification 1, the $j$-th level of classification 2, and the $k$-th level of classification 3, where $i = 1, 2, \ldots, r$, $j = 1, 2, \ldots, c$, and $k = 1, 2, \ldots, m$. Let $p_{ijk}$ be the probability that a randomly selected element falls into class $ijk$. Then the $X_{ijk}$'s are multinomial with parameters $n$ and $p_{ijk}$. If all three classifications are independent, then $p_{ijk} = w_i s_j r_k$ for all $i, j, k$, and we could use sample values to estimate the $w_i$'s, $s_j$'s, and $r_k$'s. Then, as before, we could compute the observed value of $V$ and compare this with the 100 $(1 - \alpha)$ percentile of the $\chi^2$ distribution with

$$rcm - 1 - (r - 1) - (c - 1) - (m - 1) = rcm - (r + c + m) + 2$$

degrees of freedom; if we exceed this percentile we reject the hypothesis of complete independence, otherwise we accept.

If we want to test the hypothesis that the first two modes of classification are independent of the last one (but not necessarily independent of each other), we would test that $p_{ijk} = w_{ij}s_k$. Thus, we could use the data to estimate the $w_{ij}$'s and $s_k$'s; Then we could compute the observed value of $v$ and compare this quantity with the 100 $(1 - \alpha)$ percentile of the $\chi^2$ distribution with $rcm - 1 - (rc - 1) - (m - 1) = (rc - 1)(m - 1)$ degrees of freedom.

*Example 8.5.3.* Psychological tests are available to measure both I.Q. and aptitude for specific fields of endeavor. Suppose that we also have available an index of achieved success for individuals in specific fields (in terms of salary, accomplishments, responsibility, etc.). Then, if the I.Q. and aptitude tests are measuring what is claimed, we would not expect the three classifications: I.Q., aptitude, and index of success, to be independent (this latter measurement is possibly made at a later time for the same individual than is either of the first two). Assume that for each of 1000 individuals we have available I.Q. scores, aptitude scores, and indices of success. Also assume that we have merely summarized all scores into high and low for each of the three classifications (thus $r = c = m = 2$). Then, to investigate whether the I.Q. and aptitude tests are measuring what they should, we might test the hypothesis that the three modes of classification are completely independent with $\alpha = .01$. If we accept this hypothesis, it would appear that the tests (or the measure of success) do not measure what they should; if we reject this hypothesis we should still check to make sure that the dependence is in the right direction (thus we should make sure that high aptitudes correspond with high success indices rather than low, for example).

Let us suppose that the data in Table 8.4 resulted from our measurements on the 1000 individuals. If we let I.Q. score be classification 1, aptitude score be classification 2, and success score be classification 3, then we have

$$\hat{w}_1 = \tfrac{200}{1000} = .2, \qquad \hat{w}_2 = \tfrac{800}{1000} = .8, \qquad \hat{s}_1 = \tfrac{150}{1000} = .15,$$

$$\hat{s}_2 = \tfrac{850}{1000} = .85, \qquad \hat{r}_1 = \tfrac{270}{1000} = .27, \qquad \hat{r}_2 = \tfrac{730}{1000} = .73.$$

Table 8.4

| | Low Success | | High Success | | |
|---|---|---|---|---|---|
| | Low I.Q. | High I.Q. | Low I.Q. | High I.Q. | Total |
| Low aptitude | 8 | 40 | 2 | 100 | 150 |
| High aptitude | 112 | 110 | 78 | 550 | 850 |
| Total | 120 | 150 | 80 | 650 | 1000 |

The observed value of $V$ is

$$v = \sum_{i,j,k} \frac{(x_{ijk} - 1000\hat{w}_i\hat{s}_j\hat{r}_k)^2}{1000\hat{w}_i\hat{s}_j\hat{r}_k} = 169.2.$$

The 99-th percentile of the $\chi^2$ distribution with 4 degrees of freedom is 13.3, so we would reject the hypothesis of independence. Since high values of all three and low values of all three go together much more frequently than we would expect with independence, on the basis of this sample we might conclude that the tests do seem to be measuring what is claimed. A much deeper analysis of the actual scores should be made before really definitive claims are given.

**EXERCISE 8.5.**

**1.** A sample of 2000 medical records was examined and the following data resulted.

| | Died of Cancer of Intestines | Died of All Other Causes | Totals |
|---|---|---|---|
| Smokers | 22 | 1178 | 1200 |
| Nonsmokers | 26 | 774 | 800 |
| Total | 48 | 1952 | 2000 |

Assume that these results were the outcome of a random sample from a certain population and test that the two classifications are independent, with $\alpha = .05$.

**2.** A random sample of 3000 Valencia oranges was selected from the 1967 crop. Each orange was graded for color (light, medium, or dark) and a determination of sugar content was made (this continuous variable was used to call the orange sweet or not sweet). The results are summarized in the following table.

| Color | Sweet | Not Sweet | Total |
|---|---|---|---|
| Light | 1300 | 200 | 1500 |
| Medium | 500 | 500 | 1000 |
| Dark | 200 | 300 | 500 |
| Total | 2000 | 1000 | 3000 |

Test the hypothesis that sweetness and color are independent, with $\alpha = .01$.

**3.** Records of 10,000 auto accidents were examined to determine the degree of injury to the driver and whether or not he was using a seat belt. The data is summarized below.

| | Seat Belt | No Seat Belt | Total |
|---|---|---|---|
| Minor injuries | 2500 | 1500 | 4000 |
| Major injuries | 450 | 4500 | 5000 |
| Death | 50 | 950 | 1000 |
| Total | 3000 | 7000 | 10,000 |

Test the hypothesis that severity of injury to the driver is independent of whether the driver wears a seat belt, with $\alpha = .01$.

**4.** A random sample of 4000 individuals (all male of the same age) yielded the following data.

| | Annual Income | | | |
|---|---|---|---|---|
| Highest Education Attained | Less than 5000 | 5000 to 15,000 | More than 15,000 | Total |
| Grade school | 350 | 35 | 15 | 400 |
| High school | 100 | 850 | 50 | 1000 |
| College | 40 | 1200 | 760 | 2000 |
| Graduate | 10 | 415 | 175 | 600 |
| Total | 500 | 2500 | 1000 | 4000 |

Test the hypothesis that annual salary (for males of this age) is independent of education attained, with $\alpha = .1$.

**5.** Use the data presented in Example 8.5.3 to test the hypothesis that the I.Q. score is independent of the success index-aptitude score combinations, with $\alpha = .05$.

**6.** Suppose that a sample of $n$ individuals could be categorized according to each of $k$ different classifications. How many degrees of freedom would we have for testing that all $k$ were independent? How many degrees of freedom would we have for testing that any single one was independent of the other $k - 1$?

**7.** Suppose that $X_{ij}$, $i = 1, 2, \ldots, r$, $j = 1, 2, \ldots, s$, is a random sample of a multinomial random variable with parameters $n$ and $p_{ij} = w_i s_j$. Show that the maximum likelihood estimators of the $w_i$'s and $s_j$'s are

$$\hat{W}_i = \frac{X_i}{n}, \qquad \hat{S}_j = \frac{X_j}{n}.$$

# 9

# Bayesian Methods

In Chapters 7 and 8 we discussed the definition and use of some classical methods for estimating parameters and testing hypotheses. In this chapter we shall look at techniques based on Bayes theorem (Theorem 2.6.1). A prime difference between these techniques and the classical ones already discussed is the interpretation of the probabilities used. In essentially everything we have studied so far, the probabilities used were interpreted in a frequency sense; they were referring to an experiment which could be repeated an indefinite number of times and, if the probability of occurrence of an event $A$ was .3, we meant that $A$ would be expected to occur in about 30% of these repetitions. This type of interpretation of probability is called objective or frequentist; the numbers called probabilities are measuring relative frequencies of occurrence in repetitions of the basic experiment.

In common English usage, however, probability is frequently used in another, more subjective sense. For example, we have all heard such statements as "it probably will rain tomorrow" or "the chances are 3 out of 5 the Yankees will win the pennant this year" or "most likely the bank robbery was an inside job." In each of these cases the individual making the statement is using his own experience and knowledge as the basis for the statement and is not referring to some experiment which can or will be repeated an indefinite number of times. These are all examples of uses of subjective probabilities, ones that base their validity strictly on the beliefs of the individuals making them. Thus a subjective probability is measuring a person's "degree of belief" in a proposition, which is not necessarily the same as its long-term frequency of occurrence if, indeed, he is referring to something which could be repeated.

The Bayesian techniques which we shall study in this chapter all make use of subjective probabilities measuring degrees of belief about the value or values of unknown parameters. These subjective probabilities are used to define what is called the prior distribution for the parameter. Thus, when using Bayesian methods we shall act as though an unknown parameter is a random variable and has a known prior distribution (prior to taking a sample); this prior distribution summarizes our subjective degree of belief about the unknown value of the parameter. If we are fairly certain of the parameter's value, we shall choose a prior with a small variance; if we are less certain about its value we shall choose a prior with a larger variance. After the prior is specified the sample values are observed and used to compute what is called the posterior distribution of the parameter. This posterior distribution is made up of both the subjective prior information (our degree of belief) about the parameter and the objective sample information. The posterior distribution then is used to construct an estimator of the unknown parameter or to make interval statements about the unknown parameter.

## 9.1. Prior and Posterior Distributions

In Chapter 7 we examined both the method of moments and maximum likelihood criteria for using a sample to estimate the value of an unknown parameter $\theta$. We assumed that the probability law of a random variable $X$, which we could sample, was dependent on the unknown value of $\theta$; then the sample information was used to construct a guess or estimate of the value of $\theta$. Neither of these two methods called for any information for estimating $\theta$ other than the sample values that occurred. Moreover, if such extraneous information were available, it would not have been possible to make use of it. In many situations additional information is available about the value of the unknown parameter; if this information can be used to construct a prior distribution for the parameter $\theta$, then the Bayesian methods discussed in this chapter may be used to estimate the unknown value of $\theta$.

DEFINITION 9.1.1. The *prior distribution* of a parameter $\theta$ is a probability function or probability density function expressing our degree of belief about the value of $\theta$, prior to observing a sample of a random variable $X$ whose distribution function is indexed by $\theta$.

It should be stressed that the prior distribution for a parameter $\theta$ is making use of additional assumed information above and beyond anything to be observed from a random sample of $X$. Since this information is expressed as a probability function or density function, we shall let $\Theta$ be the symbol for the corresponding random variable and then $\theta$ would represent its observed value (which would be the same as the true unknown value). Some examples should clarify this notation.

*Example 9.1.1.* Suppose that we have a brand new 50-cent piece and are interested in estimating $p$, the probability of getting a head with this coin on a single flip. We know that $p$ must lie between zero and 1. If we are not willing to assume that any values in this interval are more likely than others, we could then reasonably assume a uniform prior for $p$ on the interval (0, 1). This would be written

$$\begin{aligned} f_P(p) &= 1, \qquad 0 < p < 1 \\ &= 0, \qquad \text{otherwise.} \end{aligned}$$

This would, in a sense, correspond to an assumption of total ignorance; we feel that all possible values of $p$ are equally likely. On the other hand, we might feel justified in assuming that $p$ must certainly lie in the interval from .4 to .6, since the coin would appear to be quite symmetrical, and that each of these values is equally likely to occur. Our prior for $p$ then would be

$$\begin{aligned} f_P(p) &= 5, \qquad .4 < p < .6 \\ &= 0, \qquad \text{otherwise.} \end{aligned}$$

This is a stronger assumption, because we have ruled out values of $p$ below .4 and above .6. Or, we might again reason that the value of $p$ must certainly lie in the interval from .4 to .6 but that only .4, .5, or .6 are possible values for $p$ with .5 being twice as likely. Then we would be assuming the discrete prior

$$\begin{aligned} p_P(p) &= \tfrac{1}{4}, \qquad \text{at } p = .4 \text{ or } .6 \\ &= \tfrac{1}{2}, \qquad \text{at } p = \tfrac{1}{2} \\ &= 0, \qquad \text{otherwise.} \end{aligned}$$

It should be noted that the priors mentioned in Example 9.1.1 were quite arbitrary and dependent upon the sort of assumptions one is willing to make regarding the unknown value of $p$; many other possible assumed priors could be mentioned. As we shall see, the final answer arrived at in using a Bayes technique is generally dependent upon the particular prior assumed, so the assumption should not be taken lightly. Furthermore, two different people observing the same sample values might very well arrive at different estimates of the value of the same parameter because they assumed different prior information.

*Example 9.1.2.* Assume that the length of time a light bulb made a certain way will burn is a normal random variable with mean $\mu$ and standard deviation 100 hours. Our prior experience with this process and similar ones might lead us to assume a normal prior for $\mu$; that is, we might assume

$$f_M(\mu) = \frac{1}{\sigma_0\sqrt{2\pi}}\, e^{-(\mu-\mu_0)^2/2\sigma_0^2}.$$

The parameters $\mu_0$ and $\sigma_0^2$ are called the prior mean and prior variance for $\mu$; the smaller the assumed value of $\sigma_0^2$, the more certain we are of the unknown value of $\mu$ since the prior density then has more area close to $\mu_0$. On the other hand, the bigger the assumed value for $\sigma_0^2$, the less certain we are of the unknown value of $\mu$ since there is correspondingly less area in a fixed interval about $\mu_0$.

With the foregoing discussion in mind regarding an assumed prior for an unknown parameter $\theta$, let us switch our attention to the probability density function for a continuous random variable $X$ which is indexed by $\theta$. We previously denoted this density simply by $f_X(x)$; this notation does not express the fact that $f_X(x)$ actually is dependent on the value of $\theta$. While discussing Bayesian techniques in this chapter, we shall use a conditional probability notation for such a density to stress the fact that $f_X(x)$ can only be used to compute probabilities for a given value of $\theta$. Thus, we shall write $f_{X|\Theta}(x \mid \theta)$ for the density of $X$. (If $X$ is discrete, we shall analogously use $p_{X|\Theta}(x \mid \theta)$ for its probability function in the context of Bayesian methods.) Thus, if $X$ is an exponential random variable with parameter $\theta$, we shall write

$$\begin{aligned} f_{X|\Theta}(x \mid \theta) &= \theta e^{-\theta x}, \qquad x > 0, \qquad \theta > 0 \\ &= 0, \qquad \text{otherwise} \end{aligned}$$

to stress the conditional dependence of the density on $\theta$. Or, if $X$ is binomial with parameters $n$ and $p$, we shall denote its probability function by

$$\begin{aligned} p_{X|P}(x \mid p) &= \binom{n}{x} p^x (1-p)^{n-x}, \qquad x = 0, 1, \ldots, n, \qquad 0 \le p \le 1, \\ &= 0, \qquad \text{otherwise.} \end{aligned}$$

Generally the parameter $n$ is known, so it plays no special role in this conditional notation.

We are in essence acting as though the parameter of the probability law for $X$ is itself a random variable when we assume a prior distribution for such a parameter; thus the density and probability functions which we are used to are actually conditional densities or probability functions. The notation is meant to acknowledge this fact.

Our notation for the density function of the elements of a random sample of $n$ observations of $X$ will also be changed accordingly. Thus if $X_1, X_2, \ldots, X_n$ is a random sample of an exponential random variable with parameter $\theta$, the joint density for the sample will be denoted by

$$f_{\underline{X}|\Theta}(\underline{x} \mid \theta) = \theta^n e^{-\theta \Sigma x_i}.$$

Given a prior density for $\theta$, $f_\Theta(\theta)$, and the conditional density of the elements of a sample, $f_{\underline{X}|\Theta}(\underline{x} \mid \theta)$, the joint (unconditional) density for the sample and the parameter, is simply the product of these two functions:

$$f_{\underline{X},\Theta}(\underline{x}, \theta) = f_{\underline{X} \mid \Theta}(\underline{x} \mid \theta) f_\Theta(\theta).$$

(The reader might like to review conditional densities in Chapter 5.) Then the marginal density of the sample values, which is independent of $\theta$, is given by the integral of the joint density over the range of $\Theta$. Thus

$$f_{\underline{X}}(\underline{x}) = \int_{\text{Range of } \Theta} f_{\underline{X},\Theta}(\underline{x}, \theta)\, d\theta.$$

This will be referred to as the *marginal of the sample*. The posterior density for the parameter $\theta$ is defined as follows.

DEFINITION 9.1.2. The *posterior density* for $\theta$ is the conditional density of $\Theta$, given the sample values. Thus

$$f_{\Theta \mid \underline{X}}(\theta \mid \underline{x}) = \frac{f_{\underline{X},\Theta}(\underline{x}, \theta)}{f_{\underline{X}}(\underline{x})}.$$

This posterior density thus is simply the conditional density of $\Theta$, given the sample values; the prior density expresses our degree of belief of the location of the value of $\theta$ prior to sampling, and the posterior density expresses our degree of belief of the location of $\theta$, given the results of the sample. Let us consider some examples of these manipulations.

*Example 9.1.3.* Let us return to the coin of Example 9.1.1 and, based on the result of a single flip, see how the posterior of $p$ changes for the three priors presented there. Thus, if we flip this 50-cent piece one time and let $X = 1$ if we get a head and $X = 0$ if we get a tail, the probability function for the sample is

$$\begin{aligned} p_{X \mid P}(x \mid p) &= 1 - p, &\quad \text{at } x = 0 \\ &= p, &\quad \text{at } x = 1. \end{aligned}$$

If we assume our prior to be

$$f_P(p) = 1, \qquad 0 < p < 1,$$

then the joint density function for $X$ and $P$ is

$$\begin{aligned} f_{X,P}(x, p) &= 1 - p, &\quad \text{for } x = 0, &\quad 0 < p < 1 \\ &= p, &\quad \text{for } x = 1, &\quad 0 < p < 1. \end{aligned}$$

(Note that this is actually a mixture, corresponding to the joint function of a discrete and a continuous random variable; we shall use $f$ to denote the joint function in such cases.) The marginal for $X$ then is

$$\begin{aligned} p_X(x) &= \int_0^1 (1 - p)\, dp = \tfrac{1}{2}, &\quad \text{for } x = 0 \\ &= \int_0^1 p\, dp = \tfrac{1}{2}, &\quad \text{for } x = 1. \end{aligned}$$

The posterior for $p$ then is

$$\begin{aligned} f_{P \mid X}(p \mid x) &= 2(1-p), \qquad \text{for } x = 0, \qquad 0 < p < 1 \\ &= 2p \qquad\qquad\ \ \text{for } x = 1, \qquad 0 < p < 1. \end{aligned}$$

Prior to flipping the coin we felt that the probability that $p$ exceeded $\frac{1}{2}$ was $\frac{1}{2}$; after we flip the coin and get a head ($x = 1$), this same probability is

$$\int_{1/2}^{1} 2p \, dp = \tfrac{3}{4}.$$

Or, if we get a tail ($x = 0$), this same probability is

$$\int_{1/2}^{1} 2(1-p) \, dp = \tfrac{1}{4}.$$

Thus our posterior probability that $p > \frac{1}{2}$ is either $\frac{3}{4}$ or $\frac{1}{4}$, depending on whether we get a head when the coin is flipped. On the other hand, if we take

$$f_P(p) = 5, \qquad .4 < p < .6$$

as our prior for $p$, then the marginal for $X$ is

$$\begin{aligned} p_X(x) &= \int_{.4}^{.6} 5(1-p) \, dp = .5, \qquad \text{for } x = 0 \\ &= \int_{.4}^{.6} 5p \, dp = .5, \qquad\qquad \text{for } x = 1. \end{aligned}$$

Notice that the marginal for $X$ is the same in this case as it was earlier. The posterior for $p$ now is

$$\begin{aligned} f_{P \mid X}(p \mid x) &= 10(1-p), \qquad \text{for } x = 0, \qquad .4 < p < .6 \\ &= 10p, \qquad\qquad\ \ \text{for } x = 1, \qquad .4 < p < .6. \end{aligned}$$

Similarly, if we use the discrete prior

$$\begin{aligned} p_P(p) &= \tfrac{1}{4}, \qquad \text{for } p = .4 \text{ or } .6 \\ &= \tfrac{1}{2}, \qquad \text{for } p = .5, \end{aligned}$$

then the posterior for $p$ is discrete as well and we find

$$\begin{aligned} p_{P \mid X}(p \mid x = 0) &= .3, \qquad \text{at } p = .4 \\ &= .5, \qquad \text{at } p = .5 \\ &= .2, \qquad \text{at } p = .6 \end{aligned}$$

and

$$\begin{aligned} p_{P \mid X}(p \mid x = 1) &= .2, \qquad \text{at } p = .4 \\ &= .5, \qquad \text{at } p = .5 \\ &= .3, \qquad \text{at } p = .6. \end{aligned}$$

Let us conclude this section with an example of a normal random variable with a known variance and derive the posterior distribution of the mean $\mu$ for a normal prior.

*Example 9.1.4.* Assume that the weights of a certain species of adult hummingbird are normally distributed with unknown mean $\mu$ and known variance $\sigma^2 = \frac{1}{4}$. Furthermore, let us assume that our experience with other types of hummingbirds leads us to conclude that $\mu$ is undoubtedly very close to 4 ounces; we feel that this additional information is well represented by a normal prior on $\mu$ with mean 4 and standard deviation $\frac{1}{10}$. We randomly select a sample of $n$ hummingbirds of the given species; the density function of the sample then is

$$f_{\underline{X}\,|\,\mathrm{M}}(\underline{x} \mid \mu) = \left(\frac{2}{2\pi}\right)^{n/2} e^{-2\Sigma(x_i-\mu)^2}.$$

We recall the identity

$$\sum (x_i - \mu)^2 = \sum (x_i - \bar{x})^2 + n(\bar{x} - \mu)^2$$

where $\bar{x} = (1/n) \sum x_i$. The joint density of the sample values and M then is

$$\begin{aligned} f_{\underline{X},\mathrm{M}}(\underline{x}, \mu) &= \left(\frac{2}{2\pi}\right)^{n/2} e^{-2[\Sigma(x_i-\bar{x})^2+n(\bar{x}-\mu)^2]} \frac{100}{\sqrt{2\pi}} e^{-50(\mu-4)^2} \\ &= Ke^{-2n(\bar{x}-\mu)^2-50(\mu-4)^2} \end{aligned}$$

where $K$ involves the constant term and the sample values $x_1, x_2, \ldots, x_n$. By completing the square, we find

$$2n(\bar{x} - \mu)^2 + 50(\mu - 4)^2 = (2n + 50)\left[\mu - \frac{n\bar{x} + 100}{n + 25}\right]^2 + \frac{50n\bar{x}^2 - 400n\bar{x}}{n + 25}$$

and thus

$$f_{X,\mathrm{M}}(\underline{x}, \mu) = Ke^{-(50n\bar{x}^2-400n\bar{x})/(n+25)}e^{-(2n+50)[\mu-(n\bar{x}+100)/(n+25)]^2}.$$

Now, to get the marginal for the sample we need to integrate over $\mu$; everything except the exponential term involving $\mu$ factors outside the integral sign and the value of the integral itself is

$$\frac{2\sqrt{n + 25}}{\sqrt{2\pi}}$$

(since it is the integral over the whole line of a normal density with mean $(n\bar{x} + 100)/(n + 25)$ and variance $1/(4n + 100)$). Thus, the marginal for the sample is

$$f_{\underline{X}}(\underline{x}) = \frac{2\sqrt{n + 25}}{\sqrt{2\pi}} Ke^{-(50n\bar{x}^2-400n\bar{x})/(n+25)}$$

and the posterior for $\mu$ is

$$f_{\mathrm{M}\,|\,\underline{X}}(\mu \mid \underline{x}) = \frac{2\sqrt{n + 25}}{\sqrt{2\pi}} e^{-(2n+50)[\mu-(n\bar{x}+100)/(n+25)]^2}.$$

That is, the posterior is again normal in form with mean

$$\frac{n\bar{x} + 100}{n + 25} = \frac{4n}{4n + 100}\bar{x} + \frac{100}{4n + 100}4$$

(a weighted sum of the observed sample mean and the prior mean) and variance $1/(4n + 100)$ (which is the reciprocal of the sum of the reciprocals of the variance of $\bar{X}$ and the prior variance). Thus, if our prior on $\mu$ is normal, so is the posterior where the posterior mean is a weighted sum of the observed sample mean and the mean of the prior.

## EXERCISE 9.1.

**1.** Assume that the probability that an item coming off an assembly line is defective in some respect is a constant $p$. Suppose that we select 1 item at random from the assembly line (constituting a sample of 1 observation of a Bernoulli random variable with parameter $p$) and we assume a uniform prior density for $p$ on the interval $(0, 1)$. What is the posterior for $p$ if the selected item is defective? If it is not defective?

**2.** Repeat problem 1, assuming that $p$ is uniform on $(a, b)$ where $a \geq 0$, $b \leq 1$.

**3.** A new psychological test has been developed to measure "intelligence." Assume that the score an individual will make on the test is a normal random variable with unknown mean $\mu$ and known variance of 10. Prior information on $\mu$ (from similar tests already in use) would lead us to assume a normal prior density for $\mu$ with mean 100 and variance 5. Given a random sample of 20 individual scores on the test, compute the posterior density for $\mu$.

**4.** If in problem 3 we assume that $\mu$ has a uniform prior on the interval $(90, 110)$ (rather than the normal prior used there), show that the posterior for $\mu$ is

$$\frac{1}{\sqrt{\pi}} \frac{1}{N_Z\left(\dfrac{110 - \bar{x}}{\sqrt{10}}\right) - N_Z\left(\dfrac{90 - \bar{x}}{\sqrt{10}}\right)} e^{-(\bar{x}-\mu)^2}, \qquad 90 < \mu < 110,$$

where $\bar{x}$ is the observed mean of the 20 scores.

**5.** Assume that the length of nylon string which can be extruded by a machine with no break occurring is an exponential random variable $X$ with parameter $\lambda$. Also assume an exponential prior density on $\lambda$ with parameter $\frac{1}{1000}$. Given a sample of 1 observation on $X$ (and we find $x = 1100$ feet), compute the posterior density for $\lambda$.

**6.** The time it will take a certain track star to run a 100-yard dash is a uniform random variable on the interval $(\alpha, 10.5)$ measured in seconds. Assume that $\alpha$ is equally likely to be 9.5 or 9.7 seconds and compute the posterior for $\alpha$, given a sample of 1 time of 10 seconds.

**7.** A coin is flipped 1 time. The probability $p$ of getting a head is equally likely to be $\frac{1}{4}$, $\frac{1}{2}$, or $\frac{3}{4}$. Compute the posterior for $p$, given that we get a tail on the single flip. Compute the posterior for $p$, given that we get a head on the single flip.

**9.2. Bayes Estimators**

Before discussing the methods used to construct Bayes estimators, let us define two additional density functions which occur frequently as prior and posterior density functions; these are the densities of the gamma and beta random variables, respectively. First $X$ is called a *gamma random variable* with parameters $m$ and $\theta$ if its density function is

$$f_{X \mid \Theta}(x \mid \theta) = \frac{1}{\Gamma(m+1)} \theta^{m+1} x^m e^{-\theta x}, \qquad \text{for } x > 0, \quad m > -1, \quad \theta > 0;$$

the parameter $m$ is generally assumed to be known (it is quite analogous to the parameter $n$ of a binomial random variable). This density function is discussed in Appendix 4. Notice that if $m = 0$, then $X$ is an exponential random variable with parameter $\theta$; if $\theta = \frac{1}{2}$, then $X$ is a $\chi^2$ random variable with $2m + 2$ degrees of freedom. Thus, both the exponential and $\chi^2$ random variables are special cases of the gamma. It is easily shown that the moment-generating function for a gamma random variable is

$$m_X(t) = \frac{1}{(1 - t/\theta)^{m+1}}, \qquad \text{for } t < \theta.$$

Using this it is found that

$$\mu_X = \frac{m+1}{\theta}, \qquad \sigma_X^2 = \frac{m+1}{\theta^2}.$$

We also easily find that $f_{X \mid \Theta}(x \mid \theta)$ is maximized at $x = m/\theta$. Example 9.2.4 illustrates a case in which the gamma density occurs as a posterior for a parameter.

The beta function is also discussed in Appendix 4. $X$ is called a *beta random variable* with parameters $a$ and $b$ if its density function is

$$f_{X \mid A,B}(x \mid a, b) = \frac{\Gamma(a+b)}{\Gamma(a)\Gamma(b)} x^{a-1}(1-x)^{b-1}, \qquad 0 < x < 1,$$

where both parameters $a$ and $b$ are positive. You are asked in the problems to verify that:

$$\mu_X = \frac{a}{a+b}$$

$$\sigma_X^2 = \frac{ab}{(a+b)^2(a+b+1)}$$

$$x = \frac{a-1}{a+b-2} \text{ maximizes } f_{X \mid A,B}(x \mid a, b) \text{ (unless } a + b = 2).$$

Note that, for $a + b$ fixed, the bigger that $a$ gets, the further to the right the center of mass is shifted as well as the mode. Thus, by appropriately choosing $a$ and $b$ the density of $X$ may take on quite a variety of shapes including, for $a = b = 1$, that of a uniform random variable. Example 9.2.3 illustrates a common case in which the beta density occurs as a posterior density.

The posterior distribution of a parameter $\theta$ measures our degree of belief as to the true unknown value of $\theta$; it combines the assumed prior knowledge of $\theta$ and the information contained in the sample about $\theta$ through Bayes theorem. Then, to give a single point as our best guess of the value of $\theta$, we might logically choose the mean of the posterior distribution. Recall from our discussion of the mean of a random variable, that this number would locate the center of gravity of the posterior distribution and would thus be a likely candidate if we had to give a single number as our guess of the value of $\theta$. This posterior mean value is called the Bayes estimate of $\theta$, as we shall see in Definition 9.2.1 below. Before giving this definition, however, let us note the fact that if the posterior density for $\theta$ is not symmetric, other measures of the middle of the posterior might equally well be used as the point estimate of $\theta$. Using the mode of the posterior, the value that maximizes the posterior density, would seem quite logical; it corresponds to the maximum likelihood estimate for a uniform prior as we shall see. Or, we might prefer to use the median of the posterior as our point estimate of $\theta$, since it would be the number that splits our total degree of belief into two equal pieces (in a probability sense). Nevertheless, the commonly used estimate is the mean.

DEFINITION 9.2.1. The mean of the posterior distribution of $\theta$ is called the *Bayes estimate* for $\theta$. This estimate is denoted by $\theta^*$ and the corresponding random variable by $\Theta^*$.

Thus, as in Chapter 7, the observed value of the estimator, called the estimate, is denoted by a lower case letter and is a function of the observed sample values. The same function of the sample random variables (rather than their observed values) is denoted by a capital letter and called the estimator.

*Example 9.2.1.* In Example 9.1.3 we used three different priors for $p$, the probability of getting a head on flipping a coin, and for each we derived the posterior distribution. Let us now compute the mean value for each posterior, which would be the Bayes estimate, using the corresponding prior. For the posterior

$$\begin{aligned} f_{P \mid X}(p \mid x) &= 2(1 - p), \qquad x = 0, \qquad 0 < p < 1 \\ &= 2p, \qquad x = 1, \qquad 0 < p < 1 \end{aligned}$$

we have

$$\begin{aligned} p^* &= \tfrac{1}{3}, \qquad \text{if } x = 0 \\ &= \tfrac{2}{3}, \qquad \text{if } x = 1, \end{aligned}$$

since these are the respective means for the two possible $x$ values we might observe in the sample. For the posterior

$$\begin{aligned} f_{P \mid X}(p \mid x) &= 10(1 - p), \qquad x = 0, \qquad .4 < x < .6 \\ &= 10p, \qquad x = 1, \qquad .4 < x < .6 \end{aligned}$$

we find

$$\begin{aligned} p^* &= .4933, \qquad \text{if } x = 0 \\ &= .5067, \qquad \text{if } x = 1. \end{aligned}$$

For the discrete posterior

$$\begin{aligned} p_{P \mid X}(p \mid x = 0) &= .3, \qquad \text{at } p = .4 \\ &= .5, \qquad \text{at } p = .5 \\ &= .2, \qquad \text{at } p = .6 \end{aligned}$$

and

$$\begin{aligned} p_{P \mid X}(p \mid x = 1) &= .2, \qquad \text{at } p = .4 \\ &= .5, \qquad \text{at } p = .5 \\ &= .3, \qquad \text{at } p = .6, \end{aligned}$$

we find

$$\begin{aligned} p^* &= .49, \qquad \text{if } x = 0 \\ &= .51, \qquad \text{if } x = 1. \end{aligned}$$

Both the maximum likelihood and method of moments estimates would be 0 if $x = 0$, 1 if $x = 1$, so in this case the Bayes estimates seem to give more sensible results. The reason for this will become apparent as we proceed.

*Example 9.2.2.* In Example 9.1.4 we assumed that weights of hummingbirds were normally distributed with unknown mean $\mu$ and known variance of $\frac{1}{4}$. We also assumed a normal prior on $\mu$ with mean 4 and variance $\frac{1}{100}$. The posterior then turned out to be normal with mean $(n\bar{x} + 100)/(n + 25)$. Thus, for this example, the Bayes estimate is

$$\mu^* = \frac{n\bar{x} + 100}{n + 25}$$

and

$$\text{M}^* = \frac{n\bar{X} + 100}{n + 25}.$$

The following example shows an instance in which the beta occurs as a posterior density.

*Example 9.2.3.* A production process turns out items, each of which is or is not defective. Assume that the probability of a defective occurring is $p$ (a constant) and that the defectives occur independently. To estimate $p$, we might assume a uniform prior for $p$ on $(0, 1)$, take a random sample of $n$ items, and let $X$ be the total number of defectives in the sample. Then $X$ is binomial with parameters $n$ and $p$ and we have

$$p_{X\,|\,P}(x \mid p) = \binom{n}{x} p^x (1-p)^{n-x}, \qquad x = 0, 1, \ldots, n, \qquad 0 < p < 1$$

$$f_P(p) = 1, \qquad 0 < p < 1.$$

The joint density of $X$ and $P$ is

$$f_{X,P}(x, p) = \binom{x}{n} p^x (1-p)^{n-x}, \qquad 0 < p < 1, \qquad x = 0, 1, \ldots, n;$$

the marginal for $X$ is

$$p_X(x) = \int_0^1 \binom{n}{x} p^x (1-p)^{n-x}\, dx = \binom{n}{x} \cdot \frac{x!\,(n-x)!}{(n+1)!} = \frac{1}{n+1}$$

for $x = 0, 1, \ldots, n$. Thus, averaging over the values of $p$, all of which are assumed to be equally likely, the values of $X$ are equally likely to occur. The posterior for $P$ then is

$$f_{P\,|\,X}(p \mid x) = \frac{\Gamma(n+2)}{\Gamma(x+1)\Gamma(n-x+1)} p^x (1-p)^{n-x}, \qquad 0 < p < 1,$$

a beta density with parameters $x + 1$ and $n - x + 1$. The mean of the posterior is $(x + 1)/(n + 2)$ and the mode (maximum value) of the posterior is $x/n$; thus the Bayes estimate of $p$ (given $x$ defectives occur in the sample of $n$) is

$$p^* = \frac{x+1}{n+2}$$

and the Bayes estimator is

$$P^* = \frac{X+1}{n+2}.$$

Both the method of moments and maximum likelihood yield $X/n$ as the estimator of $p$ for this case; the Bayes estimator thus is essentially increasing the sample size by 2 and the number of defectives by 1 and forming the same ratio as maximum likelihood to estimate $p$. Or, as noted above, the mode of the posterior is $x/n$ (this locates the maximum point on the posterior) and thus, in this example, the Bayes estimate is given by the mean (by definition) and the maximum likelihood estimate by the mode of the posterior density.

Theorem 9.2.1 shows that the maximum likelihood estimate of a parameter $\theta$ always is given by the mode of the posterior, if we assume a uniform prior on $\theta$.

***Theorem 9.2.1.*** Suppose that $X$ is a random variable with density $f_{X\,|\,\Theta}(x \mid \theta)$ and assume a uniform prior for $\theta$ on the interval $(a, b)$. Then, if $\theta^+$ locates the mode of the posterior (given a random sample of $n$ observations of $X$), $\theta^{+}$ is the maximizing value of the likelihood function for $\theta$ in the interval $(a, b)$.

*Proof:* The posterior is

$$f_{\Theta \mid \underline{X}}(\theta \mid x) = \frac{f_{\underline{X} \mid \Theta}(\underline{x} \mid \theta) f_{\Theta}(\theta)}{f_{\underline{X}}(\underline{x})}.$$

But if $f_{\Theta}(\theta)$ is constant for all $\theta$, then maximizing $f_{\Theta \mid \underline{X}}(\theta \mid \underline{x})$ with respect to $\theta$ is equivalent to maximizing $f_{\underline{X} \mid \Theta}(\underline{x} \mid \theta)$ with respect to $\theta$ and the likelihood function is

$$L(\theta) = f_{X \mid \Theta}(x \mid \theta).$$

Thus the mode of the posterior is the maximum likelihood estimate of $\theta$, given a uniform prior. ◀

Theorem 9.2.1 then gives some indication of the relationship between the Bayes estimator and the maximum likelihood estimator for a special case. It can be shown that the two will, in general, differ by an amount that is small compared to $1/\sqrt{n}$. Thus, for moderately large sample sizes the two criteria will lead to essentially the same estimator.

Let us now examine a case in which the posterior density for a parameter has the gamma density function.

*Example 9.2.4.* Suppose that the time to failure $T$ of an electron tube is an exponential random variable with parameter $\lambda$. Thus, the density for a random sample of $n$ times is

$$f_{T \mid \Lambda}(t \mid \lambda) = \lambda^{n} e^{-\lambda \Sigma t_i}.$$

Assume that we are given prior information that the average time to failure of tubes of this type is most likely to be at least 100 hours; that is, since the mean of $T$ is $1/\lambda$, we are fairly sure that $1/\lambda > 100$ or $\lambda < .01$. We might then arbitrarily say that we are 80% sure that $\lambda < .01$ and adopt an exponential prior for $\lambda$ that satisfies this statement. (We choose the exponential type of density mainly because it is convenient; it enables us to represent the given information quite easily.) Thus, we assume as a prior for $\lambda$

$$f_{\Lambda}(\lambda) = 160 e^{-160\lambda}, \qquad \lambda > 0.$$

Note that

$$\int_0^{.01} 160 e^{-160\lambda}\, d\lambda \doteq .798.$$

The joint density of the sample values and $\lambda$ then is

$$f_{\underline{T},\Lambda}(t, \lambda) = 160 \lambda^{n} e^{-\lambda(\Sigma t_i + 160)};$$

the marginal for the sample values is

$$f_{\underline{T}}(\underline{t}) = \int_0^\infty 160\lambda^n e^{-\lambda(\Sigma t_i+160)}\, d\lambda$$

$$= 160 \frac{\Gamma(n+1)}{(\Sigma t_i + 160)^{n+1}}$$

and the posterior for $\lambda$ is

$$f_{\Lambda \mid \underline{T}}(\lambda \mid \underline{t}) = \frac{1}{\Gamma(n+1)} [\Sigma t_i + 160]^{n+1} \lambda^n e^{-\lambda(\Sigma t_i + 160)}.$$

Thus the posterior is that of a gamma random variable with parameters $m = n$, $\theta = \sum t_i + 160$. The mean of the posterior, which is the Bayes estimate of $\lambda$, then is $(n + 1)/(\sum t_i + 160)$. The mode of the posterior is easily seen to be $n/(\sum t_i + 160)$. We recall that the maximum likelihood estimate of $\lambda$ is $n/\sum t_i$; if, for example, we took a sample of $n = 10$ lifetimes and found $\sum t_i = 1600$, which agrees perfectly with the assumed prior mean, the maximum likelihood estimate is $\hat{\lambda} = \frac{1}{160}$ (equal to the prior mean) while the Bayes estimate is $\lambda^* = \frac{1}{176}$. Thus, if our sample agrees perfectly with the assumed prior mean, it is the maximum likelihood estimate that reproduces the prior estimate, rather than the Bayes, for an exponential random variable with an exponential prior for $\lambda$.

Table 9.1 gives some common examples of assumed priors for the parameters of some common probability laws and, given a random sample of $n$ observations of the random variable involved, also gives the resulting posterior distribution for the parameter

*Example 9.2.5.* Assume that the number of times, $X$, that a light switch can be turned to the on position until it fails is a geometric random variable with parameter $p$. Let us also assume that our prior experience with switches of this type indicates that $p$ is probably very close to .001; we shall use this information to construct a prior for $p$, using a beta distribution. A beta random variable with parameters $a$ and $b$ has mean $a/(a + b)$ and variance

$$\frac{ab}{(a+b)^2(a+b+1)},$$

as we have seen. Then, we might use our prior information about $p$ and assume the prior mean to be .001 and the prior standard deviation to be .0005. These assumptions lead to $a = 3.995$ and $b = 3991$ as values for the parameters of the prior. Then, if we have a sample of $n$ observations on $X$, we see from line 7 of Table 9.1 that the posterior for $p$ is

$$\frac{\Gamma(3995 + \sum x_i)}{\Gamma(n + 3.995)\Gamma(3991 + \sum x_i - n)} p^{2.995+n}(1 - p)^{3990+\Sigma x_i - n}.$$

The mean of this density (the Bayes estimate) is $(3.995 + n)/(3995 + \sum x_i)$. The

**Table 9.1** ***Priors and Posteriors, Given a Random Sample of Size n***

| Line | Density of Sample (given value of unknown parameter) | Prior Density for Unknown Parameter | Posterior Density for Unknown Parameter |
|---|---|---|---|
| 1 | $\left(\frac{1}{2\pi\sigma_X^2}\right)^{n/2} e^{-\Sigma(x_i-\mu)^2/2\sigma_X^2}$ ($\sigma_X^2$ known) | $\frac{1}{\sigma_0\sqrt{2\pi}} e^{-(\mu-\mu_0)^2/2\sigma_0^2}$ | $\frac{a^{1/2}}{\sqrt{2\pi}} \exp\left\{-\frac{a}{2}\left[\mu - \frac{1}{a}\left(\frac{n\bar{x}}{\sigma_X^2} + \frac{\mu_0}{\sigma_0^2}\right)\right]^2\right\}$, $a = \frac{n}{\sigma_X^2} + \frac{1}{\sigma_0^2}$ |
| 2 | $\left(\frac{1}{2\pi\sigma_X^2}\right)^{n/2} e^{-\Sigma(x_i-\mu)^2/2\sigma_X^2}$ ($\sigma_X^2$ known) | $\frac{1}{b-a}$ | $\sqrt{\frac{n}{2\pi\sigma_X^2}}\left[N_Z\left(\frac{b-\bar{x}}{\sigma_X/\sqrt{n}}\right) - N_Z\left(\frac{a-\bar{x}}{\sigma_X/\sqrt{n}}\right)\right]^{-1}$ $\times e^{-n(\mu-\bar{x}^2)/2\sigma_X^2}$ |
| 3 | $\left(\frac{1}{2\pi\sigma^2}\right)^{n/2} e^{-\Sigma(x_i-\mu_X)^2/2\sigma^2}$ ($\mu_X$ known) | $\frac{1}{m!}[m\sigma_0^2]^{m+1}\left(\frac{1}{\sigma^2}\right)^{m+2} e^{-m\sigma_0^2/\sigma^2}$ | $\frac{1}{\left(m+\frac{n}{2}\right)!} b^{n/2+m+1}\left(\frac{1}{\sigma^2}\right)^{n/2+m+2} e^{-b/\sigma^2}$, $b = m\sigma_0^2 + \Sigma(x_i - \mu_X)^2$ |
| 4 | $\lambda^n e^{-\lambda\Sigma x_i}$ | $\frac{1}{m!}\left(\frac{m+1}{\lambda_0}\right)^{m+1}\lambda^m e^{-(m+1)\lambda/\lambda_0}$ | $\frac{\{n\bar{x} + [(m+1)/\lambda_0]\}^{n+m+1}}{\Gamma(n+m+1)}\lambda^{n+m}$ $\times e^{-\lambda[n\bar{x}+(m+1)/\lambda_0]}$ |
| 5 | $\frac{\lambda^{\Sigma x_i}}{\Pi x_i!} e^{-n\lambda}$ | $\frac{1}{m!}\left(\frac{m+1}{\lambda_0}\right)^{m+1}\lambda^m e^{-(m+1)\lambda/\lambda_0}$ | $\frac{[n+(m+1)/\lambda_0]^{c+1}}{c!}\lambda^c e^{-[n+(m+1)/\lambda_0]\lambda}$, $c = m + \Sigma x_i$ |
| 6 | $p^{\Sigma x}(1-p)^{n-\Sigma x}$ | $\frac{\Gamma(a+b)}{\Gamma(a)\Gamma(b)} p^{a-1}(1-p)^{b-1}$ | $\frac{\Gamma(a+b+n)}{\Gamma(a+\Sigma x_i)\Gamma(b+n-\Sigma x_i)} p^{a+\Sigma x_i-1}$ $\times (1-p)^{b+n-\Sigma x_i-1}$ |
| 7 | $p^n(1-p)^{\Sigma x_i - n}$ | $\frac{\Gamma(a+b)}{\Gamma(a)\Gamma(b)} p^{a-1}(1-p)^{b-1}$ | $\frac{\Gamma(a+b+\Sigma x_i)}{\Gamma(a+n)\Gamma(b+\Sigma x_i - n)} p^{a+n-1}$ $\times (1-p)^{b+\Sigma x_i-n-1}$ |

maximum likelihood estimate is $n/\sum x_i$. Thus, if $n = 25$ and we observe that $\sum x_i = 25{,}295$, then

$$p^* = \frac{28.995}{29{,}290} = .000990, \qquad \hat{p} = \frac{25}{25{,}295} = .000988;$$

either of the two methods gives essentially the same answer. This is because we assume sample results which are consistent with the assumed prior information (the mean of the prior is .001 and if $p = .001$ we would expect each $x_i$ close to $1/.001 = 1000$ and the sum of 25 sample values to be close to 25,000). Generally, the less consistent the sample results are with the assumed prior, the bigger is the difference between $p^*$ and $\hat{p}$; $p^*$ always tends to "shade" the sample estimate toward the prior information. If, for example, $n = 25$ and $\sum x_i = 1000$, then

$$p^* = \frac{28.995}{4995} = .0058, \qquad \hat{p} = \frac{25}{1000} = .025,$$

and the estimates are quite widely separated. If the sample results do differ markedly from what is expected, as in this latter case, it would appear that the prior information is incorrect; if the sample results are accepted as being correct, the maximum likelihood estimate would be the better one to use.

If, in fact, we are absolutely certain that the value of a parameter $\theta$ is $\theta_0$, then the prior for $\theta$ would be

$$\begin{aligned} p_\Theta(\theta) &= 1, \qquad \text{for } \theta = \theta_0 \\ &= 0, \qquad \text{otherwise.} \end{aligned}$$

In such a case we would certainly not want to take a sample to estimate $\theta$, since we would already be willing to assume that $\theta$ is known. However, an interesting property of Bayes estimators is that the sample values cannot change our minds about such a strong assumption, no matter what happens in the sample. To see this, assume that we have a continuous random variable $X$ whose density is indexed by $\theta$; the density function for a random sample of $n$ observations of $X$ then is

$$f_{\underline{X} \mid \Theta}(\underline{x} \mid \theta)$$

and the joint density of $\underline{X}$ and $\Theta$ is

$$\begin{aligned} f_{\underline{X},\Theta}(\underline{x}, \theta) &= f_{\underline{X} \mid \Theta}(\underline{x} \mid \theta), \qquad \text{for } \theta = \theta_0 \\ &= 0, \qquad \text{otherwise,} \end{aligned}$$

and the marginal for $X$ is

$$f_{\underline{X}}(\underline{x}) = f_{\underline{X} \mid \Theta}(\underline{x} \mid \theta_0)$$

since there is only one value of $\theta(\theta_0)$ to average over. Then the posterior for $\theta$ is

$$f_{\Theta \mid \underline{X}}(\theta \mid \underline{x}) = \frac{f_{\underline{X},\Theta}(\underline{x}, \theta)}{f_{\underline{X}}(\underline{x})}$$

$$= 1, \qquad \text{for } \theta = \theta_0$$

$$= 0, \qquad \text{otherwise.}$$

Thus the posterior probability that $\theta = \theta_0$ is 1 and we are still absolutely certain that $\theta = \theta_0$, whether the sample results are consistent with this value or not.

## EXERCISE 9.2.

**1.** If $X$ is a beta random variable with parameters $a$ and $b$, verify that its mode is $(a - 1)/(a + b - 2)$, its mean is $a/(a + b)$, and that its variance is

$$\frac{ab}{(a + b)^2(a + b + 1)} .$$

**2.** Derive the moment-generating function for a gamma random variable with parameters $m$ and $\theta$.

**3.** Find the mode of a gamma random variable with parameters $m$ and $\theta$.

**4.** Find the mean and variance of $X$ if

$$f_X(x) = \frac{[m\sigma_0^2]^{m+1}}{m!}\left(\frac{1}{x}\right)^{m+1} e^{-m\sigma_0^2/x}, \qquad \text{for } x > 0.$$

(Compare with the prior for $\sigma^2$, line 3, Table 9.1.)

**5.** Find the mean and variance of $X$ if

$$f_X(x) = \frac{1}{m!}\left(\frac{m + 1}{\lambda_0}\right)^{m+1} x^m e^{-((m+1)x)/\lambda_0}, \qquad x > 0.$$

(Compare with the prior for $\lambda$, lines 4 and 5, Table 9.1.)

**6.** Verify the posterior for $\mu$ that is given on line 2, Table 9.1, given a uniform prior on $(a, b)$ with $\sigma_X^2$ known.

**7.** Given a sample result such that $a \le \bar{x} \le b$, what is the modal value of the posterior discussed in problem 6 above? If we find $\bar{x} < a$ or $\bar{x} > b$, what is the modal value?

**8.** Verify the posterior for $\sigma^2$ given on line 3, Table 9.1, for the given prior. What is the Bayes estimate for $\sigma^2$?

**9.** Verify the posterior for $\lambda$ given on line 4, Table 9.1, and compute the Bayes estimate of $\lambda$.

**10.** Verify the posterior for $\lambda$ given on line 5, Table 9.1, and compute the Bayes estimate of $\lambda$.

**11.** Compute the Bayes estimate for $\mu$, given a uniform prior on $(a, b)$ for $\mu$ with $\sigma_X^2$ known (line 2, Table 9.1). (**Hint:**

$$\int (\mu - \bar{x})e^{-(1/2)(\mu-\bar{x})^2}\, d\mu = -e^{-(1/2)(\mu-\bar{x})^2}.)$$

**12.** If we are certain that $p = .75$, where $p$ is the parameter of a Bernoulli random variable $X$, show that no matter how many successes or failures in a row we might observe, the posterior probability that $p = .75$ is 1.

### 9.3. Interval Estimation

Using Bayesian methods it is easy to construct interval estimates of parameters. If we have available the posterior distribution of the parameter, we can construct an interval, generally centered at the posterior mean, which contains 100 $(1 - \alpha)\%$ of the posterior probability. Such an interval is an example of a 100 $(1 - \alpha)\%$ Bayes interval for the unknown parameter (to contrast it with the non-Bayesian confidence interval approach we studied earlier). The formal definition follows.

DEFINITION 9.3.1. Let

$$f_{\Theta \mid \underline{X}}(\theta \mid \underline{x})$$

be the posterior density for $\theta$. If

$$\int_a^b f_{\Theta \mid \underline{X}}(\theta \mid \underline{x})\, d\theta = 1 - \alpha,$$

then the interval $(a, b)$ will be called a 100 $(1 - \alpha)\%$ *Bayes interval for* $\theta$. (Either $a$ or $b$ may be infinite for one-sided intervals.)

*Example 9.3.1.* Suppose that the "true" weights of 1-year-old trout are normally distributed with mean $\mu$ and standard deviation $\sigma = .2$ (in pounds); it is known that the standard deviation does not change with environment although the mean weight $\mu$ does. Men from the State Department of Fish and Game stocked a new, man-made lake with freshly hatched trout a year ago and are now interested in an interval estimate of $\mu$, the average weight for 1-year-old trout in this lake. Based on their experience with this fish in similar circumstances, they assume a normal prior with mean $\mu_0 = 7$ pounds and standard deviation $\sigma_0 = \frac{1}{3}$ pound. They randomly select a sample of 30 fish from the lake and then compute the posterior density for $\mu$ and use it to derive their interval estimate for $\mu$. From line 1, Table 9.1, we see that the posterior for $\mu$ is normal with mean $(750\bar{x} + 63)/750.11$ and variance $1/750.11$ where $\bar{x}$ is the observed average weight of the 30 fish in the sample. Then, a 90% Bayes interval for $\mu$ (centered at the posterior mean) would go from

$$\frac{750\bar{x} + 63}{750.11} - \frac{1.64}{\sqrt{750.11}}$$

to

$$\frac{750\bar{x} + 63}{750.11} + \frac{1.64}{\sqrt{750.11}};$$

ignoring the prior information about $\mu$, the observed value of a 90% confidence interval for $\mu$ would be from

$$\bar{x} - (1.64)\sqrt{\frac{0.2}{30}}$$

to

$$\bar{x} + (1.64)\sqrt{\frac{0.2}{30}}.$$

Notice that the Bayes interval is slightly shorter than the confidence interval because the additional assumed prior information is equivalent to a slightly larger sample size if no prior information is assumed.

*Example 9.3.2.* The number of fatal traffic accidents per week in the state of California is a Poisson random variable with parameter $\lambda$. Let us assume that there has been an average of 80 fatal accidents per week in the last 4 years. We assume a prior for $\lambda$ having mean 80 and (arbitrarily) variance equal to 20; if we use the $\gamma$ prior given on line 5 of Table 9.1, we then would choose $\lambda_0 = 80$, $m = 319$ to satisfy the assumed prior mean and variance. Assume that we take a sample of 20 weeks and want to compute a 95% Bayes interval for $\lambda$, the true rate per week. From line 5, Table 9.1, the posterior density function for $\lambda$ is

$$\frac{(24)^{320+\Sigma x_i}}{(319 + \sum x_i)!} \lambda^{319+\Sigma x_i} e^{-24\lambda};$$

once the observed number of fatal accidents is known, this density can then be used to compute the desired 95% Bayes interval.

In many cases the Bayes interval will be quite similar to an observed confidence interval with the same confidence coefficient; the interpretation of the two is quite different however. As we recall, an observed 95% confidence interval is a pair of observed values of two statistics, $L$ and $U$, whose probability, before the sample is taken, of bracketing the unknown $\theta$ is .95. A 95% Bayes interval for $\theta$, on the other hand, is simply an interval which contains 95% of the posterior probability (degree of belief, given the sample values) for $\theta$. Thus, the confidence interval probability references a relative proportion of the time that $\theta$ would be bracketed by $U$ and $L$, whereas the Bayes interval probability references a subjective degree of belief which is affected by the sample results but is not entirely determined by them.

Using the concept of Bayes intervals, we can answer questions rather analogous to those solved by the classical theory of tests of hypotheses, which we examined in Chapter 8. Suppose that we have taken a sample of size $n$

of a random variable $X$, assumed a prior for the parameter $\theta$ of $F_X$, and computed the posterior density for $\theta$. Let us also assume that $(a, b)$ is the 100 $(1 - \alpha)\%$ Bayesian interval for $\theta$; then for any particular value $\theta_0$ of $\theta$, we might say that we reject $H_0: \theta = \theta_0$ if either $\theta_0 < a$ or $\theta_0 > b$, since in these cases the hypothesized value for $\theta$ does not fall in the Bayes interval for $\theta$. That is, the hypothesized value does not seem very likely in terms of posterior degrees of belief, so we decide to reject $H_0$. It is not straightforward to compute the probabilities of type I and type II error when using such a procedure. Essentially, for the assumed prior we would have to compute the probability that the posterior interval would not include $\theta_0$, given that $\theta_0$ is in fact the true value of $\theta$, to evaluate the probability of type I error. That is, we would need to get the probability distribution of the posterior itself and use it to compute the probability of type I error. This will not in general be equal to $\alpha$ if we base the decision on a 100 $(1 - \alpha)\%$ Bayes interval. This probability will depend quite heavily on how close the prior mean is to the observed sample mean, if $\theta$ is the mean of $X$. The probability of type II error would be evaluated analogously.

Thus, it should be evident that basing the rule for accepting or rejecting $H_0: \theta = \theta_0$ on the 100 $(1 - \alpha)\%$ Bayes interval for $\theta$, as mentioned above, is not equivalent to the classical test of hypothesis method for determining the rejection rule. The classical method allows for no *a priori* information and, if such *a priori* information is available, perhaps the classical test of hypothesis method should not logically be used. Thus the merits of the Bayesian procedure possibly should not be judged through the framework of tests of hypotheses, with their probabilities of type I and type II errors. The cogent point is the fact that rejection rules can be formulated through the use of Bayesian methods; their worth should not be judged solely in the classical framework. It would seem appropriate to point out to the reader that a considerable body of literature on decision theory is available, much of which is based on statistical techniques. One well-accepted procedure in decision theory is based upon Bayesian analysis in much the same framework that we are using; we do not have the space necessary to develop these ideas, but the Bayes estimator we have defined and the above rule for accepting a hypothesis can both be well established through this theory.

Before looking at some examples of this Bayes rule for testing a hypothesis, let us make note of a technical difficulty which was actually encountered in Example 9.3.2 but was not mentioned. In Section 9.2 we saw that both the beta and the gamma functions commonly occur as posterior densities for parameters. In such a case, the Bayes interval $(a, b)$ for a parameter $\theta$ would necessitate the evaluation of either the incomplete beta or the incomplete gamma function, since we would want to find the definite limits, $a$ and $b$, such that the area under the beta or gamma was $1 - \alpha$. Unfortunately, for general values of the parameters, these integrations cannot be carried out in closed form; thus, $a$ and $b$ cannot be explicitly obtained as functions of the sample results and the prior parameters. However, quite complete tables of both the incomplete beta and incomplete gamma functions are available which can be used to determine the values of $a$ and $b$ for a given value of $1 - \alpha$. Thus, for a given sample result, $a$ and $b$ can be numerically evaluated.

*Example 9.3.3.* Let $p$ be the probability of a head on a given coin and let $X$ be the total number of heads that occur in $n = 50$ flips. Assume that we observe $x = 27$ heads; given a uniform prior for $p$ on (0, 1), let us compute a 90% Bayes interval for $p$, given the sample result, and see from this which hypothesized values, $p_0$, we would accept for $p$. We see from Table 9.1 that the posterior for $p$ is the beta density

$$f_{P|X}(p \mid x) = \frac{\Gamma(52)}{\Gamma(28)\Gamma(24)} p^{27}(1 - p)^{23};$$

the posterior mean, which is the Bayes estimate, is

$$p^* = \tfrac{28}{52} = .5384.$$

Generally, if we want a 90% interval for $p$, it would seem reasonable to center it at $p^*$ which is our point estimate of $p$. Thus, we would want to find $a$ and $b$ such that

$$\int_{.5384}^{b} f_{P|X}(p \mid x)\,dp = \int_{a}^{.5384} f_{P|X}(p \mid x)\,dp = .45;$$

as was mentioned above, it is not possible to express these in a simple closed form. From a table of values of integrals of the beta function (the incomplete beta), we find that $b = .6486$ and $a = .4224$. Thus, on the basis of this sample, we would reject $H_0{:}p = p_0$, if $p_0 < .4224$ or if $p_0 > .6486$, since such values do not fall in an interval centered at $p^*$ which contains 90% of the posterior probability.

*Example 9.3.4.* Assume that the scores that high school seniors would make on achievement test $A$ are normally distributed with unknown mean $\mu$ and known standard deviation $\sigma = 15$. Let us also assume a uniform prior for $\mu$ on the interval (90, 110). Suppose that we randomly select 100 high school seniors, measure their scores on the test, and find $\bar{x} = 105$. Then the posterior for $\mu$ is

$$f_{M|\underline{X}}(\mu \mid \underline{x}) = \sqrt{\frac{100}{2\pi(15)^2}}\left[N_Z\left(\frac{110 - 105}{1.5}\right) - N_Z\left(\frac{90 - 105}{1.5}\right)\right]^{-1} e^{-100(\mu-105)^2/2(15)^2}$$

$$= \frac{1}{.9996}\sqrt{\frac{100}{2\pi(15)^2}}\, e^{-100(\mu-105)^2/2(15)^2}, \qquad \text{for } 90 < \mu < 110.$$

Then, using the normal tables, the 99% Bayes interval for $\mu$ (centered at $\mu^* = 105$) goes from $105 - (2.63)(1.5) = 101.06$ to $105 + (2.63)(1.5) = 108.95$. We would then reject $H_0{:}\mu = \mu_0$ for any $\mu_0 < 101.06$ or $\mu_0 > 108.95$.

## EXERCISE 9.3.

**1.** The time necessary for a barber to give a haircut to a customer is a normal random variable $X$ with $\sigma_X = 2$ minutes. Assuming a normal prior on $\mu$ with mean 20 and standard deviation 1 and that the total observed length of time for him to give 25 haircuts was 8 hours, construct a 90% Bayes interval for $\mu$.

**2.** Assume that the number of cars per hour that arrive at a service station for gasoline is a Poisson random variable with parameter $\lambda$. Also assume an exponential prior for $\lambda$ with parameter .05 and that a total of 230 cars arrived during a 10-hour period. Set up the integral that would define a 99% Bayes interval for $\lambda$.

**3.** In an industrial process for making bolts it is critical that the average length of the bolts should be 2 inches. It is also very critical that the variability of lengths should be small, since bolts that are too short or too long are not usable. Assume that for a given machine setting the lengths of the bolts produced are normally distributed with mean 2 and unknown variance $\sigma^2$. We are given prior information that the average value of $\sigma^2$ should be about .0001 and the variance of $\sigma^2$ about .000001. Find the parameters of the prior for $\sigma^2$ of the form given in Table 9.1; assuming that a sample of 100 bolt lengths yields

$$\sum x_i = 198.25, \qquad \sum x_i^2 = 393.0008,$$

derive the posterior for $\sigma^2$ and set up the integral that defines the 95% Bayes interval for $\sigma^2$.

**4.** Let $p_k$ be the proportion of married couples that have $k$ children, $k = 0, 1, 2, \ldots$. If you assume that $p_k$ is uniformly distributed on $(a_k, b_k)$, $(0 \le a_k < b_k \le 1)$, what is the posterior density for $p_k$, given that in a random sample of $n$ families, $x_k$ had $k$ children? Set up the integral defining a $100(1 - \alpha)\%$ Bayes interval for $p_k$.

**5.** The time between births at a large hospital is an exponential random variable with parameter $\lambda$. Assume that the average observed time (between births) over the past month has been 45 minutes. If you assume an exponential prior for $\lambda$, what is the posterior for $\lambda$, given a random sample of $n$ birth times? Set up a 95% Bayes interval for $\lambda$.

**6.** In testing hypotheses by using Bayesian methods, it would perhaps seem illogical to test that the unknown parameter is equal to any value other than its prior mean. Suppose then that $X$ is a normal random variable with unknown mean $\mu$ and known variance $\sigma_X^2$; assume a normal prior for $\mu$ with prior mean $\mu_0$ and prior variance $\sigma_0^2$. If we take a random sample of $n$ observations of $X$, show that we shall reject $H_0: \mu = \mu_0$ if either

$$\bar{x} > \frac{n}{\sigma_X^2}\mu_0 + z_{1-\alpha/2}\sqrt{a}$$

or

$$\bar{x} < \frac{n}{\sigma_X^2}\mu_0 - z_{1-\alpha/2}\sqrt{a}$$

where $a = n/\sigma_X^2 + 1/\sigma_0^2$.

# 10

# Least Squares and Regression Theory

A commonly occurring problem in many fields of endeavor is that of estimating the relationship between two variables. For example, the weight of a baby ($Y$) is undoubtedly related to its age ($x$); the salary ($Y$) which an individual of a given age is paid may very likely be related to the number of years ($x$) of formal education he has had; the cumulative grade point average ($Y$) which a college student maintains is probably related to the score ($x$) he made on an entrance examination; the maintenance cost ($Y$) per year for an automobile is undoubtedly related to its age ($x$). The methods of least squares and regression theory address the problem of estimating the relationship between a variable $Y$ and an "independently varying" variable $x$ (or more than one $x$), given a sample of $Y$ values with their associated values of $x$. The assumed functional form of the relationship between $Y$ and $x$ can in theory be anything; we shall concentrate our effort on a linear relation but shall indicate procedures for more general models.

## 10.1. Least Squares Estimation

Let us first of all clarify what we shall mean by a relationship between a random variable $Y$ and a variable $x$. We shall assume that for any given value of an "independent" variable $x$ there exists a distribution or population of associated $y$ values; this distribution of associated $y$ values, for a given $x$, is described by a probability density function $f_Y(y \mid x)$ (or, if the population has a discrete number of possible values, by a probability function $p_Y(y \mid x)$). For the time being we shall assume that $f_Y(y \mid x)$ depends on the associated $x$ value only through its mean value. That is, $E(Y \mid x)$ is a function of $x$ but any higher moments of $Y$ do not depend on $x$. Thus, we shall assume that the mean value of a random variable $Y$ depends upon the associated $x$ value but that the variance of $Y$ does not.

The particular functional form of the dependence of $E[Y \mid x]$ is arbitrary, of course, and varies from one application to another. In this section we shall study the special case where we assume that

$$E[Y \mid x] = a + bx$$

and that the higher moments of $Y$ do not depend on $x$. The two constants, $a$ and $b$, are unknown and are to be estimated from a sample of values of $Y$ with their associated $x$ values.

Let us note that we can then define the random variable

$$E = Y - (a + bx)$$

which would have mean zero, no matter what the values of $a$ and $b$, and which would have variance $\sigma^2$ identical with the variance of $Y$. Then we can write

$$Y = a + bx + E,$$

which is a standard expression in treatments of least squares. $E$ is called the residual error and represents the deviations of the $Y$ values about their mean. Now suppose that we have available a random sample of $n$ pairs $(Y_1, x_1)$, $(Y_2, x_2), \ldots, (Y_n, x_n)$ which we have written as 2-tuples to stress the fact that each sample of $Y$ that we take has an associated value of $x$. The values of $x$ may or may not be all distinct but, as we shall see, we must have at least two different values of $x$ represented if we are to estimate both $a$ and $b$. The method of least squares estimation of $a$ and $b$ specifies that we should take as our estimates of $a$ and $b$ those values that minimize

$$\sum_{i=1}^{n} (y_i - a - bx_i)^2$$

where the lower case $y_i$'s are the observed sample $Y$ values and the $x_i$'s are the associated values of the independent variable $x$. Thus we are minimizing the sum of squares of the residuals when applying the method of least squares.

Figure 10.1 gives a graphical presentation of this procedure. We can plot the $n$ sample pairs $(x_1, y_1), (x_2, y_2), \ldots, (x_n, y_n)$ as pictured. Because of the fact that the $y_i$'s are observed values of random variables, these points will not necessarily lie on a straight line, even if the mean value of $Y$ is linearly related to $x$. The method of least squares then says that we should consider

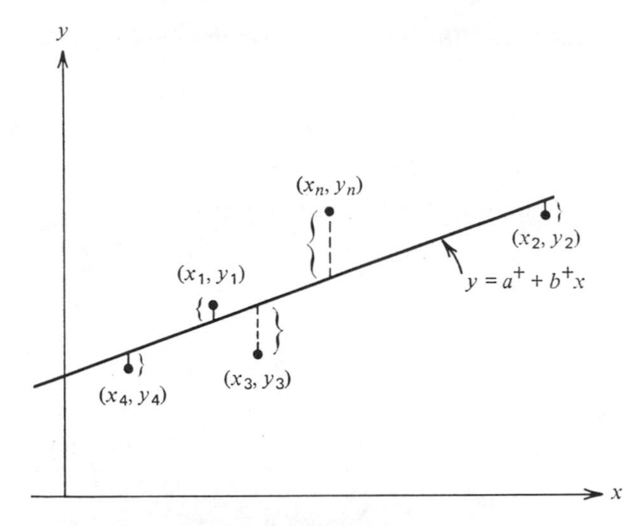


**Figure 10.1.**

all possible straight lines that we might plot in this plane and choose the one that minimizes the sum of squares of deviations between the observed sample points and the line which we draw. The line which does this minimization is called the least squares line and is denoted by $y = a^+ + b^+x$; $a^+$ and $b^+$ are called the least squares estimates of $a$ and $b$. If we choose any other line

$$y = a^* + b^*x,$$

the sum of squares of deviations of the $y_i$'s about this second line is necessarily larger. The line $y = a^+ + b^+x$ is also called the regression of $Y$ on $x$ (see Chapter 5, Section 5.3 where both $Y$ and $X$ were random variables).

***Theorem 10.1.1.*** Let $(y_1, x_1), (y_2, x_2), \ldots, (y_n, x_n)$ be the observed values of a random variable $Y$, with their associated $x$ values, where $E[Y \mid x] = a + bx$. Then the least squares line is given by $y = a^+ + b^+x$ where

$$b^+ = \frac{\sum (x_i - \bar{x})(y_i - \bar{y})}{\sum (x_i - \bar{x})^2}$$

and

$$a^+ = \bar{y} - b^+\bar{x}.$$

*Proof:* The quantity to be minimized, with respect to $a$ and $b$, is

$$Q = \sum_{i=1}^{n} (y_i - a - bx_i)^2.$$

To accomplish this we need to derive the partial derivatives of $Q$ with respect to $a$ and $b$, set them equal to zero, and solve the resulting equations simultaneously for $a^+$ and $b^+$. Then, if

$$\frac{\partial^2 Q}{\partial a^2} > 0$$

and if the determinant of second partial derivatives

$$\begin{vmatrix} \dfrac{\partial^2 Q}{\partial a^2} & \dfrac{\partial^2 Q}{\partial a \partial b} \\ \dfrac{\partial^2 Q}{\partial a \partial b} & \dfrac{\partial^2 Q}{\partial b^2} \end{vmatrix}$$

is positive at $a^+$, $b^+$, we have in fact located the minimum value for $Q$. Thus,

$$\frac{\partial Q}{\partial a} = -2 \sum_{i=1}^{n} (y_i - a - bx_i)$$

$$\frac{\partial Q}{\partial b} = -2 \sum_{i=1}^{n} x_i(y_i - a - bx_i).$$

The equations to be solved for $a^+$ and $b^+$ (called the *normal equations*) then are

$$na^+ + b^+ \sum x_i = \sum y_i$$

$$a^+ \sum x_i + b^+ \sum x_i^2 = \sum x_i y_i.$$

From the first equation we have

$$a^+ = \bar{y} - b^+\bar{x}$$

and substituting this result in the second yields

$$(\bar{y} - b^+\bar{x}) \sum x_i + b^+ \sum x_i^2 = \sum x_i y_i$$

or

$$b^+[\sum x_i^2 - \bar{x} \sum x_i] = \sum x_i y_i - \bar{y} \sum x_i.$$

Thus

$$b^+ = \frac{\sum x_i y_i - \bar{y} \sum x_i}{\sum x_i^2 - \bar{x} \sum x_i} = \frac{\sum (x_i - \bar{x})(y_i - \bar{y})}{\sum (x_i - \bar{x})^2}.$$

The second partials are

$$\frac{\partial^2 Q}{\partial a^2} = 2n, \qquad \frac{\partial^2 Q}{\partial a \partial b} = 2 \sum x_i, \qquad \frac{\partial^2 Q}{\partial b^2} = 2 \sum x_i^2.$$

Thus $\partial^2 Q/\partial a^2 = 2n > 0$ and

$$\begin{vmatrix} 2n & 2\sum x_i \\ 2\sum x_i & 2\sum x_i^2 \end{vmatrix} = 4n \sum x_i^2 - 4(\sum x_i)^2$$

$$= 4n \sum (x_i - \bar{x})^2,$$

which is positive if we have at least two different values represented in $x_1$, $x_2, \ldots, x_n$, for any values of $a$ and $b$. Thus the line $y = a^+ + b^+x$ does give the minimum value of $Q$. ◄

We now examine a numerical example.

*Example 10.1.1.* Suppose that it is medically plausible that the weight of a baby is a linear function of its age, on the average, for ages between 1 month and 6 months and that the higher moments of the weights do not depend on $x$. We are assuming then that for babies aged $x$ months there is a distribution of weights; for any fixed $x$ the mean of this distribution is given by $a + bx$, $1 \leq x \leq 6$, where $a$ and $b$ are unknown constants, and there is no further dependence between the weights and the age. The following sample of 10 weights (in pounds) and ages (in months) was observed:

| $y_i$ | $x_i$ |
|---|---|
| 14 | 3 |
| 16.5 | 5 |
| 14 | 2 |
| 10 | 1 |
| 16 | 4.5 |
| 20 | 5.5 |
| 22 | 6 |
| 15 | 3.5 |
| 16.5 | 4 |
| 12 | 2.5 |

We find that $n = 10$, $\sum x_i = 37$, $\sum x_i^2 = 160$, $\sum y_i = 156$, and $\sum x_i y_i = 625$; thus $\bar{x} = 3.7$, $\bar{y} = 15.6$, $\sum (x_i - \bar{x})^2 = 23.1$, $\sum (x_i - \bar{x})(y_i - \bar{y}) = 47.8$, $b^+ = \dfrac{47.8}{23.1} = 2.07$, $a^+ = 15.6 - (2.07)(3.7) = 7.94$, and the least squares line is $y = 7.94 + 2.07x$.

The original data and this line are plotted in Figure 10.2. By referring either to the equation or its graph, we would estimate the average weight of 2-month-old babies to be 12.08 pounds; we would estimate the average weight of 5-month-old babies to be 18.29 pounds.

We have seen what the estimates of $a$ and $b$ turn out to be as a function of the observed sample values, using the least squares criterion; as always the least squares *estimators* are defined as the same functions of the random

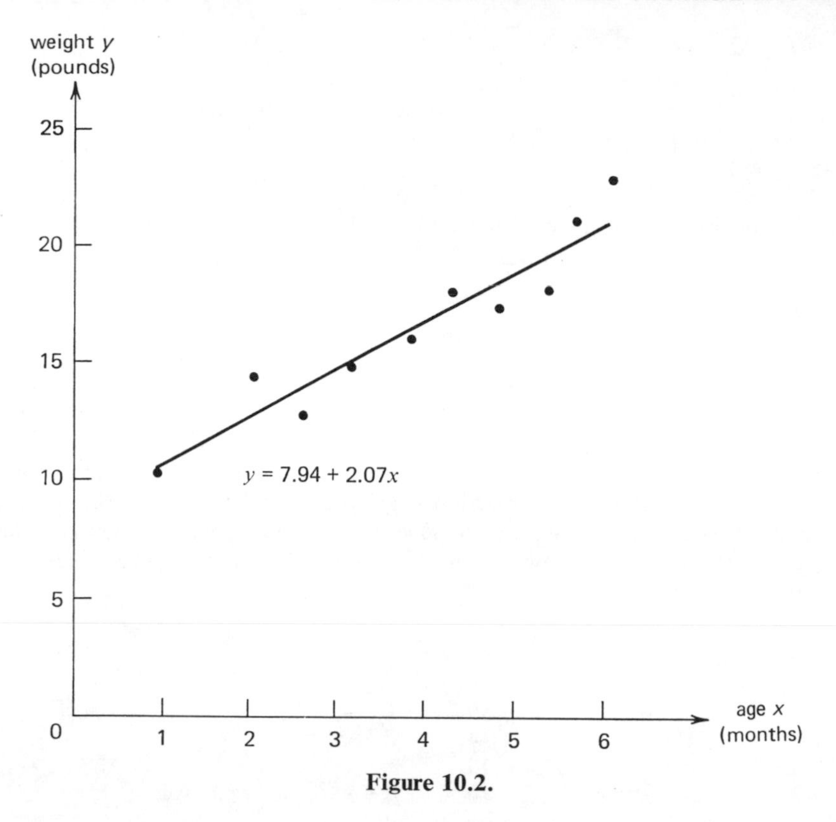


**Figure 10.2.**

variables rather than their observed values. Thus

$$B^{+} = \frac{\sum (x_i - \bar{x})(Y_i - \bar{Y})}{\sum (x_i - \bar{x})^2}$$

$$A^{+} = \bar{Y} - B^{+}\bar{x},$$

since we are assuming that there is a distribution of $y$ values for any given $x$. We would then like to investigate questions regarding the expected values of these estimators, their variances, and their covariance. The following theorems give these expected values, variances, and covariances.

***Theorem 10.1.2.*** Assume that $Y$ is a random variable with mean $a + bx$, for a given value of $x$, and let $(Y_1, x_1)$, $(Y_2, x_2)$, $\ldots$, $(Y_n, x_n)$ be a random sample of size $n$ of $Y$ (with their associated $x$ values). Define

$$B^{+} = \frac{\sum (x_i - \bar{x})(Y_i - \bar{Y})}{\sum (x_i - \bar{x})^2}$$

$$A^{+} = \bar{Y} - B^{+}\bar{x};$$

then $A^{+}$ and $B^{+}$ are unbiased estimators of $a$ and $b$.

*Proof:*

$$E[B^+] = \frac{\sum (x_i - \bar{x})E[Y_i - \bar{Y}]}{\sum (x_i - \bar{x})^2}$$
$$= \frac{\sum (x_i - \bar{x})[b(x_i - \bar{x})]}{\sum (x_i - \bar{x})^2}$$
$$= b,$$

since

$$E[\bar{Y}] = \frac{1}{n} E[\sum Y_i]$$
$$= \frac{1}{n} \sum (a + bx_i) = a + b\bar{x}$$

and $E[Y_i - \bar{Y}] = [a + bx_i] - [a + b\bar{x}] = b(x_i - \bar{x})$. We also have

$$E[A^+] = E[\bar{Y} - B^+\bar{x}]$$
$$= [a + b\bar{x}] - b\bar{x} = a$$

and the theorem is proved. ◀

***Theorem 10.1.3.*** Again, assume that $Y$ is a random variable with mean $a + bx$, for a given value of $x$, and that the variance of $Y$ is $\sigma^2$ for any $x$ (that is, each distribution of $Y$ values, for differing values of $x$, has the same variance). Let $(Y_1, x_1)$, $(Y_2, x_2)$, $\ldots$, $(Y_n, x_n)$ be a random sample of $Y$ with their associated values of $x$. Then

$$V(A^+) = \frac{\sigma^2 \sum x_i^2}{n \sum (x_i - \bar{x})^2}$$
$$V(B^+) = \frac{\sigma^2}{\sum (x_i - \bar{x})^2}$$
$$\text{Cov}(A^+, B^+) = -\frac{\sigma^2 \bar{x}}{\sum (x_i - \bar{x})^2},$$

where $A^+$ and $B^+$ are the least squares estimators.

*Proof:* Let us note first of all that

$$\sum (x_i - \bar{x})(Y_i - \bar{Y}) = \sum (x_i - \bar{x}) Y_i - \bar{Y} \sum (x_i - \bar{x})$$
$$= \sum (x_i - \bar{x}) Y_i,$$

since $\sum (x_i - \bar{x}) = 0$; then we can write

$$B^+ = \frac{\sum (x_i - \bar{x}) Y_i}{\sum (x_i - \bar{x})^2}.$$

Now we know that if $Y_1, Y_2, \ldots, Y_n$ are independent random variables, each with variance $\sigma^2$ and if

$$Z = \sum_{i=1}^{n} c_i Y_i,$$

then

$$V(Z) = \sigma^2 \sum_{i=1}^{n} c_i^2.$$

Thus

$$B^+ = \sum_{i=1}^{n} c_i Y_i$$

where

$$c_i = \frac{x_i - \bar{x}}{\sum (x_i - \bar{x})^2}$$

and then

$$V(B^+) = \sigma^2 \sum c_i^2 = \sigma^2 \frac{\sum (x_i - \bar{x})^2}{[\sum (x_i - \bar{x})^2]^2} = \frac{\sigma^2}{\sum (x_i - \bar{x})^2}.$$

(We have written $V(Z)$ as shorthand for the variance of $Z$ for any random variable $Z$.) Let us now turn to the variance of $A^+$. Since $A^+ = \bar{Y} - B^+\bar{x}$, we have

$$V(A^+) = V(\bar{Y}) + \bar{x}^2 V(B^+) - 2\bar{x} \operatorname{Cov}(\bar{Y}, B^+).$$

Now $V(\bar{Y}) = \sigma^2/n$ and we have just evaluated $V(B^+)$, so the one remaining quantity we need to get $V(A^+)$ is the covariance between $\bar{Y}$ and $B^+$. We saw in Chapter 5 that if $Y_1, Y_2, \ldots, Y_n$ are independent random variables, each with variance $\sigma^2$, and $Z_1 = \sum c_i Y_i$, $Z_2 = \sum d_i Y_i$, then $\operatorname{Cov}(Z_1, Z_2) = \sigma^2 \sum c_i d_i$. Thus

$$\operatorname{Cov}(\bar{Y}, B^+) = \sigma^2 \sum \frac{1}{n} \frac{(x_i - \bar{x})}{\sum (x_i - \bar{x})^2} = 0$$

and we have

$$\begin{aligned}
V(A^+) &= V(\bar{Y}) + \bar{x}^2 V(B^+) \\
&= \sigma^2 \left[\frac{1}{n} + \frac{\bar{x}^2}{\sum (x_i - \bar{x})^2}\right] = \frac{\sigma^2}{n} \frac{\sum x_i^2}{\sum (x_i - \bar{x})^2}. \\
\operatorname{Cov}(A^+, B^+) &= \operatorname{Cov}(\bar{Y} - B^+\bar{x}, B^+) \\
&= \operatorname{Cov}(\bar{Y}, B^+) - \bar{x} \operatorname{Cov}(B^+, B^+) \\
&= 0 - \bar{x} V(B^+) \\
&= -\sigma^2 \frac{\bar{x}}{\sum (x_i - \bar{x})^2},
\end{aligned}$$

and the theorem is proved. ◀

Notice then that the variances and the covariance of our estimators of $a$ and $b$ depend upon the values of $x$. If we are free to choose the $x$ values, we can select them in optimal ways in terms of accuracy or variability of our estimators of $a$ and $b$. Since

$$V(B^+) = \frac{\sigma^2}{\sum (x_i - \bar{x})^2},$$

it is plain that we shall get the most precise (least variable) estimate of the slope $b$ if we make $\sum (x_i - \bar{x})^2$ as large as possible. In Example 10.1.1, we assumed that the average weight of babies of age $x$ months was $a + bx$ for $x$ between 1 and 6. To get the most precise estimate of $b$, we should choose half the babies in the sample to be 1 month old and the other half to be 6 months old because these values for $x$ would maximize $\sum (x_i - \bar{x})^2$. It is intuitively plausible, regardless of the result about $V(B^+)$, that this distribution of $x$'s should give the best estimate of the slope since, in taking half the values at each of the two end points of the interval, we are getting the most precise estimates we can of the average weights at both 1 month and 6 months. Then, because we assume the mean of $Y$ to be $a + bx$ over this interval, the most precise estimates at the end points are connected by a straight line to give the most precise estimate of the slope. In general, if we can choose the $x$ values, we shall get the smallest variance when estimating $b$ by maximizing $\sum (x_i - \bar{x})^2$. Along the same lines, we should note that $V(A^+)$ is minimized if $\bar{x}$ is zero; at the same time, $A^+$ and $B^+$ are uncorrelated if $\bar{x} = 0$.

The method of least squares does not contain any rationale for estimating $\sigma^2$, the variance of $Y$. However, the following theorem presents an unbiased estimator for $\sigma^2$ whose observed value can be computed from a sample.

***Theorem 10.1.4.*** Assume that $E[Y \mid x] = a + bx$ and that $V(Y \mid x) = \sigma^2$. Then

$$(S^+)^2 = \frac{1}{n-2} \sum_{i=1}^{n} (Y_i - A^+ - B^+ x_i)^2$$

$$= \frac{1}{n-2} \left[\sum (Y_i - \bar{Y})^2 - (B^+)^2 \sum (x_i - \bar{x})^2\right]$$

is an unbiased estimator of $\sigma^2$.

*Proof:* Let us first verify the identity given above. Clearly

$$Y_i - A^+ - B^+ x_i = Y_i - (\bar{Y} - B^+ \bar{x}) - B^+ x_i$$

$$= Y_i - \bar{Y} - B^+(x_i - \bar{x});$$

thus,

$$
\begin{aligned}
&\sum (Y_i - A^+ - B^+x_i)^2 \\
&\quad = \sum (Y_i - \bar{Y} - B^+(x_i - \bar{x}))^2 \\
&\quad = \sum (Y_i - \bar{Y})^2 - 2B^+ \sum (x_i - \bar{x})(Y_i - \bar{Y}) + (B^+)^2 \sum (x_i - \bar{x})^2 \\
&\quad = \sum (Y_i - \bar{Y})^2 - (B^+)^2 \sum (x_i - \bar{x})^2
\end{aligned}
$$

since

$$\sum (x_i - \bar{x})(Y_i - \bar{Y}) = B^+ \sum (x_i - \bar{x})^2.$$

Clearly

$$Y_1 - \bar{Y} = \frac{n-1}{n} Y_1 - \frac{1}{n} \sum_2^n Y_i$$

and thus

$$
\begin{aligned}
V(Y_1 - \bar{Y}) &= \frac{(n-1)^2}{n^2} \sigma^2 + \frac{1}{n^2}(n-1)\sigma^2 \\
&= \frac{n-1}{n} \sigma^2.
\end{aligned}
$$

Similarly

$$V(Y_i - \bar{Y}) = \frac{n-1}{n} \sigma^2, \qquad i = 2, 3, \ldots, n.$$

Now $E[Y_i - \bar{Y}] = b(x_i - \bar{x})$ and thus

$$
\begin{aligned}
E[\sum (Y_i - \bar{Y})^2] &= \sum \left( \frac{n-1}{n} \sigma^2 + b^2(x_i - \bar{x})^2 \right) \\
&= (n-1)\sigma^2 + b^2 \sum (x_i - \bar{x})^2
\end{aligned}
$$

since the expected value of the square of a variable is its variance plus the square of its mean. Similarly

$$
\begin{aligned}
\sum (x_i - \bar{x})^2 E[(B^+)^2] &= \sum (x_i - \bar{x})^2 \left[ \frac{\sigma^2}{\sum (x_i - \bar{x})^2} + b^2 \right] \\
&= \sigma^2 + b^2 \sum (x_i - \bar{x})^2;
\end{aligned}
$$

thus, combining these two results, we have

$$
\begin{aligned}
E[(S^+)^2] &= \frac{1}{n-2} E[\sum (Y_i - \bar{Y})^2 - (B^+)^2 \sum (x_i - \bar{x})^2] \\
&= \frac{1}{n-2} [(n-1)\sigma^2 + b^2 \sum (x_i - \bar{x})^2 - \sigma^2 - b^2 \sum (x_i - \bar{x})^2] \\
&= \sigma^2,
\end{aligned}
$$

and $(S^+)^2$ is an unbiased estimator of $\sigma^2$ as was claimed. ◀

It will be recalled that if we had a random sample $Y_1, Y_2, \ldots, Y_n$ of a random variable $Y$, whose mean was a constant $\mu$ and not related to some additional variable $x$, then we constructed an unbiased estimator of $\sigma^2$ by taking the sum of squares of the $Y_i$'s about their mean $\bar{Y}$ (which estimates $\mu$) and dividing by $n - 1$. Now, in the least squares model where the mean value of $Y$ is dependent on another variable $x$, we estimate $\sigma^2$ by taking the sum of squares of the deviations of the $Y_i$'s about the estimated line, $A^+ + B^+x$, which for any $x$ gives the estimator of the mean of $Y$, and dividing by $n - 2$. In the former case the divisor is $n - 1$ since only one statistic ($\bar{Y}$) is needed to estimate the mean of $Y$; in the latter case the divisor is $n - 2$ since two statistics ($A^+$ and $B^+$) are needed to estimate the mean of $Y$ for a given $x$.

*Example 10.1.2.* Assume that the cumulative grade point average a student achieves at the end of the freshman year is a linear function of the score, $x$, that he made on an entrance exam. That is, we are assuming that if we look at all students scoring $x$ on the entrance exam, we have a distribution of cumulative grade point averages achieved at the end of the freshman year; furthermore the means of these distributions, for varying values of $x$, lie on the straight line $a + bx$. As before, we also assume that the higher moments are all independent of $x$. The entrance exam scores can range from 60 to 100, say, and the grade point average is computed on the basis of 4 points for an A, 3 points for a B, 2 for a C, 1 for a D, and 0 for an F.

Suppose that we want to estimate $a$ and $b$ and especially want a precise estimate of $b$; to do so we would want our sample $x$ values separated as widely as possible. When examining the entrance exam records, we find that there are relatively few scores at 60 and 100 but that for $x = 65$ and $x = 95$ there are a lot of students to choose from. Accordingly, we select at random 4 students who scored 65 and 4 students who scored 95 and find their grade point averages to be 2.25, 2.17, 2.67, 2.42 for $x = 65$ and 3.63, 2.95, 3.02, 3.49 for $x = 95$. (In actual practice, of course, we would want to take a larger sample than this to get good estimates of $a$ and $b$, but for illustrative purposes $n$ is kept small.) Then we find that $\bar{x} = 80$, $\bar{y} = 2.825$, $\sum (x_i - \bar{x})^2 = 1800$, $\sum (y_i - \bar{y})^2 = 2.0916$, $\sum (x_i - \bar{x})(y_i - \bar{y}) = 53.7$, $b^+ = \frac{53.7}{1800} = .030$, and $a^+ = 2.825 - .030(80) = .425$. Thus $y = .425 + .030x$ is the least squares line.

Then we estimate that if a student scores $x = 75$ on the entrance exam, he will on the average have a $.425 + (.03)(75) = 2.675$ grade point at the end of his freshman year; a student scoring $x = 90$ on the entrance exam should on the average have a $.425 + (.03)(90) = 3.125$ grade point at the end of his freshman year. Our estimate of $\sigma^2$ is

$$(s^+)^2 = \tfrac{1}{6}[2.0916 - (.03)^2(1800)] = .079$$

and our estimate of $\sigma$ is

$$s^+ = \sqrt{0.79} = .280.$$

The estimated variance of $B^+$ is $(.079)(\frac{1}{1800}) = .0000439$; the estimated variance of $A^+$ is $(.079)\left(\frac{1}{6} + \frac{(80)^2}{1800}\right) = .294$; the estimated covariance between $A^+$ and $B^+$ is $(.079)\left(\frac{-80}{1800}\right) = -.00351$.

When selecting only two different $x$ values, as in this example, we are relying very heavily on the assumed linearity of $E[Y \mid x]$. Since we did not observe $x$ at intermediate points between 65 and 95, we would not be able to detect from the sample whether this linear assumption appeared correct.

The Gauss-Markov theorem, which we shall now state in a restricted context, is the major reason why the method of least squares is so frequently employed when estimating the relationship between the mean of a random variable $Y$ and an independent variable $x$. We shall state the theorem only in the context of a linear relationship between $E[Y \mid x]$ and $x$, but it can be extended to more general cases.

***Theorem 10.1.4.*** Assume that $E[Y \mid x] = a + bx$, for given values of $x$, and that the variance of $Y$ is $\sigma^2$ for all $x$. Then if $(Y_1, x_1)$, $(Y_2, x_2)$, $\ldots$, $(Y_n, x_n)$ is a random sample of $Y$ with their associated $x$ values, the least squares estimators $A^+$ and $B^+$ are the best linear unbiased estimators of $a$ and $b$.

The remarkable thing about this theorem is the fact that the actual distribution of the $Y$ values does not have to be known for us to compute the best linear unbiased estimators of $a$ and $b$. It will be recalled that "best linear unbiased" means that among all linear functions of the $Y_i$'s that are unbiased estimators, none has a smaller variance. However, it is important to realize that we assume the variance of $Y$ to be a constant for all values of $x$ in order that $A^+$ and $B^+$ have this property; if this is not satisfied, then the best linear unbiased estimators are altered somewhat from the above values (see problem 6 below).

## EXERCISE 10.1.

**1.** Eight 1-foot-tall Monterey pine trees were planted in similar controlled environments but were subjected to differing amounts of irrigation to simulate the effect of differing amounts of rainfall. At the end of 1 year, their achieved heights were measured. The following table presents their measured heights (in inches) at the end of a year ($y_i$) and the amount of rainfall (in inches) simulated for each ($x_i$). Assume that $Y$, the height of a tree at the end of a year (starting from a 1-foot height), is a random variable with mean $a + bx$, where $x$ is the rainfall, and with constant variance $\sigma^2$ for all $x$. Compute the best linear unbiased estimates of $a$ and $b$ and compute an unbiased estimate of $\sigma^2$ and of $V(A^+)$ and $V(B^+)$.

| $y_i$ | $x_i$ |
|---|---|
| 19 | 10 |
| 22 | 14 |
| 25 | 18 |
| 31 | 22 |
| 33 | 26 |
| 39 | 30 |
| 44 | 34 |
| 45 | 38 |

**2.** Show that

$$\begin{aligned}\sum x_i y_i - \bar{x} \sum y_i &= \sum x_i y_i - \bar{y} \sum x_i \\ &= \sum (x_i - \bar{x})(y_i - \bar{y}).\end{aligned}$$

**3.** A random sample of 10 men aged 30 yielded the following information on their current annual salary ($y_i$) (in thousands of dollars) and the number of years of formal schooling they had ($x_i$). Assume that the mean of the distribution of salaries

| $y_i$ | $x_i$ |
|---|---|
| 7.0 | 10 |
| 6.2 | 12 |
| 8.1 | 13 |
| 7.5 | 14 |
| 6.5 | 16 |
| 10.5 | 16 |
| 8.0 | 16 |
| 13.2 | 16 |
| 12.8 | 18 |
| 16.5 | 20 |

is a linear function of the number of years of schooling and that the variance of $Y$ is independent of $x$; compute the best linear unbiased estimates of $a$ and $b$. Also estimate $\sigma^2$ and $V(A^+)$, $V(B^+)$.

**4.** Assume that $Y$ is a random variable with $E[Y \mid x] = bx$, for any $x$, and that the variance of $Y$ is $\sigma^2$, independent of $x$. Assuming $(Y_1, x_1), (Y_2, x_2), \ldots, (Y_n, x_n)$ is a random sample of $Y$, with the associated $x$ values, derive the least squares estimator of $b$. Define an unbiased estimator of $\sigma^2$ and show that it is unbiased.

**5.** Show that the least squares estimator of $b$ derived in problem 4 is the best linear unbiased estimator of $b$, in the context of the assumptions given there. (**Hint:**

$$B^+ = \frac{\sum x_i Y_i}{\sum x_i^2} = \sum c_i Y_i, \qquad \text{where } c_i = \frac{x_i}{\sum x_i^2}.$$

If $B$ is any other linear unbiased estimator of $b$, then

$$B = \sum d_i Y_i = \sum (c_i - f_i) Y_i, \qquad \text{with} \sum f_i x_i = 0.$$

Compute $V(B)$.)

**6.** Suppose that $E[Y \mid x] = bx$ and that $(Y_1, x_1)$, $(Y_2, x_2)$, ..., $(Y_n, x_n)$ is a random sample of $Y$ with their associated $x$ values. Assume that $V(Y_i) = \sigma_i^2$; that is, the variances of the $Y$ values are now conceivably dependent upon the $x$ values. Show that

$$B^+ = \frac{\sum x_i Y_i / \sigma_i^2}{\sum x_i^2 / \sigma_i^2}$$

is the best linear unbiased estimator of $b$. (**Hint:** Define

$$Z_i = \frac{Y_i}{\sigma_i}, \qquad w_i = \frac{x_i}{\sigma_i}$$

and apply the result of problem 5.)

**7.** Suppose that everything is as given in problem 6, with $x_i > 0$, $\sigma_i^2 = \sigma^2 x_i$. Evaluate the best linear unbiased estimator of $b$.

**8.** Again, suppose that everything is as given in problem 6 and now assume that $\sigma_i^2 = \sigma^2 x_i^2$; evaluate the best linear unbiased estimator of $b$.

**9.** Show that the least squares line $y = a^+ + b^+ x$ always goes through the point whose coordinates are $(\bar{x}, \bar{y})$.

**10.** Assume that $E[Y \mid x] = a + b/x$, for $x > 0$, and derive the least squares estimates of $a$ and $b$.

## 10.2. Interval Estimation and Tests of Hypotheses

We ended the last section with a theorem giving an optimum property of the least squares estimates, no matter what the form of the underlying distribution for $Y$. In this section we shall make the assumption that $Y$ is normally distributed, with mean $a + bx$ and constant variance $\sigma^2$. Under this distributional assumption, we shall find it easy to construct confidence intervals for $a$ and $b$ and for $a + bx$, the mean value of $Y$, and to test hypotheses about these quantities. We shall first investigate the maximum likelihood estimators of $a$, $b$, and $\sigma^2$ for a normal assumption on $Y$.

***Theorem 10.2.1.*** $Y$ is a normal random variable with mean $a + bx$, for a specified value of $x$, and with variance $\sigma^2$ for any $x$. The maximum likelihood estimators of $a$, $b$, and $\sigma^2$ are

$$\hat{A} = \bar{Y} - \hat{B}\bar{x}$$

$$\hat{B} = \frac{\sum (x_i - \bar{x})(Y_i - \bar{Y})}{\sum (x_i - \bar{x})^2}$$

$$\hat{\Sigma}^2 = \frac{1}{n} \sum (Y_i - \hat{A} - \hat{B} x_i)^2,$$

given a random sample $(Y_1, x_1), (Y_2, x_2), \ldots, (Y_n, x_n)$ of $Y$ with their associated $x$ values.

*Proof:* The likelihood function for the sample is

$$L(a, b, \sigma^2) = \left(\frac{1}{2\pi\sigma^2}\right)^{n/2} e^{(-1/2\sigma^2)\sum(y-a-bx_i)^2}.$$

Then

$$\ln L = -\frac{n}{2}\ln 2\pi\sigma^2 - \frac{1}{2\sigma^2}\sum (y_i - a - bx_i)^2$$

and

$$\frac{\partial \ln L}{\partial a} = \frac{1}{\sigma^2}\sum (y_i - a - bx_i)$$

$$\frac{\partial \ln L}{\partial b} = \frac{1}{\sigma^2}\sum x_i(y_i - a - bx_i)$$

$$\frac{\partial \ln L}{\partial \sigma^2} = -\frac{n}{2\sigma^2} + \frac{1}{2\sigma^4}\sum (y_i - a - bx_i)^2.$$

We note that if we set these three equations equal to zero and multiply through by $\sigma^2$, the first two equations are identical with the least squares normal equations that we solved in Section 10.1. Thus, their solution is given by

$$\hat{a} = \bar{y} - \hat{b}\bar{x}$$

$$\hat{b} = \frac{\sum (x_i - \bar{x})(y_i - \bar{y})}{\sum (x_i - \bar{x})^2};$$

once the first two equations are solved for $\hat{a}$ and $\hat{b}$, their solutions can be put into the third equation, yielding

$$\hat{\sigma}^2 = \frac{1}{n}\sum (y_i - \hat{a} - \hat{b}x_i)^2.$$

Thus we have the desired result upon replacing the sample values $y_1, y_2, \ldots, y_n$ with the random variables $Y_1, Y_2, \ldots, Y_n$. ◀

We note immediately that the maximum likelihood estimators of $a$ and $b$ are identical with the least squares estimators of these same parameters. Thus we can say immediately that $\hat{A}$ and $\hat{B}$ are the best linear unbiased estimators of $a$ and $b$, respectively. We shall use $\hat{A}$ and $\hat{B}$, rather than $A^+$ and $B^+$, to denote these estimators from now on. We also know that

$$E[n\hat{\Sigma}^2] = (n-2)\sigma^2$$

so that $\hat{\Sigma}^2$ is a biased estimator of $\sigma^2$. From now on we shall use only the unbiased estimator of $\sigma^2$ (gotten by dividing by $n - 2$ rather than $n$) and shall denote it by $S^2$ rather than $(S^+)^2$.

The estimator $\hat{B}$ of $b$ is a linear combination of normal random variables, under the assumption of normality for $Y$, and thus is itself a normal random variable; its mean is $b$ and its variance is

$$\frac{\sigma^2}{\sum (x_i - \bar{x})^2}.$$

Similarly, since $\hat{A}$ is a linear combination of normally distributed random variables, it is itself normally distributed with mean $a$ and variance

$$\frac{\sum x_i^2}{n \sum (x_i - \bar{x})^2} \sigma^2.$$

It can also be shown that $[(n - 2)/\sigma^2]\, S^2$ is a $\chi^2$ random variable with $n - 2$ degrees of freedom, under our normal assumption, and that $S^2$ is distributed independently of $\hat{A}$ and $\hat{B}$. These facts will enable us to put confidence intervals about $a$ and $b$ and to test hypotheses about $a$ and $b$. The following theorem gives the basic result for these techniques.

***Theorem 10.2.2.*** Let $Y$ be a normal random variable with mean $a + bx$, for a fixed value of $x$, and with constant variance $\sigma^2$. Let $\hat{A}$ and $\hat{B}$ be the maximum likelihood (least squares) estimators of $a$ and $b$ and let $S^2$ be the unbiased estimator of $\sigma^2$ already defined. Then

$$\frac{\hat{A} - a}{S\sqrt{\dfrac{\sum x_i^2}{n \sum (x_i - \bar{x})^2}}} \quad \text{and} \quad \frac{\hat{B} - b}{S\sqrt{\dfrac{1}{\sum (x_i - \bar{x})^2}}}$$

each have the $t$-distribution with $n - 2$ degrees of freedom.

*Proof:* We are given that

$$\frac{\hat{A} - a}{\sigma\sqrt{\dfrac{\sum x_i^2}{n \sum (x_i - \bar{x})^2}}} \quad \text{and} \quad \frac{\hat{B} - b}{\sigma\sqrt{\dfrac{1}{\sum (x_i - \bar{x})^2}}}$$

are each standard normal random variables; $[(n - 2)/\sigma^2]\, S^2$ is a $\chi^2$ random variable with $n - 2$ degrees of freedom and is independent of both of the standard normal random variables just mentioned. Thus,

$$\frac{\hat{A} - a}{\sigma\sqrt{\dfrac{\sum x_i^2}{n \sum (x_i - \bar{x})^2}}} \Bigg/ \sqrt{\frac{(n - 2)S^2}{\sigma^2(n - 2)}} = \frac{\hat{A} - a}{S\sqrt{\dfrac{\sum x_i^2}{n \sum (x_i - \bar{x})^2}}}$$

and

$$\frac{\hat{B} - b}{\sigma\sqrt{\dfrac{1}{\sum (x_i - \bar{x})^2}}} \Bigg/ \sqrt{\frac{(n-2)S^2}{\sigma^2(n-2)}} = \frac{\hat{B} - b}{S\sqrt{\dfrac{1}{\sum (x_i - \bar{x})^2}}}$$

each have the $t$-distribution with $n - 2$ degrees of freedom. (See Theorem 7.4.3.) ◀

Constructing a confidence interval for $a$ or $b$, or testing a hypothesis about either, then will proceed in exactly the same way as we saw for the mean of a normal random variable in Chapters 7 and 8. Specifically, the statistics

$$L = \hat{B} - t_{1-\alpha/2} S \frac{1}{\sum (x_i - \bar{x})^2}$$

$$U = \hat{B} + t_{1-\alpha/2} S \frac{1}{\sum (x_i - \bar{x})^2},$$

where $t_{1-\alpha/2}$ is the 100 $(1 - \alpha/2)$ percentile of the $t$-distribution with $n - 2$ degrees of freedom, form a 100 $(1 - \alpha)\%$ confidence interval for $b$. A 100 $(1 - \alpha)\%$ confidence interval for $a$ is defined similarly.

We can also construct 100 $(1 - \alpha)\%$ confidence bands about the regression line $\hat{Y}_x = \hat{A} + \hat{B}x$; that is, for any given constant value of $x$, $\hat{Y}_x$ is normally distributed with mean $a + bx$ and variance

$$\begin{aligned} V(\bar{Y}_x) &= V(\hat{A} + \hat{B}x) \\ &= V(\hat{A}) + x^2 V(\hat{B}) + 2x \operatorname{Cov}(\hat{A}, \hat{B}) \\ &= \frac{\sigma^2}{\sum (x_i - \bar{x})^2}\left[\frac{1}{n}\sum x_i^2 + x^2 - 2x\bar{x}\right] \\ &= \frac{\sigma^2}{\sum (x_i - \bar{x})^2}\left[\frac{1}{n}\sum (x_i - \bar{x})^2 + (x - \bar{x})^2\right] \\ &= \sigma^2\left[\frac{1}{n} + \frac{(x - \bar{x})^2}{\sum (x_i - \bar{x})^2}\right]. \end{aligned}$$

Thus,

$$\frac{\hat{Y}_x - (a + bx)}{S\sqrt{\dfrac{1}{n} + \dfrac{(x - \bar{x})^2}{\sum (x_i - \bar{x})^2}}}$$

has the $t$-distribution with $n - 2$ degrees of freedom and

$$L_x = \hat{Y}_x - t_{1-\alpha/2} S \sqrt{\frac{1}{n} + \frac{(x - \bar{x})^2}{\sum (x_i - \bar{x})^2}}$$

$$U_x = \hat{Y}_x + t_{1-\alpha/2} S \sqrt{\frac{1}{n} + \frac{(x - \bar{x})^2}{\sum (x_i - \bar{x})^2}}$$

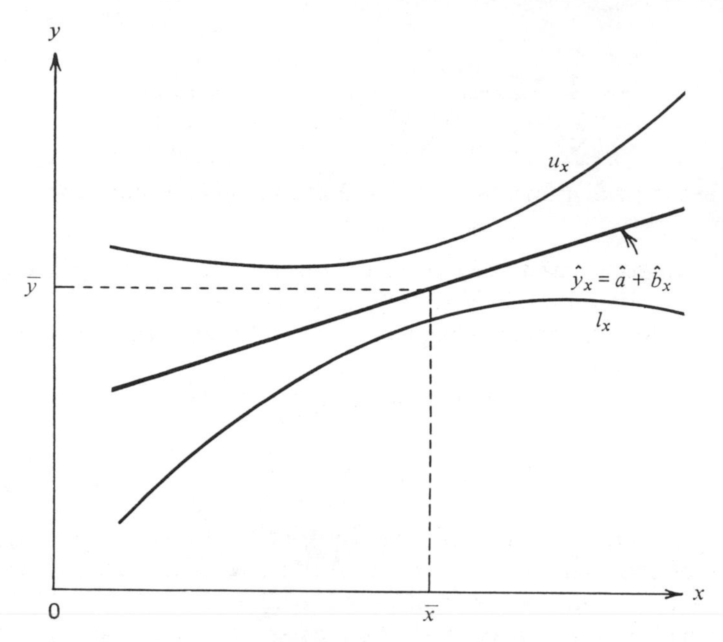


**Figure 10.3.**

form a 100 $(1 - \alpha)\%$ confidence interval for $a + bx$, for any given value of $x$. The observed values of these statistics can be plotted as functions of $x$, along with the observed regression line $\hat{y}_x$, and it is obvious that they will form a band about the regression line; this is the reason they are referred to as confidence bands. Note that the interval is shortest for $x = \bar{x}$ and that as $x$ moves away from $\bar{x}$, in either direction, the confidence interval for $a + bx$ gets wider. (See Figure 10.3.)

*Example 10.2.1.* For the regression of weights of babies on their ages (Example 10.1.1), we had $n = 10$, $\bar{x} = 3.7$, $\sum (x_i - \bar{x})^2 = 23.1$, $\hat{b} = 2.07$, and $\hat{a} = 7.94$. By assuming that weights of babies of any given age $x$ are normally distributed, we can put confidence intervals about $a$, $b$, and $a + bx$. To do this we need to compute $s^2$ from the data given in Example 10.1.1; we find that

$$\begin{aligned} s^2 &= \tfrac{1}{8}[\sum (y_i - \bar{y})^2 - \hat{b}^2 \sum (x_i - \bar{x})^2] \\ &= \tfrac{1}{8}[111.9 - 98.91] = 1.624 \\ s &= \sqrt{1.624} = 1.274. \end{aligned}$$

Then the observed 95% confidence limits for $b$ are $2.07 - (2.306)(1.274)\left(\frac{1}{4.806}\right) = 1.49$ and $2.07 + (2.306)(1.274)\left(\frac{1}{4.806}\right) = 2.68$. The estimated standard deviation of our estimate of $a$ is

$$s\sqrt{\frac{\sum x_i^2}{n\sum(x_i-\bar{x})^2}} = (1.274)(.931) = 1.186;$$

thus the observed 95% confidence limits for $a$ are $7.94 - (2.306)(1.186) = 5.190$ and $7.94 + (2.306)(1.186) = 10.689$. The observed 95% confidence bands for the average weight of a baby age $x$ months (for $1 \le x \le 6$) are

$$7.94 + 2.07x - 2.306\sqrt{\frac{1}{10} + \frac{(x-3.7)^2}{23.1}}$$

and

$$7.94 + 2.07x + 2.306\sqrt{\frac{1}{10} + \frac{(x-3.7)^2}{23.1}};$$

thus we are 95% sure that the interval (8.52, 11.50) covers the average weight of $x = 1$-month-old babies. Similarly, we are 95% sure that the interval (14.52, 16.00) covers the average weight of 4-month-old babies and that the interval (19.04, 21.68) covers the average weight of 6-month-old babies.

Tests of hypotheses about the values of $a$ and $b$ are easily carried out, based on the distributional result given in Theorem 10.2.2. Suppose that we wanted to test $H_0{:}b = b_0$ versus $H_1{:}b \ne b_0$, where $b_0$ is some specified value. The likelihood ratio test criterion (Chapter 8) then says we should reject $H_0$ if $|\hat{b} - b_0|$ exceeds

$$t_{1-\alpha/2}s\sqrt{\frac{1}{\sum(x_i-\bar{x})^2}},$$

where $t_{1-\alpha/2}$ is the $100\,(1-\alpha/2)$ percentile of the $t$-distribution with $n-2$ degrees of freedom (since we have the two-sided alternative). Similarly, we can easily construct the likelihood ratio test criterion that $a$ takes on any specified value, using the fact that on the basis of a sample of $n$ observations the estimator $\hat{A}$ is normally distributed with mean $a$ and variance that can be independently estimated from the sample values (this also turns out to be a $t$-test, of course). A frequently tested hypothesis for least-squares models is that $b = 0$; this would test that there is no relation between the mean of $Y$ and the independent variable $x$ (since if the slope is zero, the mean of $Y$ is a constant for all $x$). The following example is of this type.

*Example 10.2.2.* In Example 10.1.2, we assumed that the cumulative grade point average for students at the end of their freshman year was linearly related to the score $x$ which they made on an entrance exam. We shall now use that data to check the assumption that these two variables are related. Specifically, it would not seem logical that there exists a decreasing relation between $y$ and $x$ (the higher the entrance exam score $x$, the lower the resulting grade point average); thus let us test $H_0:b = 0$ versus $H_1:b > 0$ with $\alpha = .1$. Since $\hat{B}$ is normally distributed with mean $b$ and standard deviation

$$\sigma\sqrt{\frac{1}{\sum (x_i - \bar{x})^2}},$$

we know that the uniformly most powerful test of this hypothesis versus this one-sided alternative is given by rejecting $H_0$ only if $\hat{b}$ exceeds

$$t_{.9}s\sqrt{\frac{1}{\sum (x_i - \bar{x})^2}},$$

where $t_{.9}$ is the 90-th percentile of the $t$-distribution with 6 degrees of freedom (and equals 1.440). From the data given in Example 10.1.2, we find that $\hat{b} = .030$,

$$s\sqrt{\frac{1}{\sum (x_i - \bar{x})^2}} = \sqrt{.0000439} = .0066;$$

since $.030 > (1.44)(.0066) = .0095$, we reject $H_0$. The probability that we are wrong in so rejecting is .1.

*Example 10.2.3.* A new type of scale has been produced which is designed to give weights accurately (in thousandths of pounds) of objects whose true weights can be as large as half a pound. We have available 5 objects whose true weights are known to be .1, .2, .3, .4, and .5 pounds, respectively. We put each of these objects on the scale twice and record the reading obtained. If we assume that the recorded weight on the scale is a normal random variable $Y$ with mean $bx$, where $x$ is the true weight, and constant variance $\sigma^2$, we can use these 10 weighings to test the hypothesis $H_0:b = 1$ versus $H_1:b \neq 1$, with $\alpha = .05$, for example. If $b \neq 1$, then the scale has a bias in its readings. (The scale would certainly not be of much use if we rejected $H_0$.) The 10 readings are as recorded below. We find that $\sum x_i = 3.0$,

| $x_i$ | $y_i$ |
|---|---|
| .1 | .098 |
| .1 | .099 |
| .2 | .208 |
| .2 | .200 |
| .3 | .302 |
| .3 | .298 |
| .4 | .405 |
| .4 | .401 |
| .5 | .502 |
| .5 | .495 |

$\sum x_i^2 = 1.1$, $\sum y_i = 3.008$, $\sum y_i^2 = 1.105$, and $\sum x_iy_i = 1.102$. From problem 4, Section 10.1, the maximum likelihood (least-squares) estimate of $b$ is

$$\hat{b} = \frac{\sum x_iy_i}{\sum x_i^2} = 1.002;$$

the unbiased estimate of $\sigma^2$ is

$$\hat{s}^2 = \frac{1}{n-1}\sum (y_i - bx_i)^2 = \frac{1}{n-1}\left[\sum y_i^2 - b^2 \sum x_i^2\right] = .000014.$$

Then

$$\hat{B} = \frac{\sum x_i Y_i}{\sum x_i^2}, \qquad V(\hat{B}) = \sigma^2 \sum \frac{x_i^2}{(\sum x_i^2)^2} = \frac{\sigma^2}{\sum x_i^2};$$

now

$$\frac{\hat{B} - b}{\sqrt{\dfrac{1}{\sum x_i^2}}}$$

has the $t$-distribution with 9 degrees of freedom and, using the likelihood ratio criterion for a two-sided test about the mean of a normal, we should reject $H_0$ if

$$|\hat{b} - 1| \geq 2.262s\sqrt{\frac{1}{\sum x_i^2}}$$

(2.262 is the 97.5 percentile of the $t$-distribution with 9 degrees of freedom). We find that

$$2.262s\sqrt{\frac{1}{\sum x_i^2}} = .0081$$

and $|\hat{b} - 1| = .002$; thus we accept $H_0$.

**EXERCISE 10.2.**

**1.** Assume that the running time to failure of an electron tube is an exponential random variable $Y$ with mean $a + bx$, where $x$ is the number of times the tube has been switched on. Assume that we have a random sample of lives of $m$ tubes $(y_1, y_2, \ldots, y_m)$, each of which has been switched on $x_1$ times, and a random sample of lives of $n$ tubes $(z_1, z_2, \ldots, z_n)$, each of which has been switched on $x_2$ times. Compute the maximum likelihood estimates of $a$ and $b$ and compare with the least squares estimates. Why might these be preferred to the least squares estimates?

**2.** Using the data on average salaries versus education in problem 3, Section 10.1, test the hypothesis that $b = 0$ versus the alternative that $b > 0$, with $\alpha = .01$.

**3.** Compute a 90% confidence interval for $a$ and a 90% confidence interval for $b$ for the height versus rainfall data of problem 1, Section 10.1.

**4.** Assume that the number of sales of a given item in a large department store per week is a Poisson random variable with mean $a + bx$, where $x$ is the amount of money spent on advertising the item during the preceding week. Assume that for $m$ weeks the advertising budget for this item was constant at $x_1$ dollars and that the weekly numbers of sales made with the budget at this level were $y_1, y_2, \ldots, y_m$.

Then some time ago the advertising budget was increased to $x_2$ dollars; the numbers of sales made during $n$ weeks after the new budget was adopted were $z_1, z_2, \ldots, z_n$. Compute the maximum likelihood estimates of $a$ and $b$ and compare with the least squares estimates. Which of these would you prefer and why?

**5.** For the scale data presented in Example 10.2.3, assume that $Y$ is normal with mean $a + bx$ and constant variance $\sigma^2$ and estimate $a$, $b$, and $\sigma^2$; test the hypothesis that $a = 0$ versus the alternative that $a \neq 0$, with $\alpha = .1$.

**6.** The question of precision of a scale is related to the $\sigma^2$ of the distribution of $Y$. If $\sigma^2$ gets too large, the individual readings would vary a great deal about the true weight, assuming no bias; thus we would not put much faith in the readings. Suppose that it has been decided that the scale in Example 10.2.3 is precise enough if $\sigma^2 \leq .00001$. Use the data presented in the example to test $H_0: \sigma^2 \leq .00001$ versus $H_1: \sigma^2 > .00001$ with $\alpha = .05$.

**7.** Assume that $Y$ is normal with mean $a + bx$ and variance $x\sigma^2$ for $x > 0$. Given a random sample of $n$ observations of $Y$, with their associated $x$ values, compute the maximum likelihood estimators of $a$, $b$, and $\sigma^2$.

### 10.3. Nonlinear Models

We have examined problems of estimation and hypothesis testing when a normal random variable $Y$ has mean $a + bx$, for a given value of $x$, and variance $\sigma^2$. If the mean value of $Y$ is a nonlinear function of $x$, then the least-squares criteria or the maximum likelihood criteria can still be employed, but the equations to be solved for the estimates may not be as easy to handle. No matter whether the mean of $Y$ is linearly related to $x$, however, the least squares estimates and the maximum likelihood estimates of the parameters in the mean will be identical. To see that this is the case, suppose that $Y$ is normally distributed with mean $m(x)$ and variance $\sigma^2$, where $m(x)$ is some function of $x$ which includes unknown parameters $b_0, b_1, \ldots, b_k$. Then, assuming that $(y_1, x_1), (y_2, x_2), \ldots, (y_n, x_n)$ are the observed values of a random sample of $Y$, with their associated $x$ values, the least squares estimates of $b_0, b_1, \ldots, b_k$ would be determined by minimizing

$$\sum (y_i - m(x_i))^2$$

simultaneously with respect to $b_0, b_1, \ldots, b_k$. The maximum likelihood estimators of $b_0, b_1, \ldots, b_k$ would be determined by maximizing

$$L(b_0, b_1, \ldots, b_k, \sigma^2) = \left(\frac{1}{2\pi\sigma^2}\right)^{n/2} e^{-\Sigma(y_i - m(x_i))^2/2\sigma^2};$$

$L$ will clearly be maximized, with respect to $b_0, b_1, \ldots, b_k$, by the values that minimize the exponent on $e$. But these are the least squares values, by definition, so the two will always coincide.

If $E[Y \mid x] = m(x)$ is linear in the unknown parameters, regardless of whether it is linear in $x$ or not, then we have what is called a linear statistical model. Examples of linear statistical models are

$$m_1(x) = a + bx$$

$$m_2(x) = b_0 + b_1 x + b_2 x^2$$

$$m_3(x_1, x_2) = b_0 + b_1 x_1 + b_2 x_2 + b_3 x_1 x_2.$$

All of these cases (or any other linear model), employing least squares criteria, are easily handled by using matrix theory which we have not assumed in this book. To see why more advanced mathematical techniques might be called for when treating linear models involving more than two parameters, let us examine the normal (least squares) equations for the quadratic model $m_2(x)$ given above. We assume that $(y_1, x_1)$, $(y_2, x_2)$, . . . , $(y_n, x_n)$ are the observed values of a random sample of a random variable $Y$, with their associated $x$ values, that

$$E[Y \mid x] = b_0 + b_1 x + b_2 x^2,$$

and that the variance of $Y$ is $\sigma^2$ for any $x$. The quantity to be minimized, with respect to $b_0$, $b_1$, $b_2$, is

$$Q = \sum (y_i - b_0 - b_1 x - b_2 x^2)^2.$$

Taking the partial derivatives of $Q$ with respect to $b_0$, $b_1$, and $b_2$ leads to the normal equations

$$n\hat{b}_0 + \hat{b}_1 \sum x_i + \hat{b}_2 \sum x_i^2 = \sum y_i$$

$$\hat{b}_0 \sum x_i + \hat{b}_1 \sum x_i^2 + \hat{b}_2 \sum x_i^3 = \sum x_i y_i$$

$$\hat{b}_0 \sum x_i^2 + \hat{b}_1 \sum x_i^3 + \hat{b}_2 \sum x_i^4 = \sum x_i^2 y_i.$$

Thus we would need to solve three linear equations in the three unknowns $\hat{b}_0$, $\hat{b}_1$, and $\hat{b}_2$ to find the least squares or maximum likelihood estimates. Without the use of matrix theory, it becomes very tedious to exhibit the solutions to such equations as functions of the $x_i$'s and $y_i$'s. We shall say nothing more about such higher order linear models. The interested reader is invited to consult the references, especially the books by Graybill and by Zehna; with the use of matrix theory the least squares solutions take extremely simple form and the higher order linear models do not present any special difficulty.

If $E[Y \mid x] = m(x)$ is a nonlinear function of the parameters, then we have what would be called a nonlinear statistical model. Examples of nonlinear statistical models are

$$m_1(x) = b + b^2x$$
$$m_2(x) = \cos a + x \sin b$$
$$m_3(x) = ae^{bx};$$

the normal equations are generally nonlinear in the unknowns and thus do not admit easy solutions for the estimates. In the case of $m_1(x)$, above, we would want to minimize

$$\sum (y_i - b - b^2x_i)^2.$$

By taking the derivative with respect to $b$ and setting it equal to zero, we find the equation we need to solve:

$$\hat{b}^3 \sum x_i^2 + \tfrac{3}{2}\hat{b}^2 \sum x_i + \left(\frac{n}{2} - \sum x_i y_i\right)\hat{b} - \tfrac{1}{2} \sum y_i = 0.$$

Such a cubic equation can be solved by standard methods, of course, but the point is that the normal equations themselves become nonlinear, if $m(x)$ is not linear in the parameters, and this frequently leads to technical difficulties. The possible variety of nonlinear models is almost endless and, partly because of this, such models have not been investigated to any great extent in the past. With the advent of modern computers and the related technology, however, the stumbling block of nonlinear equations is disappearing and, where appropriate, such models are bound to be assumed.

# References

The following is a selected list of texts which contain information on various aspects of probability theory and statistics.


**Probability**

Feller, William. *An Introduction to Probability Theory and its Applications*, Vol. 1 (1968, third edition) and Vol. 2 (1966). John Wiley, New York.

Parzen, E. *Modern Probability Theory and its Applications* (1960). John Wiley, New York.

**Probability and Mathematical Statistics**

Anderson, R. L. and T. A. Bancroft. *Statistical Theory in Research* (1952). McGraw-Hill, New York.

Fisz, Marek. *Probability Theory and Mathematical Statistics* (1963, third edition). John Wiley, New York.

Graybill, F. A. *An Introduction to Linear Statistical Models*, Vol. I (1961). McGraw-Hill, New York.

Hoel, Paul G. *Introduction to Mathematical Statistics* (1962, third edition). John Wiley, New York.

Hogg, R. V. and A. T. Craig. *Introduction to Mathematical Statistics* (1965, second edition). Macmillan, New York.

Mood, A. M. and F. A. Graybill. *Introduction to the Theory of Statistics* (1963, second edition). McGraw-Hill, New York.

Rao, C. R. *Linear Statistical Inference and its Applications* (1965). John Wiley, New York.

Wilks, S. S. *Mathematical Statistics* (1962). John Wiley, New York.

Zehna, Peter W. *Probability Distributions and Statistics* (1969). Allyn and Bacon, Inc., Boston.

**Methods of Statistics**

Brownlee, K. A. *Statistical Theory and Methodology in Science and Engineering* (1965, second edition). John Wiley, New York.

Ostle, Bernard. *Statistics in Research* (1964, second edition). Iowa State Univ. Press, Ames, Iowa.

# I

# Appendix: Summation and Product Notations

### A1.1. Summation Notation

We shall frequently want a shorthand notation both to indicate sums of numbers and to allow manipulation of such sums easily. The Greek letter $\sum$ is used to indicate the summation operation. The symbol representing the quantities to be summed occurs immediately to the right of the $\sum$, properly indexed to represent all of the terms to be summed together. Below the $\sum$ is given the lowest index value to be represented in the sum, the convention being to increase the index by 1 successively until it is equal to the value given on top of the $\sum$ and to add together all terms corresponding to these values of the index. Thus

$$\sum_{i=1}^{4} x_i = x_1 + x_2 + x_3 + x_4$$

$$\sum_{i=5}^{8} y_i = y_5 + y_6 + y_7 + y_8$$

$$\sum_{k=m}^{n} x_k = x_m + x_{m+1} + x_{m+2} + \cdots + x_n.$$

It is not necessary that the index of summation be used only as a subscript. For example, if we want to add together the first through fifth powers of 2, we can represent this operation by $\sum_{i=1}^{5} 2^i$. More generally,

$$\sum_{i=3}^{5} x^i = x^3 + x^4 + x^5 = \sum_{i=1}^{3} x^{i+2} = \sum_{j=2}^{4} x^{j+1}.$$

The number of terms to be added together is clearly given by the largest value of the index represented in the sum less the smallest value of the index in the sum plus 1. Thus there are $n - 1 + 1 = n$ terms in $\sum_{i=1}^{n} x_i$ and $n - m + 1$ terms in $\sum_{i=m}^{n} x_i$. We can easily derive the following simple results regarding the summation notation (any quantity not depending on the index of summation in any way is a constant with respect to the summation).

$$\sum_{i=1}^{n} cx_i = cx_1 + cx_2 + \cdots + cx_n$$

$$= c(x_1 + x_2 + \cdots + x_n) = c \sum_{i=1}^{n} x_i$$

$$\sum_{i=1}^{n} c = c + c + \cdots + c = nc$$

$$\sum_{i=1}^{n} (x_i + y_i) = (x_1 + y_1) + (x_2 + y_2) + \cdots + (x_n + y_n)$$

$$= (x_1 + x_2 + \cdots + x_n) + (y_1 + y_2 + \cdots + y_n)$$

$$= \sum_{i=1}^{n} x_i + \sum_{i=1}^{n} y_i.$$

We shall indicate the summation over the full range of the index by not indicating any range. Thus

$$\sum x_i = x_1 + x_2 + \cdots + x_n$$

when it is understood that the range of $i$ is the integers 1, 2, 3, . . . , $n$.

We shall also have need of double sums occasionally. The rules that govern them are exactly the same as those just discussed. Thus

$$\sum_{i=1}^{4} \sum_{j=1}^{2} x_{ij} = \sum_{i=1}^{4} (x_{i1} + x_{i2})$$

$$= (x_{11} + x_{12}) + (x_{21} + x_{22}) + (x_{31} + x_{32}) + (x_{41} + x_{42})$$

$$\sum_{i=1}^{3} \sum_{j=1}^{2} x_i y_j = \sum_{i=1}^{3} x_i(y_1 + y_2) = x_1(y_1 + y_2) + x_2(y_1 + y_2) + x_3(y_1 + y_2)$$

$$= (x_1 + x_2 + x_3)(y_1 + y_2)$$

$$= \left(\sum_{i=1}^{3} x_i\right)\left(\sum_{j=i}^{2} y_j\right).$$

A double sum is most easily read or understood (when the two limits of summation are independent of each other) if you read from the inside to the outside; that is, perform the most interior summation first and then the most exterior summation last, as was done in both examples above. If the two limits of summation are completely independent (that is, the limits of summation of $j$ do not depend on the summation index $i$ and vice versa), then it clearly does not matter whether the interior or exterior summation is done first. If the limits of summation of $j$ depend upon $i$, then the summation on $i$ must of course be accomplished first. For example,

$$\sum_{i=1}^{3}\sum_{j=i}^{4} x_{ij} = \sum_{j=1}^{4} x_{1j} + \sum_{j=2}^{4} x_{2j} + \sum_{j=3}^{4} x_{3j},$$

$$\sum_{i=4}^{5}\sum_{j=i-1}^{i+1} x_i y_j = x_4 \sum_{j=3}^{5} y_j + x_5 \sum_{j=4}^{6} y_j.$$

It would not make any sense for the limits of $i$ to involve the index $j$ and, at the same time, the limits of $j$ to involve $i$, so this case will never occur. The extension to three or more summations is easily made.

*Example A1.1.1.* The sum of the first $n$ integers is

$$\frac{n(n+1)}{2}; \qquad \text{i.e., } \sum_{i=1}^{n} i = \frac{n(n+1)}{2}.$$

To see that this is in fact correct, notice that

$$\begin{aligned}
2\sum_{i=1}^{n} i &= \sum_{i=1}^{n} i + \sum_{i=1}^{n} i \\
&= (1 + 2 + 3 + \cdots + n) + (1 + 2 + 3 + \cdots + n) \\
&= (1 + 2 + 3 + \cdots + n) + (n + (n-1) + (n-2) + \cdots + 1) \\
&= [1 + n] + [2 + (n-1)] + [3 + (n-2)] + \cdots + [n + 1] \\
&= [n + 1] + [n + 1] + [n + 1] + \cdots + [n + 1] \\
&= \sum_{i=1}^{n} [n + 1] = n(n + 1).
\end{aligned}$$

Thus, dividing by 2,

$$\sum_{i=1}^{n} i = \frac{n(n+1)}{2}.$$

Then the sum of the first 5 integers is $\frac{5 \cdot 6}{2} = 15$, the sum of the first 50 integers is $\frac{50 \cdot 51}{2} = 1275$, etc.

*Example A1.1.2.* Suppose that we record the heights and the weights of 10 boys. The height of the first boy is $x_1$, of the second is $x_2, \ldots,$ and of the tenth is $x_{10}$. The weight of the first boy is $y_1$, of the second is $y_2, \ldots,$ and of the tenth boy is $y_{10}$. The arithmetic average or mean of the 10 heights is the sum of the heights divided by 10; if we let $\bar{x}$ be this arithmetic average, then $\bar{x} = \frac{1}{10} \sum x_i$ where the range of $i$ is $1, 2, \ldots, 10$. If we let $\bar{y}$ be the arithmetic average of the weights, we also have $\bar{y} = \frac{1}{10} \sum y_i$. No matter what the particular values of $x_1, x_2, \ldots, x_{10}$ are, we have

$$\begin{aligned} \sum (x_i - \bar{x}) &= \sum x_i - \sum \bar{x} \\ &= \sum x_i - 10\bar{x} \\ &= \sum x_i - \sum x_i = 0. \end{aligned}$$

Thus the sum of deviations of the individual heights about their mean value must be zero. Similarly, $\sum (y_i - \bar{y}) = 0$. The sum of squares of the deviations of the heights about their mean value is

$$\begin{aligned} \sum (x_i - \bar{x})^2 &= \sum (x_i^2 - 2x_i\bar{x} + \bar{x}^2) \\ &= \sum x_i^2 + \sum (-2x_i\bar{x}) + \sum \bar{x}^2 \\ &= \sum x_i^2 - 2\bar{x} \sum x_i + 10\bar{x}^2 \\ &= \sum x_i^2 - 2\bar{x}(10\bar{x}) + 10\bar{x}^2 \\ &= \sum x_i^2 - 10\bar{x}^2. \end{aligned}$$

This says that the sum of the squares of the deviations of the original heights about their mean value is equal to the sum of the squares of the original heights less 10 times the square of their mean. Similarly we would have

$$\sum (y_i - \bar{y})^2 = \sum y_i^2 - 10\bar{y}^2.$$

### A1.2. Product Notation

Capital pi ($\prod$) is used to indicate products in a manner quite analogous to the summation notation just discussed. The symbol for a general term in the product occurs to the right of the $\prod$, the first value of the product index is given below the $\prod$, and the final value of the index is given above the $\prod$. Thus

$$\prod_{i=1}^{n} X_i = X_1 \cdot X_2 \cdot X_3 \cdots X_n$$

$$\prod_{j=m}^{n} Y_j = Y_m \cdot Y_{m+1} \cdots Y_n.$$

As with the summation notation, any quantity that does not depend on the product index is a constant with respect to the product. We note the following results:

$$\prod_{i=1}^{n} cX_i = (cX_1)\cdot(cX_2)\cdots(cX_n)$$
$$= c^n \prod_{i=1}^{n} X_i$$

$$\prod_{i=1}^{n} c = c^n$$

$$\left[\prod_{i=1}^{n} X_i\right]\left[\prod_{j=1}^{n} Y_j\right] = \prod_{i=1}^{n} X_i Y_i$$

$$\prod_{i=1}^{n} c^{X_i} = c^{X_1}\cdot c^{X_2}\cdots c^{X_n} = c^{\sum_{i=1}^{n} X_i}$$

$$\prod_{i=1}^{n} \frac{X_i}{Y_i} = \frac{X_1}{Y_1}\cdot\frac{X_2}{Y_2}\cdots\frac{X_n}{Y_n} = \frac{\prod X_i}{\prod Y_i}.$$

## EXERCISE A1.1.

**1.** Let $x_1 = 4$, $x_2 = 3$, $x_3 = 5$, $x_4 = 7$, $x_5 = 0$, $x_6 = -2$, $x_7 = 8$, and $x_8 = 6$, and define $y_i = x_i + 1$. Evaluate each of the sums

(a) $\sum_{i=1}^{8} x_i$

(b) $\sum_{j=1}^{8} y_j$

(c) $\sum_{k=1}^{8} \frac{1}{y_k}$

(d) $\sum_{i=1}^{8} \frac{x_i}{y_i}$

(e) $\sum_{i=5}^{7} x_i$

(f) $\sum_{j=6}^{8} x_j^{j-3}$

(g) $\sum_{i=1}^{4} x_i y_i$

(h) $\sum_{i=3}^{5} (x_i^2 + y_{i-1}^2)$.

**2.** Suppose that we measure the heights and weights of 4 boys and find that $x_1 = 70$, $x_2 = 68$, $x_3 = 65$, $x_4 = 72$, $y_1 = 160$, $y_2 = 140$, $y_3 = 170$, and $y_4 = 150$ where the $x_i$'s are the heights (in inches) and the $y_i$'s are the weights (in pounds). Compute the average height $\bar{x}$, the average weight $\bar{y}$, and $\sum (x_i - \bar{x})^2$ and $\sum (y_i - \bar{y})^2$. Verify that

$$\sum (x_i - \bar{x}) = \sum (y_i - \bar{y}) = 0.$$

**3.** Give an example to show that

$$\sum \frac{x_i}{y_i} \neq \frac{\sum x_i}{\sum y_i}.$$

**4.** Give an example to show that $\sum x_i^2 \neq \sum x_i$.

**5.** Give an example to show that

$$\sum \frac{1}{x_i} \neq \frac{1}{\sum x_i}$$

**6.** Show that, for a general $n$,

$$\sum (x_i - \bar{x}) = 0, \qquad \text{where } \bar{x} = \frac{1}{n} \sum x_i.$$

**7.** Show that, for a general $n$,

$$\sum (x_i - \bar{x})^2 = \sum x_i^2 - n\bar{x}^2, \qquad \text{where } \bar{x} = \frac{1}{n} \sum x_i.$$

**8.** Suppose that we have $2n$ numbers $x_1, x_2, \ldots, x_n, y_1, y_2, \ldots, y_n$. Show that

(a) $\sum (x_i - \bar{x})(y_i - \bar{y}) = \sum x_i(y_i - \bar{y})$
(b) $\sum (x_i - \bar{x})(y_i - \bar{y}) = \sum x_i y_i - n\bar{x}\bar{y}$,

where

$$\bar{x} = \frac{1}{n} \sum x_i, \qquad \bar{y} = \frac{1}{n} \sum y_i.$$

# 2

# Appendix: The Binomial Theorem

Most courses in high school algebra prove that for any $x$ and any $y$

$$(x + y)^2 = x^2 + 2xy + y^2$$
$$(x + y)^3 = x^3 + 3x^2y + 3xy^2 + y^3.$$

The binomial theorem is concerned with proving the expansion of $(x + y)^n$ in sums of terms of the form $x^k y^{n-k}$ for $n$ equalling any positive integer. We shall prove by induction that

$$(x + y)^n = \sum_{i=0}^{n} \binom{n}{i} x^i y^{n-i}, \qquad \text{where } \binom{n}{i} = \frac{n!}{(n-i)!\, i!}.$$

We note first of all that

$$(x + y)^1 = x + y = \binom{1}{0} x^0 y^{1-0} + \binom{1}{1} x^1 y^{1-1} = \sum_{i=0}^{1} \binom{1}{i} x^i y^{1-i}.$$

Now we assume that

$$(x + y)^k = \sum_{i=0}^{k} \binom{k}{i} x^i y^{k-i}$$

and wish to use only this assumption to establish that

$$(x + y)^{k+1} = \sum_{i=0}^{k+1} \binom{k+1}{i} x^i y^{k+1-i}.$$

Now

$$(x + y)^{k+1} = (x + y)(x + y)^k$$
$$= (x + y) \sum_{i=0}^{k} \binom{k}{i} x^i y^{k-i}$$
$$= \sum_{i=0}^{k} \binom{k}{i} x^{i+1} y^{k-i} + \sum_{i=0}^{k} \binom{k}{i} x^i y^{k+1-i}.$$

Combining like terms in these two sums, we have

$$(x + y)^{k+1} = y^{k+1} + \sum_{i=1}^{k} \left[ \binom{k}{i-1} + \binom{k}{i} \right] x^i y^{k+1-i} + x^{k+1}.$$

Now

$$\binom{k}{i-1} + \binom{k}{i} = \frac{k!}{(k+1-i)!\,(i-1)!} + \frac{k!}{(k-i)!\,i!}$$
$$= \frac{i \cdot k! + (k+1-i)k!}{(k+1-i)!\,i!}$$
$$= \frac{(k+1)k!}{(k+1-i)!\,i!} = \binom{k+1}{i}.$$

Since $\binom{k+1}{0} = \binom{k+1}{k+1} = 1$, we have

$$(x + y)^{k+1} = \sum_{i=0}^{k+1} \binom{k+1}{i} x^i y^{k+1-i}$$

and we thus have established that

$$(x + y)^n = \sum_{i=0}^{n} \binom{n}{i} x^i y^{n-i}, \tag{A2.1}$$

for $n = 2, 3, 4, 5, \ldots$.

We note here also that

$$0 = (1 - 1)^n = \sum_{i=0}^{n} \binom{n}{i} (1)^i (-1)^{n-i}$$
$$= \binom{n}{0} - \binom{n}{1} + \binom{n}{2} - \binom{n}{3} + \cdots + \binom{n}{n},$$

if $n$ is even, a result that is occasionally of use.

Two results which are of interest in studying the binomial random variable (see Section 4.1) are:

$$\begin{aligned}
\sum_{i=0}^{n} i \binom{n}{i} x^i y^{n-i} &= \sum_{i=1}^{n} i \frac{n!}{(n-i)!\, i!} x^i y^{n-i} \\
&= \sum_{i=1}^{n} \frac{n(n-1)!}{(n-i)!\,(i-1)!} x^i y^{n-i} \\
&= nx \sum_{i=1}^{n} \frac{(n-1)!}{(n-i)!\,(i-1)!} x^{i-1} y^{n-i} \\
&= nx \sum_{i-1=0}^{n-1} \binom{n-1}{i-1} x^{i-1} y^{n-i} \\
&= nx(x+y)^{n-1} \qquad \text{(A2.2)}
\end{aligned}$$

and

$$\begin{aligned}
\sum_{i=0}^{n} i(i-1) \binom{n}{i} x^i y^{n-i} &= \sum_{i=2}^{n} i(i-1) \frac{n!}{(n-i)!\, i!} x^i y^{n-i} \\
&= \sum_{i=2}^{n} \frac{n(n-1)(n-2)!}{(n-i)!\,(i-2)!} x^i y^{n-i} \\
&= n(n-1)x^2 \sum_{i=2}^{n} \frac{(n-2)!}{(n-i)!\,(i-2)!} x^{i-2} y^{n-i} \\
&= n(n-1)x^2 \sum_{i-2=0}^{n-2} \binom{n-2}{i-2} x^{i-2} y^{n-i} \\
&= n(n-1)x^2(x+y)^{n-2}. \qquad \text{(A2.3)}
\end{aligned}$$

# 3

# Appendix: Infinity, Infinite Sums, and Geometric Progressions

The student becomes acquainted with the concept of infinity in his calculus course. We say that a set has an infinite number of elements if there is no integer $M$ larger than the number of elements in the set. Perhaps the easiest example of an infinite set is the set of positive integers themselves. Rather obviously there is no integer $M$ larger than the number of integers since, if we are given any such $M$, we know that $M + 1$, $M + 2$, etc., are also integers. Thus the number of integers that there are must exceed $M$, no matter how large $M$ might be.

We shall have need to distinguish between two different orders of infinity—a countable infinity and an uncountable infinity.

DEFINITION A3.1. A set $S$ contains a *countable infinity* of elements if we can define a 1 to 1 correspondence between the elements of $S$ and the set of positive integers.

A set containing a countable infinity (also called a denumerable infinity) of elements is a particular example of a discrete set and is called a countable or denumerable set. It can be shown that each of the following sets is countable: the set of all integers, the set of even integers, the set of rational numbers, and the set of rational numbers between 0 and 1.

The second order of infinity which we shall need to refer to occasionally is called an uncountable (or nondenumerable) infinity. This can be defined in the following way.

DEFINITION A3.2. A set $S$ contains an *uncountable infinity* of elements if: (1) $S$ is an infinite set and (2) we cannot set up a 1 to 1 correspondence between the elements of $S$ and the set of positive integers.

Generally, sets with an uncountable number of elements will be all the points in some continuous interval on the real line. It can be shown that each of the following sets is uncountably infinite: the set of all real positive numbers, the set of all real numbers, the set of irrational numbers, and the set of all real numbers in any finite length interval from $a$ to $b$.

As is usual in calculus, if we have an infinite sequence of quantities $a_1$, $a_2$, $a_3, \ldots$, and say that the sum of these quantities is $A$

$$\left(\sum_{i=1}^{\infty} a_i = A\right),$$

we mean that the sequence of partial sums converges to $A$; that is, terms in the sequence

$$a_1, a_1 + a_2, a_1 + a_2 + a_3, \ldots, \sum_{i=1}^{n} a_i, \ldots$$

get closer and closer to $A$ the farther along in the sequence you go.

A particular infinite series we shall have use for is the *geometric progression*. For any real number $r$, we can easily show that

$$\sum_{i=0}^{n} r^i = 1 + r + r^2 + \cdots + r^n = \frac{1 - r^{n+1}}{1 - r} = \frac{1}{1 - r} - \frac{r^{n+1}}{1 - r}.$$

If $|r| < 1$, then $\lim_{n \to \infty} r^{n+1} = 0$ and thus

$$\sum_{i=0}^{\infty} r^i = \lim_{n \to \infty} \sum_{i=0}^{n} r^i = \lim_{n \to \infty} \left[\frac{1}{1 - r} - \frac{r^{n+1}}{1 - r}\right] = \frac{1}{1 - r}, \qquad \text{for } |r| < 1.$$

If we take the first and second derivatives on both sides of this equation, with respect to $r$, we arrive at the values of two additional infinite sums which are of interest in studying the geometric random variable. Thus

$$\sum_{i=0}^{\infty} i r^{i-1} = \frac{1}{(1 - r)^2}, \qquad \text{if } |r| < 1 \tag{A3.1}$$

$$\sum_{i=0}^{\infty} i(i - 1) r^{i-2} = \frac{2}{(1 - r)^3}, \qquad \text{if } |r| < 1. \tag{A3.2}$$

Taylor series expansions of functions can be used to expand many functions in infinite series. The Taylor series expansion of a function $f(x)$ about an arbitrary point $a$ is given by

$$f(x) = f(a) + (x - a)f'(a) + \frac{(x - a)^2}{2!} f''(a) + \cdots \tag{A3.3}$$

where $f'(a), f''(a), \ldots$, are the first, second, ..., derivatives of $f(x)$ evaluated at $a$. In particular, taking $a$ to be zero gives

$$f(x) = f(0) + xf'(0) + \frac{x^2}{2!} f''(0) + \cdots. \tag{A3.4}$$

One of the most useful applications of Taylor series, for our purposes, is the expansion of $e^x$ about the origin ($e$ is the base of natural logarithms). Define $f(x) = e^x$. Then

$$f'(x) = e^x, \qquad f''(x) = e^x, \qquad f'''(x) = e^x, \text{ etc.}$$

and

$$f'(0) = e^0 = 1, \qquad f''(0) = 1, \qquad f'''(0) = 1, \text{ etc.}$$

Thus we have

$$e^x = 1 + x + \frac{x^2}{2!} + \frac{x^3}{3!} + \cdots = \sum_{i=0}^{\infty} \frac{x^i}{i!}. \tag{A3.5}$$

It can be shown that this infinite series converges for any value of $x$ whatsoever.

# 4

# Appendix: Improper Integrals the Gamma Function, and the Beta Function

We shall frequently have need of integration of functions over infinite ranges; let us review a few facts concerning this. By definition,

$$\int_a^\infty f(x)\,dx = \lim_{A\to\infty}\int_a^A f(x)\,dx$$

$$\int_{-\infty}^b f(x)\,dx = \lim_{B\to-\infty}\int_B^b f(x)\,dx$$

$$\int_{-\infty}^\infty f(x)\,dx = \lim_{A\to\infty}\lim_{B\to-\infty}\int_B^A f(x)\,dx.$$

An improper integral

$$\int_{-\infty}^\infty f(x)\,dx$$

is said to be absolutely convergent if

$$\int_{-\infty}^\infty |f(x)|\,dx$$

exists.

The gamma function will be of use to us in some applications. The gamma function is defined to be

$$\Gamma(n) = \int_0^\infty x^{n-1} e^{-x}\, dx. \tag{A4.1}$$

It can be shown that $\Gamma(n)$ exists for any $n > 0$. If we use integration by parts, we can see that $\Gamma(n + 1) = n\Gamma(n)$; thus, if $n$ is a positive integer, then

$$\Gamma(n) = (n - 1)!. \tag{A4.2}$$

Note, in particular, that

$$\Gamma(1) = \int_0^\infty e^{-x}\, dx = 1$$

and thus $\Gamma(1) = (1 - 1)! = 0! = 1$. It can be shown that

$$\Gamma(\tfrac{1}{2}) = \sqrt{\pi} \tag{A4.3}$$

An additional function which we shall find useful in some statistical applications is called the beta function; it is defined to be

$$\mathrm{B}(p, q) = \int_0^1 x^{p-1}(1 - x)^{q-1}\, dx. \tag{A4.4}$$

For $p > 0$, $q > 0$, it can be shown that

$$\mathrm{B}(p, q) = \frac{\Gamma(p)\Gamma(q)}{\Gamma(p + q)}. \tag{A4.5}$$

# 5

# Appendix: Tables

**Table 5.A** ***Binomial Distribution Function:***

$$F_X(t) = \sum_{x=0}^{[t]} \binom{n}{x} p^x (1 - p)^{n-x}$$

| $n$ | $[t]$ | $p$ 0.10 | 0.20 | 0.30 | 0.40 | 0.50 |
|---|---|---|---|---|---|---|
| 2 | 0 | .8100 | .6400 | .4900 | .3600 | .2500 |
| | 1 | .9900 | .9600 | .9100 | .8400 | .7500 |
| 3 | 0 | .7290 | .5120 | .3430 | 2160 | .1250 |
| | 1 | .9720 | .8960 | .7840 | .6480 | .5000 |
| | 2 | .9990 | .9920 | .9730 | .9360 | .8750 |
| 4 | 0 | .6561 | .4096 | .2401 | .1296 | .0625 |
| | 1 | .9477 | .8192 | .6517 | .4752 | .3125 |
| | 2 | .9963 | .9728 | .9163 | .8208 | .6875 |
| | 3 | .9999 | .9984 | .9919 | .9744 | .9375 |
| 5 | 0 | .5905 | .3277 | .1681 | .0778 | .0312 |
| | 1 | .9185 | .7373 | .5282 | .3370 | .1875 |
| | 2 | .9914 | .9421 | .8369 | .6826 | .5000 |
| | 3 | .9995 | .9933 | .9692 | .9130 | .8125 |
| | 4 | 1.0000 | .9997 | .9976 | .9898 | .9688 |
| 6 | 0 | .5314 | .2621 | .1176 | .0467 | .0156 |
| | 1 | .8857 | .6554 | .4202 | .2333 | .1094 |

*(continued)*

Table 5.A ***Binomial Distribution Function*** (*continued*)

| $n$ | $[t]$ | $p$ 0.10 | 0.20 | 0.30 | 0.40 | 0.50 |
|---|---|---|---|---|---|---|
| 6 | 2 | .9842 | .9011 | .7443 | .5443 | .3438 |
| | 3 | .9987 | .9830 | .9295 | .8208 | .6562 |
| | 4 | .9999 | .9984 | .9891 | .9590 | .8906 |
| | 5 | 1.0000 | .9999 | .9993 | .9959 | .9844 |
| 7 | 0 | .4783 | .2097 | .0824 | .0280 | .0078 |
| | 1 | .8503 | .5767 | .3294 | .1586 | .0625 |
| | 2 | .9743 | .8520 | .6471 | .4199 | .2266 |
| | 3 | .9973 | .9667 | .8740 | .7102 | .5000 |
| | 4 | .9998 | .9953 | .9712 | .9037 | .7734 |
| | 5 | 1.0000 | .9996 | .9962 | .9812 | .9375 |
| | 6 | 1.0000 | 1.0000 | .9998 | .9984 | .9922 |
| 8 | 0 | .4305 | .1678 | .0576 | .0168 | .0039 |
| | 1 | .8131 | .5033 | .2553 | .1064 | .0352 |
| | 2 | .9619 | .7969 | .5518 | .3154 | .1445 |
| | 3 | .9950 | .9437 | .8059 | .5941 | .3633 |
| | 4 | .9996 | .9896 | .9420 | .8263 | .6367 |
| | 5 | 1.0000 | .9988 | .9887 | .9502 | .8555 |
| | 6 | 1.0000 | .9999 | .9987 | .9915 | .9648 |
| | 7 | 1.0000 | 1.0000 | .9999 | .9993 | .9961 |
| 9 | 0 | .3874 | .1342 | .0404 | .0101 | .0020 |
| | 1 | .7748 | .4362 | .1960 | .0705 | .0195 |
| | 2 | .9470 | .7382 | .4628 | .2318 | .0898 |
| | 3 | .9917 | .9144 | .7297 | .4826 | .2539 |
| | 4 | .9991 | .9804 | .9012 | .7334 | .5000 |
| | 5 | .9999 | .9969 | .9747 | .9006 | .7461 |
| | 6 | 1.0000 | .9997 | .9957 | .9750 | .9102 |
| | 7 | 1.0000 | 1.0000 | .9996 | .9962 | .9805 |
| | 8 | 1.0000 | 1.0000 | 1.0000 | .9997 | .9980 |
| 10 | 0 | .3487 | .1074 | .0282 | .0060 | .0010 |
| | 1 | .7361 | .3758 | .1493 | .0464 | .0107 |
| | 2 | .9298 | .6778 | .3828 | .1673 | .0547 |
| | 3 | .9872 | .8791 | .6496 | .3823 | .1719 |
| | 4 | .9984 | .9672 | .8497 | .6331 | .3770 |

**Table 5.A** ***Binomial Distribution Function*** *(continued)*

| $n$ | $[t]$ | $p$ = 0.10 | 0.20 | 0.30 | 0.40 | 0.50 |
|---|---|---|---|---|---|---|
| 10 | 5 | .9999 | .9936 | .9527 | .8338 | .6230 |
| | 6 | 1.0000 | .9991 | .9894 | .9452 | .8281 |
| | 7 | 1.0000 | .9999 | .9984 | .9877 | .9453 |
| | 8 | 1.0000 | 1.0000 | .9999 | .9983 | .9893 |
| | 9 | 1.0000 | 1.0000 | 1.0000 | .9999 | .9990 |
| 11 | 0 | .3138 | .0859 | .0198 | .0036 | .0005 |
| | 1 | .6974 | .3221 | .1130 | .0302 | .0059 |
| | 2 | .9104 | .6174 | .3127 | .1189 | .0327 |
| | 3 | .9815 | .8389 | .5696 | .2963 | .1133 |
| | 4 | .9972 | .9496 | .7897 | .5328 | .2744 |
| | 5 | .9997 | .9883 | .9218 | .7535 | .5000 |
| | 6 | 1.0000 | .9980 | .9784 | .9006 | .7256 |
| | 7 | 1.0000 | .9998 | .9957 | .9707 | .8867 |
| | 8 | 1.0000 | 1.0000 | .9994 | .9941 | .9673 |
| | 9 | 1.0000 | 1.0000 | 1.0000 | .9993 | .9941 |
| | 10 | 1.0000 | 1.0000 | 1.0000 | 1.0000 | .9995 |
| 12 | 0 | .2824 | .0687 | .0138 | .0022 | .0002 |
| | 1 | .6590 | .2749 | .0850 | .0196 | .0032 |
| | 2 | .8891 | .5583 | .2528 | .0834 | .0193 |
| | 3 | .9744 | .7946 | .4925 | .2253 | .0730 |
| | 4 | .9957 | .9274 | .7237 | .4382 | .1938 |
| | 5 | .9995 | .9806 | .8822 | .6652 | .3872 |
| | 6 | .9999 | .9961 | .9614 | .8418 | .6128 |
| | 7 | 1.0000 | .9994 | .9905 | .9427 | .8062 |
| | 8 | 1.0000 | .9999 | .9983 | .9847 | .9270 |
| | 9 | 1.0000 | 1.0000 | .9998 | .9972 | .9807 |
| | 10 | 1.0000 | 1.0000 | 1.0000 | .9997 | .9968 |
| | 11 | 1.0000 | 1.0000 | 1.0000 | 1.0000 | .9998 |
| 13 | 0 | .2542 | .0550 | .0097 | .0013 | .0001 |
| | 1 | .6213 | .2336 | .0637 | .0126 | .0017 |
| | 2 | .8661 | .5017 | .2025 | .0579 | .0112 |
| | 3 | .9658 | .7473 | .4206 | .1686 | .0461 |
| | 4 | .9935 | .9009 | .6543 | .3530 | .1334 |
| | 5 | .9991 | .9700 | .8346 | .5744 | .2905 |
| | 6 | .9999 | .9930 | .9376 | .7712 | .5000 |
| | 7 | 1.0000 | .9988 | .9818 | .9023 | .7095 |

*(continued)*

**Table 5.A** ***Binomial Distribution Function*** (*continued*)

| $n$ | $[t]$ | $p$ 0.10 | 0.20 | 0.30 | 0.40 | 0.50 |
|---|---|---|---|---|---|---|
| 13 | 8 | 1.0000 | .9998 | .9960 | .9679 | .8666 |
| | 9 | 1.0000 | 1.0000 | .9993 | .9922 | .9539 |
| | 10 | 1.0000 | 1.0000 | .9999 | .9987 | .9888 |
| | 11 | 1.0000 | 1.0000 | 1.0000 | .9999 | .9983 |
| | 12 | 1.0000 | 1.0000 | 1.0000 | 1.0000 | .9999 |
| 14 | 0 | .2288 | .0440 | .0068 | .0008 | .0001 |
| | 1 | .5846 | .1979 | .0475 | .0081 | .0009 |
| | 2 | .8416 | .4481 | .1608 | .0398 | .0065 |
| | 3 | .9559 | .6982 | .3552 | .1243 | .0287 |
| | 4 | .9908 | .8702 | .5842 | .2793 | .0898 |
| | 5 | .9985 | .9561 | .7805 | .4859 | .2120 |
| | 6 | .9998 | .9884 | .9067 | .6925 | .3953 |
| | 7 | 1.0000 | .9976 | .9685 | .8499 | .6047 |
| | 8 | 1.0000 | .9996 | .9917 | .9417 | .7880 |
| | 9 | 1.0000 | 1.0000 | .9983 | .9825 | .9102 |
| | 10 | 1.0000 | 1.0000 | .9998 | .9961 | .9713 |
| | 11 | 1.0000 | 1.0000 | 1.0000 | .9994 | .9935 |
| | 12 | 1.0000 | 1.0000 | 1.0000 | .9999 | .9991 |
| | 13 | 1.0000 | 1.0000 | 1.0000 | 1.0000 | .9999 |
| 15 | 0 | .2059 | .0352 | .0047 | .0005 | .0000 |
| | 1 | .5490 | .1671 | .0353 | .0052 | .0005 |
| | 2 | .8159 | .3980 | .1268 | .0271 | .0037 |
| | 3 | .9444 | .6482 | .2969 | .0905 | .0176 |
| | 4 | .9873 | .8358 | .5155 | .2173 | .0592 |
| | 5 | .9978 | .9389 | .7216 | .4032 | .1509 |
| | 6 | .9997 | .9819 | .8689 | .6098 | .3036 |
| | 7 | 1.0000 | .9958 | .9500 | .7869 | .5000 |
| | 8 | 1.0000 | .9992 | .9848 | .9050 | .6964 |
| | 9 | 1.0000 | .9999 | .9963 | .9662 | .8491 |
| | 10 | 1.0000 | 1.0000 | .9993 | .9907 | .9408 |
| | 11 | 1.0000 | 1.0000 | .9999 | .9981 | .9824 |
| | 12 | 1.0000 | 1.0000 | 1.0000 | .9997 | .9963 |
| | 13 | 1.0000 | 1.0000 | 1.0000 | 1.0000 | .9995 |
| | 14 | 1.0000 | 1.0000 | 1.0000 | 1.0000 | 1.0000 |

Table 5.A ***Binomial Distribution Function*** (*continued*)

| $n$ | $[t]$ | $p$ 0.10 | 0.20 | 0.30 | 0.40 | 0.50 |
|---|---|---|---|---|---|---|
| 16 | 0 | .1853 | .0281 | .0033 | .0003 | .0000 |
| | 1 | .5147 | .1407 | .0261 | .0033 | .0003 |
| | 2 | .7892 | .3518 | .0994 | .0183 | .0021 |
| | 3 | .9316 | .5981 | .2459 | .0651 | .0106 |
| | 4 | .9830 | .7982 | .4499 | .1666 | .0384 |
| | 5 | .9967 | .9183 | .6598 | .3288 | .1051 |
| | 6 | .9995 | .9733 | .8247 | .5272 | .2272 |
| | 7 | .9999 | .9930 | .9256 | .7161 | .4018 |
| | 8 | 1.0000 | .9985 | .9743 | .8577 | .5982 |
| | 9 | 1.0000 | .9998 | .9929 | .9417 | .7728 |
| | 10 | 1.0000 | 1.0000 | .9984 | .9809 | .8949 |
| | 11 | 1.0000 | 1.0000 | .9997 | .9951 | .9616 |
| | 12 | 1.0000 | 1.0000 | 1.0000 | .9991 | .9894 |
| | 13 | 1.0000 | 1.0000 | 1.0000 | .9999 | .9979 |
| | 14 | 1.0000 | 1.0000 | 1.0000 | 1.0000 | .9997 |
| | 15 | 1.0000 | 1.0000 | 1.0000 | 1.0000 | 1.0000 |
| 17 | 0 | .1668 | .0225 | .0023 | .0002 | .0000 |
| | 1 | .4818 | .1182 | .0193 | .0021 | .0001 |
| | 2 | .7618 | .3096 | .0774 | .0123 | .0012 |
| | 3 | .9174 | .5489 | .2019 | .0464 | .0064 |
| | 4 | .9779 | .7582 | .3887 | .1260 | .0245 |
| | 5 | .9953 | .8943 | .5968 | .2639 | .0717 |
| | 6 | .9992 | .9623 | .7752 | .4478 | .1662 |
| | 7 | .9999 | .9891 | .8954 | .6405 | .3145 |
| | 8 | 1.0000 | .9974 | .9597 | .8011 | .5000 |
| | 9 | 1.0000 | .9995 | .9873 | .9081 | .6855 |
| | 10 | 1.0000 | .9999 | .9968 | .9652 | .8338 |
| | 11 | 1.0000 | 1.0000 | .9993 | .9894 | .9283 |
| | 12 | 1.0000 | 1.0000 | .9999 | .9975 | .9755 |
| | 13 | 1.0000 | 1.0000 | 1.0000 | .9995 | .9936 |
| | 14 | 1.0000 | 1.0000 | 1.0000 | .9999 | .9988 |
| | 15 | 1.0000 | 1.0000 | 1.0000 | 1.0000 | .9999 |
| | 16 | 1.0000 | 1.0000 | 1.0000 | 1.0000 | 1.0000 |

(*continued*)

**Table 5.A** ***Binomial Distribution Function*** *(continued)*

| $n$ | $[t]$ | $p$ 0.10 | 0.20 | 0.30 | 0.40 | 0.50 |
|---|---|---|---|---|---|---|
| 18 | 0 | .1501 | .0180 | .0016 | .0001 | .0000 |
| | 1 | .4503 | .0991 | .0142 | .0013 | .0001 |
| | 2 | .7338 | .2713 | .0600 | .0082 | .0007 |
| | 3 | .9018 | .5010 | .1646 | .0328 | .0038 |
| | 4 | .9718 | .7164 | .3327 | .0942 | .0154 |
| | 5 | .9936 | .8671 | .5344 | .2088 | .0481 |
| | 6 | .9988 | .9487 | .7217 | .3743 | .1189 |
| | 7 | .9998 | .9837 | .8593 | .5634 | .2403 |
| | 8 | 1.0000 | .9957 | .9404 | .7368 | .4073 |
| | 9 | 1.0000 | .9991 | .9790 | .8653 | .5927 |
| | 10 | 1.0000 | .9998 | .9939 | .9424 | .7597 |
| | 11 | 1.0000 | 1.0000 | .9986 | .9797 | .8811 |
| | 12 | 1.0000 | 1.0000 | .9997 | .9942 | .9519 |
| | 13 | 1.0000 | 1.0000 | 1.0000 | .9987 | .9846 |
| | 14 | 1.0000 | 1.0000 | 1.0000 | .9998 | .9962 |
| | 15 | 1.0000 | 1.0000 | 1.0000 | 1.0000 | .9993 |
| | 16 | 1.0000 | 1.0000 | 1.0000 | 1.0000 | .9999 |
| | 17 | 1.0000 | 1.0000 | 1.0000 | 1.0000 | 1.0000 |
| 19 | 0 | .1351 | .0144 | .0011 | .0001 | .0000 |
| | 1 | .4203 | .0829 | .0104 | .0008 | .0000 |
| | 2 | .7054 | .2369 | .0462 | .0055 | .0004 |
| | 3 | .8850 | .4551 | .1332 | .0230 | .0022 |
| | 4 | .9648 | .6733 | .2822 | .0696 | .0096 |
| | 5 | .9914 | .8369 | .4739 | .1629 | .0318 |
| | 6 | .9983 | .9324 | .6655 | .3081 | .0835 |
| | 7 | .9997 | .9767 | .8180 | .4878 | .1796 |
| | 8 | 1.0000 | .9933 | .9161 | .6675 | .3238 |
| | 9 | 1.0000 | .9984 | .9674 | .8139 | .5000 |
| | 10 | 1.0000 | .9997 | .9895 | .9115 | .6762 |
| | 11 | 1.0000 | 1.0000 | .9972 | .9648 | .8204 |
| | 12 | 1.0000 | 1.0000 | .9994 | .9884 | .9165 |
| | 13 | 1.0000 | 1.0000 | .9999 | .9969 | .9682 |
| | 14 | 1.0000 | 1.0000 | 1.0000 | .9994 | .9904 |
| | 15 | 1.0000 | 1.0000 | 1.0000 | .9999 | .9978 |
| | 16 | 1.0000 | 1.0000 | 1.0000 | 1.0000 | .9996 |
| | 17 | 1.0000 | 1.0000 | 1.0000 | 1.0000 | 1.0000 |

Table 5.A ***Binomial Distribution Function*** (*continued*)

| $n$ | $[t]$ | $p$ 0.10 | 0.20 | 0.30 | 0.40 | 0.50 |
|---|---|---|---|---|---|---|
| 20 | 0 | .1216 | .0115 | .0008 | .0000 | .0000 |
| | 1 | .3917 | .0692 | .0076 | .0005 | .0000 |
| | 2 | .6769 | .2061 | .0355 | .0036 | .0002 |
| | 3 | .8670 | .4114 | .1071 | .0160 | .0013 |
| | 4 | .9568 | .6296 | .2375 | .0510 | .0059 |
| | 5 | .9887 | .8042 | .4164 | .1256 | .0207 |
| | 6 | .9976 | .9133 | .6080 | .2500 | .0577 |
| | 7 | .9996 | .9679 | .7723 | .4159 | .1316 |
| | 8 | .9999 | .9900 | .8867 | .5956 | .2517 |
| | 9 | 1.0000 | .9974 | .9520 | .7553 | .4119 |
| | 10 | 1.0000 | .9994 | .9829 | .8725 | .5881 |
| | 11 | 1.0000 | .9999 | .9949 | .9435 | .7483 |
| | 12 | 1.0000 | 1.0000 | .9987 | .9790 | .8684 |
| | 13 | 1.0000 | 1.0000 | .9997 | .9935 | .9423 |
| | 14 | 1.0000 | 1.0000 | 1.0000 | .9984 | .9793 |
| | 15 | 1.0000 | 1.0000 | 1.0000 | .9997 | .9941 |
| | 16 | 1.0000 | 1.0000 | 1.0000 | 1.0000 | .9987 |
| | 17 | 1.0000 | 1.0000 | 1.0000 | 1.0000 | .9998 |
| | 18 | 1.0000 | 1.0000 | 1.0000 | 1.0000 | 1.0000 |

**Table 5.B** ***Poisson Distribution Function:***

$$F_X(t) = \sum_{x=0}^{[t]} \frac{\lambda^x}{x!} e^{-\lambda}$$

| | $\lambda$ | | | | | | | | | |
|---|---|---|---|---|---|---|---|---|---|---|
| $[t]$ | .50 | 1.0 | 2.0 | 3.0 | 4.0 | 5.0 | 6.0 | 7.0 | 8.0 | 9.0 |
| 0 | .607 | .368 | .135 | .050 | .018 | .007 | .002 | .001 | .000 | .000 |
| 1 | .910 | .736 | .406 | .199 | .092 | .040 | .017 | .007 | .003 | .001 |
| 2 | .986 | .920 | .677 | .423 | .238 | .125 | .062 | .030 | .014 | .006 |
| 3 | .998 | .981 | .857 | .647 | .433 | .265 | .151 | .082 | .042 | .021 |
| 4 | 1.000 | .996 | .947 | .815 | .629 | .440 | .285 | .173 | .100 | .055 |
| 5 | 1.000 | .999 | .983 | .961 | .785 | .616 | .446 | .301 | .191 | .116 |
| 6 | 1.000 | 1.000 | .995 | .966 | .889 | .762 | .606 | .450 | .313 | .207 |
| 7 | 1.000 | 1.000 | .999 | .988 | .949 | .867 | .744 | .599 | .453 | .324 |
| 8 | 1.000 | 1.000 | 1.000 | .996 | .979 | .932 | .847 | .729 | .593 | .456 |
| 9 | 1.000 | 1.000 | 1.000 | .999 | .992 | .968 | .916 | .830 | .717 | .587 |
| 10 | 1.000 | 1.000 | 1.000 | 1.000 | .997 | .986 | .957 | .901 | .816 | .706 |
| 11 | 1.000 | 1.000 | 1.000 | 1.000 | .999 | .995 | .980 | .947 | .888 | .803 |
| 12 | 1.000 | 1.000 | 1.000 | 1.000 | 1.000 | .998 | .991 | .973 | .936 | .876 |
| 13 | 1.000 | 1.000 | 1.000 | 1.000 | 1.000 | .999 | .996 | .987 | .966 | .926 |
| 14 | 1.000 | 1.000 | 1.000 | 1.000 | 1.000 | 1.000 | .999 | .994 | .983 | .959 |
| 15 | 1.000 | 1.000 | 1.000 | 1.000 | 1.000 | 1.000 | .999 | .998 | .992 | .978 |
| 16 | 1.000 | 1.000 | 1.000 | 1.000 | 1.000 | 1.000 | 1.000 | .999 | .996 | .989 |
| 17 | 1.000 | 1.000 | 1.000 | 1.000 | 1.000 | 1.000 | 1.000 | 1.000 | .998 | .995 |
| 18 | 1.000 | 1.000 | 1.000 | 1.000 | 1.000 | 1.000 | 1.000 | 1.000 | .999 | .998 |
| 19 | 1.000 | 1.000 | 1.000 | 1.000 | 1.000 | 1.000 | 1.000 | 1.000 | 1.000 | .999 |
| 20 | 1.000 | 1.000 | 1.000 | 1.000 | 1.000 | 1.000 | 1.000 | 1.000 | 1.000 | 1.000 |

Table 5.B *Poisson Distribution Function* (*continued*)

| | $\lambda$ | | | | | |
|---|---|---|---|---|---|---|
| [$t$] | 10.0 | 11.0 | 12.0 | 13.0 | 14.0 | 15.0 |
| 2 | .003 | .001 | .001 | .000 | .000 | .000 |
| 3 | .010 | .005 | .002 | .001 | .000 | .000 |
| 4 | .029 | .015 | .008 | .004 | .002 | .001 |
| 5 | .067 | .038 | .020 | .011 | .006 | .003 |
| 6 | .130 | .079 | .046 | .026 | .014 | .008 |
| 7 | .220 | .143 | .090 | .054 | .032 | .018 |
| 8 | .333 | .232 | .155 | .100 | .062 | .037 |
| 9 | .458 | .341 | .242 | .166 | .109 | .070 |
| 10 | .583 | .460 | .347 | .252 | .176 | .118 |
| 11 | .697 | .579 | .462 | .353 | .260 | .185 |
| 12 | .792 | .689 | .576 | .463 | .358 | .268 |
| 13 | .864 | .781 | .682 | .573 | .464 | .363 |
| 14 | .917 | .854 | .772 | .675 | .570 | .466 |
| 15 | .951 | .907 | .844 | .764 | .669 | .568 |
| 16 | .973 | .944 | .899 | .835 | .756 | .664 |
| 17 | .986 | .968 | .937 | .890 | .827 | .749 |
| 18 | .993 | .982 | .963 | .930 | .883 | .819 |
| 19 | .997 | .991 | .979 | .957 | .923 | .875 |
| 20 | .998 | .995 | .988 | .975 | .952 | .917 |
| 21 | .999 | .998 | .994 | .986 | .971 | .947 |
| 22 | 1.000 | .999 | .997 | .992 | .983 | .967 |
| 23 | 1.000 | 1.000 | .999 | .996 | .991 | .981 |
| 24 | 1.000 | 1.000 | .999 | .998 | .995 | .989 |
| 25 | 1.000 | 1.000 | 1.000 | .999 | .997 | .994 |
| 26 | 1.000 | 1.000 | 1.000 | 1.000 | .999 | .997 |
| 27 | 1.000 | 1.000 | 1.000 | 1.000 | .999 | .998 |
| 28 | 1.000 | 1.000 | 1.000 | 1.000 | 1.000 | .999 |
| 29 | 1.000 | 1.000 | 1.000 | 1.000 | 1.000 | 1.000 |

Abridged with permission from E. C. Molina, *Poisson's Exponential Binomial Limit*, D. Van Nostrand, 1949.

**Table 5.C** ***Standard Normal Distribution Function:***

$$N_Z(t) = \int_{-\infty}^{t} \frac{1}{\sqrt{2\pi}} e^{-x^2/2}\, dx$$

| $t$ | 0 | 1 | 2 | 3 | 4 | 5 | 6 | 7 | 8 | 9 |
|---|---|---|---|---|---|---|---|---|---|---|
| −3. | .0013 | | | | | | | | | |
| −2.9 | .0019 | .0018 | .0017 | .0017 | .0016 | .0016 | .0015 | .0015 | .0014 | .0014 |
| −2.8 | .0026 | .0025 | .0024 | .0023 | .0023 | .0022 | .0021 | .0021 | .0020 | .0019 |
| −2.7 | .0035 | .0034 | .0033 | .0032 | .0031 | .0030 | .0029 | .0028 | .0027 | .0026 |
| −2.6 | .0047 | .0045 | .0044 | .0043 | .0041 | .0040 | .0039 | .0038 | .0037 | .0036 |
| −2.5 | .0062 | .0060 | .0059 | .0057 | .0055 | .0054 | .0052 | .0051 | .0049 | .0048 |
| −2.4 | .0082 | .0080 | .0078 | .0075 | .0073 | .0071 | .0069 | .0068 | .0066 | .0064 |
| −2.3 | .0107 | .0104 | .0102 | .0099 | .0096 | .0094 | .0091 | .0089 | .0087 | .0084 |
| −2.2 | .0139 | .0136 | .0132 | .0129 | .0125 | .0122 | .0119 | .0116 | .0113 | .0110 |
| −2.1 | .0179 | .0174 | .0170 | .0166 | .0162 | .0158 | .0154 | .0150 | .0146 | .0143 |
| −2.0 | .0227 | .0222 | .0217 | .0212 | .0207 | .0202 | .0197 | .0192 | .0188 | .0183 |
| −1.9 | .0287 | .0281 | .0274 | .0268 | .0262 | .0256 | .0250 | .0244 | .0239 | .0233 |
| −1.8 | .0359 | .0351 | .0344 | .0336 | .0329 | .0322 | .0314 | .0307 | .0300 | .0294 |
| −1.7 | .0446 | .0436 | .0427 | .0418 | .0409 | .0401 | .0392 | .0384 | .0375 | .0367 |
| −1.6 | .0548 | .0537 | .0526 | .0516 | .0505 | .0495 | .0485 | .0475 | .0465 | .0455 |
| −1.5 | .0668 | .0655 | .0643 | .0630 | .0618 | .0606 | .0594 | .0582 | .0571 | .0559 |
| −1.4 | .0808 | .0793 | .0778 | .0764 | .0749 | .0735 | .0721 | .0708 | .0694 | .0681 |
| −1.3 | .0968 | .0951 | .0934 | .0918 | .0901 | .0885 | .0869 | .0853 | .0838 | .0823 |
| −1.2 | .1151 | .1131 | .1112 | .1093 | .1075 | .1056 | .1038 | .1020 | .1003 | .0985 |
| −1.1 | .1357 | .1335 | .1314 | .1292 | .1271 | .1251 | .1230 | .1210 | .1190 | .1170 |
| −1.0 | .1587 | .1562 | .1539 | .1515 | .1492 | .1469 | .1446 | .1423 | .1401 | .1379 |
| −.9 | .1841 | .1814 | .1788 | .1762 | .1736 | .1711 | .1685 | .1660 | .1635 | .1611 |
| −.8 | .2119 | .2090 | .2061 | .2033 | .2005 | .1977 | .1949 | .1921 | .1894 | .1867 |
| −.7 | .2420 | .2389 | .2358 | .2326 | .2297 | .2266 | .2236 | .2206 | .2177 | .2148 |
| −.6 | .2743 | .2709 | .2676 | .2643 | .2611 | .2578 | .2546 | .2514 | .2483 | .2451 |
| −.5 | .3085 | .3050 | .3015 | .2981 | .2946 | .2912 | .2877 | .2843 | .2810 | .2776 |
| −.4 | .3446 | .3409 | .3372 | .3336 | .3300 | .3264 | .3228 | .3192 | .3156 | .3121 |
| −.3 | .3821 | .3783 | .3745 | .3707 | .3669 | .3632 | .3594 | .3557 | .3520 | .3483 |
| −.2 | .4207 | .4168 | .4129 | .4090 | .4052 | .4013 | .3974 | .3936 | .3897 | .3859 |
| −.1 | .4602 | .4562 | .4522 | .4483 | .4443 | .4404 | .4364 | .4325 | .4286 | .4247 |
| −.0 | .5000 | .4960 | .4920 | .4880 | .4840 | .4801 | .4761 | .4721 | .4681 | .4641 |

**Table 5.C** ***Standard Normal Distribution Function*** *(continued)*

| $t$ | 0 | 1 | 2 | 3 | 4 | 5 | 6 | 7 | 8 | 9 |
|---|---|---|---|---|---|---|---|---|---|---|
| .0 | .5000 | .5040 | .5080 | .5120 | .5160 | .5199 | .5239 | .5279 | .5319 | .5359 |
| .1 | .5398 | .5438 | .5478 | .5517 | .5557 | .5596 | .5636 | .5675 | .5714 | .5753 |
| .2 | .5793 | .5832 | .5871 | .5910 | .5948 | .5987 | .6026 | .6064 | .6103 | .6141 |
| .3 | .6179 | .6217 | .6255 | .6293 | .6331 | .6368 | .6406 | .6443 | .6480 | .6517 |
| .4 | .6554 | .6591 | .6628 | .6664 | .6700 | .6736 | .6772 | .6808 | .6844 | .6879 |
| .5 | .6915 | .6950 | .6985 | .7019 | .7054 | .7088 | .7123 | .7157 | .7190 | .7224 |
| .6 | .7257 | .7291 | .7324 | .7357 | .7389 | .7422 | .7454 | .7486 | .7517 | .7549 |
| .7 | .7580 | .7611 | .7642 | .7673 | .7704 | .7734 | .7764 | .7794 | .7823 | .7852 |
| .8 | .7881 | .7910 | .7939 | .7967 | .7995 | .8023 | .8051 | .8079 | .8106 | .8133 |
| .9 | .8159 | .8186 | .8212 | .8238 | .8264 | .8289 | .8315 | .8340 | .8365 | .8389 |
| 1.0 | .8413 | .8438 | .8461 | .8485 | .8508 | .8531 | .8554 | .8577 | .8599 | .8621 |
| 1.1 | .8643 | .8665 | .8686 | .8708 | .8729 | .8749 | .8770 | .8790 | .8810 | .8830 |
| 1.2 | .8849 | .8869 | .8888 | .8907 | .8925 | .8944 | .8962 | .8980 | .8997 | .9015 |
| 1.3 | .9032 | .9049 | .9066 | .9082 | .9099 | .9115 | .9131 | .9147 | .9162 | .9177 |
| 1.4 | .9192 | .9207 | .9222 | .9236 | .9251 | .9265 | .9279 | .9292 | .9306 | .9319 |
| 1.5 | .9332 | .9345 | .9357 | .9370 | .9382 | .9394 | .9406 | .9418 | .9429 | .9441 |
| 1.6 | .9452 | .9463 | .9474 | .9484 | .9495 | .9505 | .9515 | .9525 | .9535 | .9545 |
| 1.7 | .9554 | .9564 | .9573 | .9582 | .9591 | .9599 | .9608 | .9616 | .9625 | .9633 |
| 1.8 | .9641 | .9649 | .9656 | .9664 | .9671 | .9678 | .9686 | .9693 | .9700 | .9706 |
| 1.9 | .9713 | .9719 | .9726 | .9732 | .9738 | .9744 | .9750 | .9756 | .9761 | .9767 |
| 2.0 | .9773 | .9778 | .9783 | .9788 | .9793 | .9798 | .9803 | .9808 | .9812 | .9817 |
| 2.1 | .9821 | .9826 | .9830 | .9834 | .9838 | .9842 | .9846 | .9850 | .9854 | .9857 |
| 2.2 | .9861 | .9864 | .9868 | .9871 | .9875 | .9878 | .9881 | .9884 | .9887 | .9890 |
| 2.3 | .9893 | .9896 | .9898 | .9901 | .9904 | .9906 | .9909 | .9911 | .9913 | .9916 |
| 2.4 | .9918 | .9920 | .9922 | .9925 | .9927 | .9929 | .9931 | .9932 | .9934 | .9936 |
| 2.5 | .9938 | .9940 | .9941 | .9943 | .9945 | .9946 | .9948 | .9949 | .9951 | .9952 |
| 2.6 | .9953 | .9955 | .9956 | .9957 | .9959 | .9960 | .9961 | .9962 | .9963 | .9964 |
| 2.7 | .9965 | .9966 | .9967 | .9968 | .9969 | .9970 | .9971 | .9972 | .9973 | .9974 |
| 2.8 | .9974 | .9975 | .9976 | .9977 | .9977 | .9978 | .9979 | .9979 | .9980 | .9981 |
| 2.9 | .9981 | .9982 | .9982 | .9983 | .9984 | .9984 | .9985 | .9985 | .9986 | .9986 |
| 3. | .9987 | | | | | | | | | |

**Table 5.D** $\chi^2$ ***Distribution Function:***

$$F_{\chi^2}(t) = \int_0^t \frac{1}{2^{d/2}\,\Gamma(d/2)}\, x^{(d-2)/2}\, e^{-x/2}\, dx = 1 - \alpha$$

Degrees of freedom = $d$

| $d$ | $1-\alpha$: .005 | .010 | .025 | .050 | .100 | .250 | .500 | .750 | .900 | .950 | .975 | .990 | .995 |
|---|---|---|---|---|---|---|---|---|---|---|---|---|---|
| 1 | .0000393 | .000157 | .000982 | .00393 | .0158 | .102 | .455 | 1.32 | 2.71 | 3.84 | 5.02 | 6.63 | 7.88 |
| 2 | .0100 | .0201 | .0506 | .103 | .211 | .575 | 1.39 | 2.77 | 4.61 | 5.99 | 7.38 | 9.21 | 10.6 |
| 3 | .0717 | .115 | .216 | .352 | .584 | 1.21 | 2.37 | 4.11 | 6.25 | 7.81 | 9.35 | 11.3 | 12.8 |
| 4 | .207 | .297 | .484 | .711 | 1.06 | 1.92 | 3.36 | 5.39 | 7.78 | 9.49 | 11.1 | 13.3 | 14.9 |
| 5 | .412 | .554 | .831 | 1.15 | 1.61 | 2.67 | 4.35 | 6.63 | 9.24 | 11.1 | 12.8 | 15.1 | 16.7 |
| 6 | .676 | .872 | 1.24 | 1.64 | 2.20 | 3.45 | 5.35 | 7.84 | 10.6 | 12.6 | 14.4 | 16.8 | 18.5 |
| 7 | .989 | 1.24 | 1.69 | 2.17 | 2.83 | 4.25 | 6.35 | 9.04 | 12.0 | 14.1 | 16.0 | 18.5 | 20.3 |
| 8 | 1.34 | 1.65 | 2.18 | 2.73 | 3.49 | 5.07 | 7.34 | 10.2 | 13.4 | 15.5 | 17.5 | 20.1 | 22.0 |
| 9 | 1.73 | 2.09 | 2.70 | 3.33 | 4.17 | 5.90 | 8.34 | 11.4 | 14.7 | 16.9 | 19.0 | 21.7 | 23.6 |
| 10 | 2.16 | 2.56 | 3.25 | 3.94 | 4.87 | 6.74 | 9.34 | 12.5 | 16.0 | 18.3 | 20.5 | 23.2 | 25.2 |
| 11 | 2.60 | 3.05 | 3.82 | 4.57 | 5.58 | 7.58 | 10.3 | 13.7 | 17.3 | 19.7 | 21.9 | 24.7 | 26.8 |
| 12 | 3.07 | 3.57 | 4.40 | 5.23 | 6.30 | 8.44 | 11.3 | 14.8 | 18.5 | 21.0 | 23.3 | 26.2 | 28.3 |
| 13 | 3.57 | 4.11 | 5.01 | 5.89 | 7.04 | 9.30 | 12.3 | 16.0 | 19.8 | 22.4 | 24.7 | 27.7 | 29.8 |
| 14 | 4.07 | 4.66 | 5.63 | 6.57 | 7.79 | 10.2 | 13.3 | 17.1 | 21.1 | 23.7 | 26.1 | 29.1 | 31.3 |
| 15 | 4.60 | 5.23 | 6.26 | 7.26 | 8.55 | 11.0 | 14.3 | 18.2 | 22.3 | 25.0 | 27.5 | 30.6 | 32.8 |

| | | | | | | | | | | | | | |
|---|---|---|---|---|---|---|---|---|---|---|---|---|---|
| 16 | 5.14 | 5.81 | 6.91 | 7.96 | 9.31 | 11.9 | 15.3 | 19.4 | 23.5 | 26.3 | 28.8 | 32.0 | 34.3 |
| 17 | 5.70 | 6.41 | 7.56 | 8.67 | 10.1 | 12.8 | 16.3 | 20.5 | 24.8 | 27.6 | 30.2 | 33.4 | 35.7 |
| 18 | 6.26 | 7.01 | 8.23 | 9.39 | 10.9 | 13.7 | 17.3 | 21.6 | 26.0 | 28.9 | 31.5 | 34.8 | 37.2 |
| 19 | 6.84 | 7.63 | 8.91 | 10.1 | 11.7 | 14.6 | 18.3 | 22.7 | 27.2 | 30.1 | 32.9 | 36.2 | 38.6 |
| 20 | 7.43 | 8.26 | 9.59 | 10.9 | 12.4 | 15.5 | 19.3 | 23.8 | 28.4 | 31.4 | 34.2 | 37.6 | 40.0 |
| 21 | 8.03 | 8.90 | 10.3 | 11.6 | 13.2 | 16.3 | 20.3 | 24.9 | 29.6 | 32.7 | 35.5 | 38.9 | 41.4 |
| 22 | 8.64 | 9.54 | 11.0 | 12.3 | 14.0 | 17.2 | 21.3 | 26.0 | 30.8 | 33.9 | 36.8 | 40.3 | 42.8 |
| 23 | 9.26 | 10.2 | 11.7 | 13.1 | 14.8 | 18.1 | 22.3 | 27.1 | 32.0 | 35.2 | 38.1 | 41.6 | 44.2 |
| 24 | 9.89 | 10.9 | 12.4 | 13.8 | 15.7 | 19.0 | 23.3 | 28.2 | 33.2 | 36.4 | 39.4 | 43.0 | 45.6 |
| 25 | 10.5 | 11.5 | 13.1 | 14.6 | 16.5 | 19.9 | 24.3 | 29.3 | 34.4 | 37.7 | 40.6 | 44.3 | 46.9 |
| 26 | 11.2 | 12.2 | 13.8 | 15.4 | 17.3 | 20.8 | 25.3 | 30.4 | 35.6 | 38.9 | 41.9 | 45.6 | 48.3 |
| 27 | 11.8 | 12.9 | 14.6 | 16.2 | 18.1 | 21.7 | 26.3 | 31.5 | 36.7 | 40.1 | 43.2 | 47.0 | 49.6 |
| 28 | 12.5 | 13.6 | 15.3 | 16.9 | 18.9 | 22.7 | 27.3 | 32.6 | 37.9 | 41.3 | 44.5 | 48.3 | 51.0 |
| 29 | 13.1 | 14.3 | 16.0 | 17.7 | 19.8 | 23.6 | 28.3 | 33.7 | 39.1 | 42.6 | 45.7 | 49.6 | 52.3 |
| 30 | 13.8 | 15.0 | 16.8 | 18.5 | 20.6 | 24.5 | 29.3 | 34.8 | 40.3 | 43.8 | 47.0 | 50.9 | 53.7 |

Abridged with permission from Catherine M. Thompson, "Tables of percentage points of the incomplete Beta function and of the chi-square distribution," *Biometrika*, **32**, 1941.

**Table 5.E** ***Student's-t Distribution Function:***

$$F_t(x) = \int_{-\infty}^{x} \frac{\Gamma((d+1)/2)}{\sqrt{d\pi}\,\Gamma(d/2)} (1 + u^2)^{-(d+1)/2}\, du = 1 - \alpha$$

| | Degrees of freedom = $d$ | | | | | | | |
|---|---|---|---|---|---|---|---|---|
| | $1-\alpha$ | | | | | | | |
| $d$ | .60 | .75 | .90 | .95 | .975 | .99 | .995 | .9995 |
| 1 | .325 | 1.000 | 3.078 | 6.314 | 12.706 | 31.821 | 63.657 | 636.619 |
| 2 | .289 | .816 | 1.886 | 2.920 | 4.303 | 6.965 | 9.925 | 31.598 |
| 3 | .277 | .765 | 1.638 | 2.353 | 3.182 | 4.541 | 5.841 | 12.941 |
| 4 | .271 | .741 | 1.533 | 2.132 | 2.776 | 3.747 | 4.604 | 8.610 |
| 5 | .267 | .727 | 1.476 | 2.015 | 2.571 | 3.365 | 4.032 | 6.859 |
| 6 | .265 | .718 | 1.440 | 1.943 | 2.447 | 3.143 | 3.707 | 5.959 |
| 7 | .263 | .711 | 1.415 | 1.895 | 2.365 | 2.998 | 3.499 | 5.405 |
| 8 | .262 | .706 | 1.397 | 1.860 | 2.306 | 2.896 | 3.355 | 5.041 |
| 9 | .261 | .703 | 1.383 | 1.833 | 2.262 | 2.821 | 3.250 | 4.781 |
| 10 | .260 | .700 | 1.372 | 1.812 | 2.228 | 2.764 | 3.169 | 4.587 |
| 11 | .260 | .697 | 1.363 | 1.796 | 2.201 | 2.718 | 3.106 | 4.437 |
| 12 | .259 | .695 | 1.356 | 1.782 | 2.179 | 2.681 | 3.055 | 4.318 |
| 13 | .259 | .694 | 1.350 | 1.771 | 2.160 | 2.650 | 3.012 | 4.221 |
| 14 | .258 | .692 | 1.345 | 1.761 | 2.145 | 2.624 | 2.977 | 4.140 |
| 15 | .258 | .691 | 1.341 | 1.753 | 2.131 | 2.602 | 2.947 | 4.073 |
| 16 | .258 | .690 | 1.337 | 1.746 | 2.120 | 2.583 | 2.921 | 4.015 |
| 17 | .257 | .689 | 1.333 | 1.740 | 2.110 | 2.567 | 2.898 | 3.965 |
| 18 | .257 | .688 | 1.330 | 1.734 | 2.101 | 2.552 | 2.878 | 3.922 |
| 19 | .257 | .688 | 1.328 | 1.729 | 2.093 | 2.539 | 2.861 | 3.883 |
| 20 | .257 | .687 | 1.325 | 1.725 | 2.086 | 2.528 | 2.845 | 3.850 |
| 21 | .257 | .686 | 1.323 | 1.721 | 2.080 | 2.518 | 2.831 | 3.819 |
| 22 | .256 | .686 | 1.321 | 1.717 | 2.074 | 2.508 | 2.819 | 3.792 |
| 23 | .256 | .685 | 1.319 | 1.714 | 2.069 | 2.500 | 2.807 | 3.767 |
| 24 | .256 | .685 | 1.318 | 1.711 | 2.064 | 2.492 | 2.797 | 3.745 |
| 25 | .256 | .684 | 1.316 | 1.708 | 2.060 | 2.485 | 2.787 | 3.725 |
| 26 | .256 | .684 | 1.315 | 1.706 | 2.056 | 2.479 | 2.779 | 3.707 |
| 27 | .256 | .684 | 1.314 | 1.703 | 2.052 | 2.473 | 2.771 | 3.690 |
| 28 | .256 | .683 | 1.313 | 1.701 | 2.048 | 2.467 | 2.763 | 3.674 |
| 29 | .256 | .683 | 1.311 | 1.699 | 2.045 | 2.462 | 2.756 | 3.659 |
| 30 | .256 | .683 | 1.310 | 1.697 | 2.042 | 2.457 | 2.750 | 3.646 |
| 40 | .255 | .681 | 1.303 | 1.684 | 2.021 | 2.423 | 2.704 | 3.551 |
| 60 | .254 | .679 | 1.296 | 1.671 | 2.000 | 2.390 | 2.660 | 3.460 |
| 120 | .254 | .677 | 1.289 | 1.658 | 1.980 | 2.358 | 2.617 | 3.373 |
| $\infty$ | .253 | .674 | 1.282 | 1.645 | 1.960 | 2.326 | 2.576 | 3.291 |

Abridged with permission from R. A. Fisher and Frank Yates, *Statistical Tables*, Oliver and Boyd, Ltd., Edinburgh, 1938.

# Answers to Problems

## CHAPTER 1

*Exercise 1.1*

**3.** $E = \{1, 2\}$, $D = \{2, 3\}$, $F = \{1, 2, 3\}$ **5.** F, F, T, T, T, T, F, T, T, F, F, F, F, F **6.** T, F, F, T, F, T, F **7.** Yes, No, No

*Exercise 1.2*

**2.** $A \cup B = \{x : 0 \leq x \leq 1\}$, $A \cup C = \{x : x = 0, \frac{1}{2}, 1\}$, $B \cup C = B$, $A \cap B = A \cap C = \phi$, $B \cap C = C$ **5.** $F \subset E$ **6.** $F \subset E$

*Exercise 1.3*

**2.** $\bar{A} = \{4, 5, 6, \ldots, n\}$, $\bar{B} = \{1, 4, 5, 6, \ldots, n - 1\}$, $\bar{A} \cup \bar{B} = \{1, 4, 5, 6, \ldots, n\}$, $\bar{A} \cup \bar{B} = \overline{A \cap B}$ **3.** No, No **8.** $A \times B = \{(1, 2), (1, 1), (2, 2), (2, 1)\}$, $A \times C = \{(1, 10), (1, 12), (2, 10), (2, 12)\} = B \times C$, $B \times A = A \times B$, $C \times A = C \times B = \{(10, 1), (10, 2), (12, 1), (12, 2)\}$, $A \times A = A \times B = B \times B$, $C \times C = \{(10, 10), (10, 12), (12, 10), (12, 12)\}$, $A \times B \times C = \{(1, 2, 10), (1, 2, 12), (1, 1, 10), (1, 1, 12), (2, 2, 10), (2, 2, 12), (2, 1, 10), (2, 1, 12)\}$, $C \times B \times A = C \times A \times B = \{(10, 2, 1), (10, 2, 2), (10, 1, 1), (10, 1, 2), (12, 2, 1), (12, 2, 2), (12, 1, 1), (12, 1, 2)\}$ **9.** $A = B$ **10.** No **11.** No

*Exercise 1.4*

**2.** $\{2, 3, \ldots, 12\}$ **3.** $\{-1, 0, 1, 2\}$ **4.** $A_7 = \{(1, 6), (2, 5), (3, 4), (4, 3), (5, 2), (6, 1)\}$, $A_3 = \{(1, 2), (2, 1)\}$, $A_{10} = \{(4, 6), (5, 5), (6, 4)\}$ **5.** $\{0, 1, 2, \ldots, k\}$ **6.** $\{0, 1/k, 2/k, \ldots, 1\}$

## CHAPTER 2

*Exercise 2.1*

For problems 1 through 5 the white balls are numbered 1, 2, 3, 4 and red balls are numbered 5, 6, 7, 8, 9, 10

**1.** $S = \{1, 2, \ldots, 10\}$ **2.** $S = \{(x_1, x_2) : x_i = 1, 2, \ldots, 10;\ i = 1, 2\}$ **3.** $S = \{(x_1, x_2) : x_i = 1, 2, 3, \ldots, 10;\ i = 1, 2;\ x_1 \neq x_2\}$ **4.** $A = \{1, 2, 3, 4\}$, $B = \{5, 6, 7, 8, 9, 10\}$ **5.** $C = \{(x_1, x_2) : x_1 = 1, 2, 3, 4;\ x_2 = 1, 2, \ldots, 10\}$, $D = \{(x_1, x_2) : x_1 = 1, 2, \ldots, 10;\ x_2 = 1, 2, 3, 4\}$, $E = \{(x_1, x_2) : x_i = 1, 2, 3, 4;\ i = 1, 2\}$,

Yes **6.** $S = \{(x_1, x_2): x_i = a, b, c, d, e;\ i = 1, 2\}$ **7.** $S = \{(x_1, x_2): x_1 =$ bald, brown, black; $x_2 =$ blue, brown$\}$, $A = \{$(bald, blue), (bald, brown)$\}$, $B = \{$(bald, blue), (brown, blue), (black, blue)$\}$, $C = \{$(brown, brown)$\}$ **8.** $S = \{(x_1, x_2): x_i =$ Marie, Sandy, Tina; $i = 1, 2;\ x_1 \neq x_2\}$, $A = \{$(Marie, Sandy), (Marie, Tina)$\}$, $B = \{$(Sandy, Marie), (Tina, Marie)$\}$, $C = \{$(Tina, Sandy), (Sandy, Tina)$\}$ **9.** $S = \{(x_1, x_2, x_3, x_4): x_i = 0, 1, 2, 3;\ i = 1, 2, 3, 4$ and $x_1 + x_2 + x_3 + x_4 = 3\}$, (Thus, $x_1$ is the number of red cards, $x_2$ the number of green cards, etc.), $A = \{(3, 0, 0, 0)\}$, $B = \{(1, 1, 1, 0)\}$, $C = \{(1, 1, 1, 0), (1, 1, 0, 1), (1, 0, 1, 1), (0, 1, 1, 1)\}$, $D = \phi$ **10.** $S = \{(x_1, x_2, x_3): x_i = 0, 1, 2, 3, 4;\ i = 1, 2, 3$ and $x_1 + x_2 + x_3 + x_4 = 4\}$ (Thus, $x_1$ gives the number of ladies choosing store 1, $x_2$ the number of ladies choosing store 2, etc.), $A = \{(4, 0, 0)\}$, $B = \{(2, 2, 0)\}$, $C = \{(1, 1, 2), (1, 2, 1), (2, 1, 1)\}$

### *Exercise 2.2*

**1.** $\frac{1}{3}, \frac{2}{3}, \frac{2}{3}, \frac{1}{3}, 1, \frac{2}{3}$ **2.** $\frac{3}{10}, \frac{7}{10}, \frac{5}{10}, \frac{3}{10}, 1, \frac{5}{10}$ **3.** 0, 1, 0, 0, 1, 0, No **7.** $\frac{1}{2}, \frac{1}{2}, \frac{1}{3}, \frac{1}{3}, \frac{2}{3}, \frac{1}{6}, \frac{1}{6}, \frac{2}{3}$ **8.** $\frac{7}{12}, \frac{7}{12}, \frac{1}{12}, \frac{1}{4}, \frac{5}{12}, \frac{3}{4}$ **9.** No **10.** No

### *Exercise 2.3*

**1.** $\frac{1}{2}$ **2.** $\frac{1}{2}, \frac{1}{4}, \frac{1}{13}, \frac{1}{52}$ **3.** $\frac{1}{5}, \frac{1}{5}, \frac{1}{25}$

**4.**

| sum | 2 | 3 | 4 | 5 | 6 | 7 | 8 | 9 | 10 | 11 | 12 |
|---|---|---|---|---|---|---|---|---|---|---|---|
| probability | $\frac{1}{36}$ | $\frac{2}{36}$ | $\frac{3}{36}$ | $\frac{4}{36}$ | $\frac{5}{36}$ | $\frac{6}{36}$ | $\frac{5}{36}$ | $\frac{4}{36}$ | $\frac{3}{36}$ | $\frac{2}{36}$ | $\frac{1}{36}$ |

**5.**

| sum | 3 | 4 | 5 | 6 |
|---|---|---|---|---|
| probability | $\frac{1}{8}$ | $\frac{3}{8}$ | $\frac{3}{8}$ | $\frac{1}{8}$ |

**6.** $\frac{18}{40}, \frac{6}{40}, \frac{7}{40}, \frac{17}{40}, \frac{4}{40}$ **7.** $\frac{1}{21}, \frac{2}{21}, \frac{3}{21}, \frac{4}{21}, \frac{5}{21}, \frac{6}{21}, \frac{12}{21}, \frac{11}{21}$

### *Exercise 2.4*

**1.** $3! = 6$ **2.** $3!$ **3.** $6! = 720$ **4.** $3^3 = 27$ **5.** $2^4 = 16$ **6.** ${}_4P_3 = 24$ **7.** $2^3 = 8$ **8.** No **9.** $\binom{28}{5}$ **10.** $\binom{8}{2}$ **11.** $\binom{10}{3}$ **12.** $\binom{20}{5}$ **13.** $\binom{15}{2}$ **14.** $\binom{15}{3}$, $\binom{15}{k}$, $k = 4, 5, \ldots, 15$ **15.** $\binom{9}{3}$ **16.** 7, 35 **17.** 24

### *Exercise 2.5*

**1.** $\binom{30}{12}$ **2.** $6 \cdot 5!$ (assuming red balls distinguishable) **3.** $2(5!)$ **4.** $\binom{20}{15}$, $\binom{20}{10}$ **5.** $\binom{9}{2}\Big/\binom{10}{3}$, $\binom{8}{1}\Big/\binom{10}{3}$ **6.** $\binom{99}{1}\Big/\binom{100}{2}$, $\binom{98}{0}\Big/\binom{100}{2}$ **7.** $(\frac{1}{2})^{30}$ **8.** $\frac{1}{2}, \frac{1}{3}$ **9.** $\frac{1}{32}, \frac{6}{32}$ **10.** $\binom{4}{2}\binom{4}{2}\binom{13}{2}\binom{44}{1}\Big/\binom{52}{5}$, $\binom{13}{2}\binom{2}{1}\binom{4}{3}\binom{4}{2}\Big/\binom{52}{5}$, $\binom{13}{5}\binom{4}{1}\Big/\binom{52}{5}$, $\binom{4}{1}^5\binom{10}{1}\Big/\binom{52}{5}$ **11.** $1 - {}_{12}P_n/(12)^n$, $n = 1, 2, \ldots, 12$

**12.** $\frac{56}{1024}$ **13.** $\frac{1}{5}, \frac{1}{45}$ **14.** $\frac{1}{20}, 0, \frac{9}{20}$

### *Exercise 2.6*

**1.** $\frac{1}{3}, \frac{1}{4}$ **2.** $\frac{1}{4}, \frac{1}{4}$ **3.** $\frac{1}{2}$ **4.** $\frac{1}{2}, \frac{1}{2}$ **5.** $\frac{5}{12}, \frac{7}{12}$ **6.** $\frac{1}{4}, \frac{3}{4}$ **7.** $\binom{26}{5} \Big/ \binom{52}{5}, \binom{13}{5} \Big/ \binom{52}{5}$
**8.** $\frac{1}{3}$ **9.** $\frac{50}{63}$ **10.** $\frac{7}{10}$ **11.** .087 **12.** .199, $\frac{1}{199}$

### *Exercise 2.7*

**1.** Yes **2.** Yes **3.** No **4.** Yes **5.** No **8.** $(.7)^{15}, \binom{15}{4}(.7)^{14}(.3)$ **9.** .07, .24, .51

### *Exercise 2.8*

**1.** $(\frac{5}{6})^9(\frac{1}{6}), \frac{91}{216}, \frac{6}{11}$ **2.** $\frac{5}{36}, \frac{5}{11}$ **3.** .01, $\frac{1}{111}$ **4.** $(.9)^4, (.9)^{10}$ **5.** $\frac{3}{10}, \frac{2}{10}$ **6.** $\frac{1}{12}, \frac{1}{12}, \frac{5}{12}$ **7.** $\frac{1}{2}(2)^{1/2}$ 8. $\frac{2}{3}$

## CHAPTER 3

### *Exercise 3.1*

**1.**.$p_Y(y) = \frac{1}{4}, y = 1, 2, 3, 4$ **2.** $p_Z(z) = \begin{cases} \frac{1}{6}, z = 3, 4, 6, 7 \\ \frac{2}{6}, z = 5 \end{cases}$

**3.**

| $z$ | 2 | 3 | 4 | 5 | 6 | 7 | 8 |
|---|---|---|---|---|---|---|---|
| $p_Z(z)$ | $\frac{1}{16}$ | $\frac{2}{16}$ | $\frac{3}{16}$ | $\frac{4}{16}$ | $\frac{3}{16}$ | $\frac{2}{16}$ | $\frac{1}{16}$ |

**4.**

| $x$ | 2 | 5 | 8 | 10 | 13 | 17 | 18 | 20 | 25 | 32 |
|---|---|---|---|---|---|---|---|---|---|---|
| $p_X(x)$ | $\frac{1}{16}$ | $\frac{2}{16}$ | $\frac{1}{16}$ | $\frac{2}{16}$ | $\frac{2}{16}$ | $\frac{2}{16}$ | $\frac{1}{16}$ | $\frac{2}{16}$ | $\frac{2}{16}$ | $\frac{1}{16}$ |

**5.**

| $x$ | 19 | $19\frac{1}{2}$ | 20 | $20\frac{1}{2}$ | $21\frac{1}{2}$ | 22 | $22\frac{1}{2}$ | 23 | $23\frac{1}{2}$ | 25 |
|---|---|---|---|---|---|---|---|---|---|---|
| $p_X(x)$ | $\frac{6}{90}$ | $\frac{24}{90}$ | $\frac{18}{90}$ | $\frac{8}{90}$ | $\frac{6}{90}$ | $\frac{8}{90}$ | $\frac{8}{90}$ | $\frac{8}{90}$ | $\frac{2}{90}$ | $\frac{2}{90}$ |

### *Exercise 3.2*

**1.** $p_X(x) = \frac{1}{3}, x = -3, -1, 0$ **2.** $p_Z(z) = \frac{1}{2}, z = -2, 0$

**3.**

| $w$ | 3 | 4 | 5 | 6 |
|---|---|---|---|---|
| $p_W(w)$ | $\frac{1}{3}$ | $\frac{1}{6}$ | $\frac{1}{6}$ | $\frac{1}{3}$ |

, $P(3 < W \leq 5) = \frac{1}{3}$

**4.**

| $y$ | 0 | 5 | 7 | 100 | 102 |
|---|---|---|---|---|---|
| $p_Y(y)$ | $\frac{1}{4}$ | $\frac{1}{12}$ | $\frac{1}{6}$ | $\frac{1}{3}$ | $\frac{1}{6}$ |

, $P(Y \leq 100) = \frac{5}{6}$

**5.** $F_Z(x) = 0, x < 0$
$= \frac{1}{3}, 0 \leq x < 1$
$= \frac{2}{3}, 1 \leq x < 2$
$= 1, x \geq 2$

**6.** $F_U(t) = 0, \quad t < -3$
$= \frac{1}{2}, \quad -3 \leq t < 0$
$= \frac{2}{3}, \quad 0 \leq t < 4$
$= 1, \quad t \geq 4$

**7.** $f_X(t) = \frac{1}{2}, \quad -1 < t < 1$
**8.** $f_Y(t) = 1/2(t)^{1/2}, \quad 0 < t < 1, \quad P(\frac{1}{4} < Y < \frac{3}{4}) = \frac{1}{2}(3)^{1/2} - \frac{1}{2}$

**9.** $f_Z(t) = 2t, \quad 0 < t < \frac{1}{2}$
$\quad = 6(1-t), \frac{1}{2} < t < 1$

**10.** No

**12.** $F_X(t) = 0, \quad t < 99$
$\quad = t - 99, 99 \leq t \leq 100$
$\quad = 1, \quad t > 100$

**13.** $F_Y(t) = 0, \quad t < 0$
$\quad = 1 - (1-t)^2, 0 \leq t \leq 1$
$\quad = 1, \quad t > 1$

**14.** $F_Z(t) = 0, \quad t < 0$
$\quad = 1 - \exp(-10t), t \geq 0$

*Exercise 3.3*

**1.** $\frac{15}{2}$, 85, $\frac{15}{64}$, $\frac{139}{2}$, $\frac{115}{4}$, $\frac{1}{2}(115)^{1/2}$ **2.** 0, $\frac{1}{3}$, 2, 7, $\frac{1}{3}$, $(\frac{1}{3})^{1/2}$ **3.** $\frac{1}{3}$, $\frac{1}{6}$, $\frac{641}{6}$, 2, $\frac{1}{18}$, $\frac{1}{3}(2)^{1/2}$ **5.** 10, 20, 55, 99, 47, 80, 50 **6.** $\exp(.5)$, $\exp(.75) - \exp(.25)$, $e - 1$, $e^2 - \frac{1}{2}$ **7.** $\log_2 1.5$, $\log_2 \frac{1.75}{1.25}$, $2 - (\log 2)^{-1}$ **8.** 2, 2, 3, 0, 1, 2 **10.** 5′ 8½″ **11.** −.70, −69, 900.51, 5229

*Exercise 3.4*

**1.** $m_1 = m_2 = m_3 = \frac{1}{6}$ **2.** $(e^{40t} - e^{30t})/10t, e^{-35t}(e^{40t} - e^{30t})/10t$ **3.** $500/(500 - t)$, $t < 500$, $\frac{1}{500}$, $1/(500)^2$ **4.** $\frac{1}{8} + (\frac{3}{8})t + (\frac{3}{8})t^2 + (\frac{1}{8})t^3, \frac{3}{4}$ **5.** $(1/n)t + (1/n)t^2 + \cdots + (1/n)t^n$ **6.** $m_1, m_2 - m_1, m_3 - 3m_2 + 2m_1$ **7.** $\xi_1, \xi_2 + \xi_1, \xi_3 + 3\xi_2 + \xi_1$ **8.** $9(t + (.1)t^2 + (.1)^2t^3 + \cdots), \frac{10}{9}, \frac{10}{81}$

*Exercise 3.5*

**1.** $1 - F_X\left(\frac{t-a}{b}\right) + p_X\left(\frac{t-a}{b}\right)$

**2.** $F_Y(t) = 0, t < a$
$\quad = 1, t \geq a$

**3.** $F_Y(t) = 0, \quad t < 13$
$\quad = \frac{1}{4}(t - 13), 13 \leq t \leq 17$
$\quad = 1, \quad t > 17$
$f_Y(t) = \frac{1}{4}, \quad 13 < t < 17$

**4.** $F_Z(t) = 0, \quad t < -1$
$\quad = (t+1)^3, \quad -1 \leq t \leq 0$
$\quad = 1, \quad t > 0$
$f_Z(t) = 3(t+1)^2, -1 < t < 0$

**5.** $F_U(t) = 0, \quad t < -120$
$\quad = \frac{1}{4}, -120 \leq t < -50$
$\quad = \frac{3}{4}, \quad -50 \leq t < 20$
$\quad = 1, \quad t \geq 20$

$p_U(t) = \frac{1}{4}, t = -120$
$\quad = \frac{1}{2}, t = -50$
$\quad = \frac{1}{4}, t = 20$

**6.** $F_X(t) = 0, \quad t < -7$
$\quad = 1 - \exp(-\frac{1}{2}(t+7)), t \geq -7$

$f_X(t) = \frac{1}{2}\exp(-\frac{1}{2}(t+7)), t > -7$

**7.** $F_{(Z-\mu)/\sigma}(t) = 0, \quad t < -\sqrt{3}$
$\quad = \sqrt{3}t/6 + \frac{1}{2}, -\sqrt{3} \leq t \leq \sqrt{3}$
$\quad = 1, \quad t > \sqrt{3}$

$f_{(Z-\mu)/\sigma}(t) = \sqrt{3}/6, -\sqrt{3} < t < \sqrt{3}$

**8.** $F_{(Z-\mu)/\sigma}(t) = 0, \quad t < -1 \qquad f_{(Z-\mu)/\sigma}(t) = \exp(-t-1),\ t > -1$
$= 1 - \exp(-t-1),\ t \geq -1$

**9.** $F_{(Z-\mu)/\sigma}(t) = 0, \quad t < -\sqrt{\frac{3}{2}} \qquad p_{(Z-\mu)/\sigma}(t) = \frac{1}{3},\ t = -\sqrt{\frac{3}{2}},\ 0,\ \sqrt{\frac{3}{2}}$
$= \frac{1}{3}, \quad -\sqrt{\frac{3}{2}} \leq t < 0$
$= \frac{2}{3}, \quad 0 \leq t < \sqrt{\frac{3}{2}}$
$= 1, \quad t > \sqrt{\frac{3}{2}}$

**10.** $F_Z(t) = 0, \quad t < 0 \qquad f_Z(t) = \dfrac{1}{2\sqrt{t}},\ 0 < t < 1$
$= \sqrt{t},\ 0 \leq t \leq 1$
$= 1, \quad t > 1$

## CHAPTER 4

### *Exercise 4.1*

**1.** $\frac{5}{6}, \frac{25}{36}, \frac{200}{243}, \frac{131}{243}$ **2.** 16, $\frac{16}{5}$, .2182, .0867, .2067 **3.** $\frac{1}{2}$, 10, 5, .2461 **4.** .01, 10, .1, .9044, .1 **7.** $\frac{1}{3}$, 18 **8.** 99, No

### *Exercise 4.2*

**1.** $\frac{3}{4}, \frac{7}{8}$ **2.** $(\frac{20}{38})^{k-1}(\frac{18}{38}), \frac{38}{18}$ **3.** $(\frac{36}{38})^{k-1}(\frac{2}{38})$, 19, 342 **4.** $(\frac{9}{10})^{k-1}(\frac{1}{10})$, 10 **5.** $\frac{1}{10}$ for $k = 1, 2, \ldots, 10$, $5\frac{1}{2}$ **6.** $\binom{3}{k}\binom{2}{2-k}\Big/\binom{5}{2}$ **7.** $\binom{26}{k}\binom{26}{13-k}\Big/\binom{52}{13}, \frac{13}{2}, \frac{169}{68}$ **8.** $\binom{8}{k}\binom{2}{5-k}\Big/\binom{10}{5}, \binom{2}{k}\binom{8}{5-k}\Big/\binom{10}{5}$ **9.** \$5 **10.** $\frac{22}{35}$ **11.** $\binom{k-1}{k-r}q^{k-r}p^r$, for $k = r, r+1, r+2, \ldots$ **12.** $\binom{k-1}{k-3}(.999)^{k-3}(.001)^3$, $k = 3, 4, \ldots$

### *Exercise 4.3*

**1.** .2231, .4422 **2.** .0003, .1353, .0003 **3.** .811, $(.811)^3$ **4.** 6, $\sqrt{6}$, .6065 **7.** $\exp(\lambda(e^t - 1))$ **8.** .6065, .3033, .0902 **9.** .9802 **10.** .0498, .2240, 149

### *Exercise 4.4*

**1.** $1/2\sqrt{t}$, $2 - \sqrt{2}$ **2.** $2t/3$ **3.** $(e^{tb} - e^{ta})/t(b-a)$ **4.** $(t^b - t^a)/(b-a)\log t$ **5.** .316, .422 **6.** .69 **7.** .2865, .5654, 0 **8.** .1882, .1882 **9.** $\frac{1}{2}\sqrt{3}$ **10.** $-\log(1-p)$ **11.** $\lambda^2 s e^{-\lambda s}$

### *Exercise 4.5*

**1.** .3085, .9987 **2.** .9773, .8413, .1587 **3.** $2N_Z(t) - 1$, $t > 0$ **4.** .3174, .3830 **5.** 0, 1.35 **6.** $\mu$, $1.35\sigma$ **8.** .455, 1.22 **9.** $\exp(-t/2)/\Gamma(n/2)\sqrt{2t}$ **10.** $\exp(\frac{1}{2}t^2)$ **11.** $(1-2t)^{-1/2}$

## CHAPTER 5

### *Exercise 5.1*

**1.**

| $y$ \ $x$ | 2 | 3 | 4 | |
|---|---|---|---|---|
| $-1$ | 0 | $\frac{1}{4}$ | 0 | $\frac{1}{4}$ |
| 0 | $\frac{1}{4}$ | 0 | $\frac{1}{4}$ | $\frac{1}{2}$ |
| 1 | 0 | $\frac{1}{4}$ | 0 | $\frac{1}{4}$ |
| | $\frac{1}{4}$ | $\frac{1}{2}$ | $\frac{1}{4}$ | |

**2.**

| $y$ \ $x$ | 2 | 3 | 4 | |
|---|---|---|---|---|
| 2 | $\frac{1}{8}$ | $\frac{1}{8}$ | 0 | $\frac{1}{4}$ |
| 3 | $\frac{1}{8}$ | $\frac{1}{4}$ | $\frac{1}{8}$ | $\frac{1}{2}$ |
| 4 | 0 | $\frac{1}{8}$ | $\frac{1}{8}$ | $\frac{1}{4}$ |
| | $\frac{1}{4}$ | $\frac{1}{2}$ | $\frac{1}{4}$ | |

**3.**

| $y$ \ $x$ | 0 | 1 | 2 | |
|---|---|---|---|---|
| 0 | 1260 | 216 | 6 | 1482 |
| 1 | 864 | 144 | 6 | 1014 |
| 2 | 132 | 24 | 0 | 156 |
| | 2256 | 384 | 12 | 2652 |

**4.**

| $y$ \ $x$ | 0 | 1 | 2 | |
|---|---|---|---|---|
| 0 | 0 | 0 | $\frac{6}{90}$ | $\frac{6}{90}$ |
| 1 | 0 | $\frac{42}{90}$ | 0 | $\frac{42}{90}$ |
| 2 | $\frac{42}{90}$ | 0 | 0 | $\frac{42}{90}$ |
| | $\frac{42}{90}$ | $\frac{42}{90}$ | $\frac{6}{90}$ | |

, $\frac{84}{90}$

**5.** $f_{X,Y}(x, y) = 1/\pi,\ x^2 + y^2 \leq 1,\ f_X(t) = f_Y(t) = 2(1 - t^2)^{1/2}/\pi$ **6.** $\dfrac{1}{2 \log 2}$ **7.** $\frac{9}{40}$

### *Exercise 5.2*

**1.**

| $y$ \ $x$ | $-1$ | 1 | |
|---|---|---|---|
| $-1$ | $\frac{1}{4}$ | $\frac{1}{4}$ | $\frac{1}{2}$ |
| 1 | $\frac{1}{4}$ | $\frac{1}{4}$ | $\frac{1}{2}$ |
| | $\frac{1}{2}$ | $\frac{1}{2}$ | |

**2.** 62.5, 3, $195\frac{5}{6}$ **3.** 260.42, $\frac{1}{3}$, $\frac{25}{3}$, .89 **4.** 1, 1, $(n + 1) \times (2n + 1)/3$, $(n + 1)/2$ **5.** $\frac{1}{4}$, $\frac{1}{4}$, $\frac{1}{20}$, $-\frac{1}{80}$ **7.** .9923 **8.** .9971

### *Exercise 5.3*

**2.** $(x + 1)/3,\ y + (n - y)/3$ **4.** $0,\ 1 - 2\sqrt{|y|} + |y|$ **6.** $\mu_Y$, No, No **8.** Either $X$ or $Y$ is a constant **11.** $\exp\,(t_1\mu_X + t_2\mu_Y + \frac{1}{2}(t_1^2\sigma_X^2 + t_2^2\sigma_Y^2))$ **12.** $\frac{1}{2}$, 0 **13.** 0, 0, $(a_1^2 + a_2^2)\sigma^2$, $(b_1^2 + b_2^2)\sigma^2$, $(a_1b_1 + a_2b_2)\sigma^2$, $(a_1b_1 + a_2b_2)/\sqrt{(a_1^2 + a_2^2)(b_1^2 + b_2^2)}$

### *Exercise 5.4*

**1.** .5881, .0207 **3.** .6915

### *Exercise 5.6*

**1.** .8643, .6976 **2.** .2451 **3.** .1379 **4.** .6915 **5.** 0 **6.** .1712 **7.** 0 **8.** .3300 **9.** .5558 **10.** .0262

## CHAPTER 6

### *Exercise 6.1*

**1.** .2146 **4.** $1 - N_Z\left(\frac{200}{\sigma} - \frac{\mu}{\sigma}\right)$ **5.** $\frac{1}{2}$ **6.** $(.6554)^{10}$, $1 - (.6554)^{10}$

### *Exercise 6.2*

**1.** $118\frac{1}{3}$, $14{,}024\frac{1}{3}$, $1{,}664{,}602\frac{1}{3}$, $118\frac{1}{3}$, 25.867, 119, 120, 15 **2.** 9.79, .047, 9.6, 9.6, 9.6, 9.7, 9.8, 9.8, 10.0, 10.2 **3.** 163.6, 146.3, 12.1, 162, 30 **4.** 3.18, .63, 3.16, .88 **5.** 56.2, 219.2, 55, 38, 46, 55, 70, 72

### *Exercise 6.3*

**1.** .3185, .3572 **2.** .9876 **3.** 0, 6250 **4.** .121, 2.775 **5.** $1 - (.1)^{10}$, $(.5)^{10}$ **6.** $1 - (.5)^{10}$, $(.1)^{10}$ **10.** $n/(n + 1)$, $1/(n + 1)$

## CHAPTER 7

### *Exercise 7.1*

**1.** .62 **2.** 3, 1.5 **3.** $63\frac{1}{3}$ **4.** $(\bar{X})^{-1}$ **5.** .192 **6.** $1/\bar{X}$ **7.** $\bar{X}$ **8.** $\bar{X}$ **9.** $\sum (X_i - 5)^2/n$ **10.** $\frac{1}{2}\bar{X} - 2.29$ **11.** $\tilde{p} = \bar{X}/\tilde{n}$, $\tilde{n} = \bar{X}^2/(\bar{X} - \sum (X_i - \bar{X})^2/N)$

### *Exercise 7.2*

**1.** $\sum (X_i - 10)^2/n$ **2.** 5.5, 1.83 **3.** $\bar{X}$ **4.** 1.5 **5.** $1/\bar{X}$ **6.** Any value between $x_{(1)} + \frac{1}{2}$ and $x_{(n)} - \frac{1}{2}$ **7.** $1/\bar{X}$ **8.** .000911 **9.** $\bar{X}/N$

### *Exercise 7.3*

**1.** $\frac{1}{6}$ **2.** $2\sigma^4/3$, No **3.** $n_1/(n_1 + n_2)$ **4.** Yes **5.** .4066 **6.** .9671, .9319 **7.** Normal, mean $\lambda$, variance $\lambda^2/n$ **8.** $\hat{\Gamma}: n\gamma/(n + 1)$, $n\gamma^2/(n + 2)(n + 1)^2$, $\tilde{\Gamma}: \gamma$, $\gamma^2/12n$, Yes **9.** $nT_{(1)}$, $\bar{T}$

### *Exercise 7.4*

**1.** (159.0, 171.4) **2.** (158.2, 172.2) **3.** $(n - 1)S^2/\chi_\alpha{}^2$, $n - 1$ d.f.
**4.** $\left(\sqrt{\sum (X_i - \bar{X})^2/\chi^2_{1-\alpha/2}}, \sqrt{\sum (X_i - \bar{X})^2/\chi^2_{\alpha/2}}\right)$ **5.** $\bar{X} - \bar{Y} \pm z_{1-\alpha/2}\sigma\sqrt{2/n}$ **6.** $\bar{X} - \bar{Y} \pm t_{1-\alpha/2}S\sqrt{2/n}$ **7.** $(\exp(-1000L_2), 1)$ **8.** 984, 6147 **9.** about .45, about .92 **10.** $\bar{X} \pm z_{1-\alpha/2}\sqrt{\bar{X}/n}$ **11.** $\frac{2\sigma}{\sqrt{n(n-1)}}\sqrt{2}\frac{\Gamma((n+1)/2)}{\Gamma(n/2)}z_{1-\alpha/2}$
**12.** $\left(\sum (X_i - \mu)^2/\chi^2_{1-\alpha/2}, \sum (X_i - \mu)^2/\chi^2_{\alpha/2}\right)$, n d.f.
**13.** $z_{1-\alpha_1/2}/z_{1-\alpha_2/2}$, $t_{1-\alpha_1/2}/t_{1-\alpha_2/2}$

## CHAPTER 8

### *Exercise 8.1*

**2.** $\mu > 2$, $\mu < 2$, none **3.** .025, $N_Z(11.608 - \mu/10)$, .5438 **4.** .05 **5.** .01, $1 - .99/\theta$

***Exercise 8.2***

**1.** .0039, .3164 **2.** .02, .510 **5.** $P(\chi^2 \leq \frac{1}{4}\chi^2_{1-\alpha})$ **6.** Reject if $\sum X_i^2 > C$ **7.** $\bar{X} > C$ **8.** $\bar{X} > C$ **9.** $(\bar{X} - \bar{Y}) > C$

***Exercise 8.3***

**1.** Reject **5.** Reject **6.** Accept **7.** $\bar{X} < C$ **8.** Yes **9.** Yes **10.** Yes

***Exercise 8.4***

**1.** Reject **2.** Accept **4.** Accept **5.** Reject **6.** Accept **7.** Accept

***Exercise 8.5***

**1.** Reject **2.** Reject **3.** Reject **4.** Reject **5.** Reject

**6.** $n_1 n_2 \cdots n_k - \sum n_i + k,\ (n_i - 1)\left(\prod_{\substack{j=1 \\ j \neq i}}^{k} n_j - 1\right)$

## CHAPTER 9

***Exercise 9.1***

**1.** $2p,\ 2(1-p)$ **2.** Given $X = 1$, $2p/(b^2 - a^2)$; given $X = 0$, $2(1-p)/(2(b-a) - (b^2 - a^2))$ **3.** $(\sqrt{2.2/2\pi}) \exp(-1.1(\mu - (2\bar{X} + .2\mu_0)/2.2)^2)$ **5.** $\lambda((x + .001)^2/\Gamma(2)) \exp(-\lambda(x + .001))$ **6.** $\frac{4}{9}$ at $\alpha = 9.5$, $\frac{5}{9}$ at $\alpha = 9.7$ **7.** Given tail, $\frac{1}{2}$ at $p = \frac{1}{4}$, $\frac{1}{3}$ at $p = \frac{1}{2}$, $\frac{1}{6}$ at $p = \frac{3}{4}$; given head, $\frac{1}{6}$ at $p = \frac{1}{4}$, $\frac{1}{3}$ at $p = \frac{1}{2}$, $\frac{1}{2}$ at $p = \frac{3}{4}$

***Exercise 9.2***

**3.** $m/\theta = x$ **4.** $\sigma_0^2,\ \sigma_0^4/(m-1)$ **5.** $\lambda_0,\ \lambda_0^2/(m+1)$ **7.** $\bar{x}$, $a$ or $b$
**8.** $\left(m\sigma_0^2 + \sum (x_i - \mu_X)^2\right)/(n/2 + m)$ **9.** $\lambda_0(n + m + 1)/(n\bar{x}\lambda_0 + m + 1)$
**10.** $\left(m + \sum x_i + 1\right)\lambda_0/(n\lambda_0 + m + 1)$
**11.** $\bar{x} + \sqrt{\sigma^2/2n\pi}\{\exp(-n(a - \bar{x})^2/2\sigma_X^2) - \exp(-n(b - \bar{x})^2/2\sigma_X^2)\}/[N_Z(\sqrt{n}(b - \bar{x})/\sigma_X) - N_Z(\sqrt{n}(a - \bar{x})/\sigma_X)]$

***Exercise 9.3***

**1.** (18.70, 19.92) **2.** $\int_a^b \frac{(10.05)^{231}}{(231)!} \lambda^{230} e^{-10.05\lambda}\, d\lambda = .99$ **3.** $\sigma_0^2 = 10^{-4}$, $m = 1.01$,
$\int_a^b \frac{(.000901)^{52.01}}{\Gamma(52.01)} (\sigma^2)^{-53.01} e^{-51.01/\sigma^2}\, d\sigma^2 = .95$

**4.** $\int_a^b \frac{\Gamma(n+2)}{\Gamma(x_k + 1)\Gamma(n + 1 - x_k)} p_k^{x_k}(1 - p_k)^{n - x_k}\, dp_k = 1 - \alpha$

**5.** $\int_a^b \frac{(n\bar{x} + 45)^{n+1}}{\Gamma(n+1)} \lambda^n e^{-\lambda(n\bar{x}+45)}\, d\lambda = .95$

# CHAPTER 10

### *Exercise 10.1*

**1.** $b^+ = 1$, $a^+ = 8\frac{1}{4}$, $\hat{\sigma}^2 = 1.583$, $v(b^+) = .00236$, $v(a^+) = 1.555$ **3.** $b^+ = .938$, $a^+ = -4.541$, $\hat{\sigma}^2 = 5.133$, $v(b^+) = .066$, $v(a^+) = 15.733$ **4.** $B^+ = \sum x_i Y_i / \sum x_i^2$, $S^2 = \left(\sum Y_i^2 - B^+ \sum x_i Y_i\right)/(n - 1)$ **7.** $\bar{Y}/\bar{x}$ **8.** $(1/n) \sum Y_i/x_i$ **10.** $A^+ = \bar{Y} - (B^+/n) \sum 1/x_i$, $B^+ = \sum (Y_i - \bar{Y})(c_i - \bar{c}) / \sum (c_i - \bar{c})^2$, where $c_i = 1/x_i$

### *Exercise 10.2*

**1.** $\hat{B} = (\bar{Z} - \bar{Y})/(x_2 - x_1)$, $\hat{A} = \bar{Z} - bx_2$ **2.** Reject $H_0$ **3.** $5.986 \leq a \leq 10.665$, $.908 \leq b \leq 1.092$ **4.** $\hat{B} = (\bar{Z} - \bar{Y})/(x_2 - x_1)$, $\hat{A} = \bar{Z} - \hat{b}x_2$, Maximum likelihood estimates maximize the probability of occurrence of the sample. **5.** $\hat{b} = .998$, $\hat{a} = .0014$, $\hat{s}^2 = .000125$, Accept $H_0: a = 0$. **7.** $\hat{\Sigma}^2 = \sum (Y_i - \hat{A} - \hat{B}x_i)^2/nx_i$, $\hat{A} = \bar{Y} - \hat{B}\bar{x}$, $\hat{B} = \left(\sum Y_i/x_i - \bar{Y} \sum 1/x_i\right)/\left(n - \bar{x} \sum 1/x_i\right)$

# Index